Springer Monographs in Mathematics

M. Banagl

Topological Invariants of Stratified Spaces

Springer

M. Banagl
Mathematisches Institut
Universität Heidelberg
Im Neuenheimer Feld 288
69120 Heidelberg, Germany
e-mail: banagl@mathi.uni-heidelberg.de

Library of Congress Control Number: 2006693517

Mathematics Subject Classification (2000): 55N33

ISSN 1439-7382

ISBN-10 3-540-38585-1 Springer Berlin Heidelberg New York
ISBN-13 978-3-540-38585-1 Springer Berlin Heidelberg New York

Springer is a part of Springer Science+Business Media
springer.com

Typesetting by the author and VTEX using a Springer LaTeX macro package
Cover design: Erich Kirchner, Heidelberg, Germany

Printed on acid-free paper SPIN: 11839545 VA 44/3100/VTEX - 5 4 3 2 1 0

Preface

The homology of manifolds enjoys a remarkable symmetry: Poincaré duality. If the manifold is triangulated, then this duality can be established by associating to a simplex its dual block in the barycentric subdivision. In a manifold, the dual block is a cell, so the chain complex based on the dual blocks computes the homology of the manifold. Poincaré duality then serves as a cornerstone of manifold classification theory. One reason is that it enables the definition of a fundamental bordism invariant, the signature. Classifying manifolds via the surgery program relies on modifying a manifold by executing geometric surgeries. The trace of the surgery is a bordism between the original manifold and the result of surgery. Since the signature is a bordism invariant, it does not change under surgery and is thus a basic obstruction to performing surgery. Inspired by Hirzebruch's signature theorem, a method of Thom constructs characteristic homology classes using the bordism invariance of the signature. These classes are not in general homotopy invariants and consequently are fine enough to distinguish manifolds within the same homotopy type.

Singular spaces do not enjoy Poincaré duality in ordinary homology. After all, the dual blocks are not cells anymore, but cones on spaces that may not be spheres. This book discusses when, and how, the invariants for manifolds described above can be established for singular spaces. By singular space we mean here a space which has points whose neighborhoods are not Euclidean, but which can still be decomposed into subsets, each of which *is* a manifold. Such a decomposition is called a stratification, provided certain conditions are satisfied where the subsets (the "strata") meet. Thus we will not be concerned with spaces that possess an intricate local structure, such as fractals. Spaces that can be stratified include all triangulated spaces, as well as algebraic varieties. Once these invariants have been constructed, we will furthermore describe tools for their computation. For instance, we will see how the invariants behave under maps.

The germ for this book was a topology seminar on sheaf theory and intersection homology which I gave in the spring of 2000 at the University of Wisconsin, Madison, and my spring 2001 special topics course on characteristic classes of stratified spaces and maps, held at the same institution, whose syllabus included Witt space bordism, the Goresky–MacPherson L-class, Verdier self-dual complexes of sheaves, perverse sheaves and t-structures, algebraic bordism of self-dual sheaves and the Cappell–Shaneson L-class formulae for stratified maps. I would like to take this opportunity to thank the Mathematics Department of the University of Wisconsin, Madison, for giving me the chance to teach these courses. The original lecture notes were subsequently expanded and, to some extent, enriched with detail.

The text has then been used as the basis for a seminar on stratified spaces and intersection homology which I taught in the winter quarter 2004 at the University of Cincinnati, as well as for the Oberseminar Topologie at the University of Heidelberg, which in the fall of 2004 was devoted to intersection homology and topological invariants of singular spaces. My colleagues' and students' feedback during these seminars led to several improvements of the text.

I have made an effort to make the book accessible to a second year graduate student in topology. The prerequisites are thus modest: (Co)homology theory, simplicial complexes, basic differential topology such as regular values and transversality, the language of categories and functors. As far as intersection homology theory is concerned, I start *ab initio*—no prior knowledge of sheaf theory, triangulated categories, derived categories or Verdier duality is required. In addition, I hope that the book may be useful for the research mathematician who wishes to learn about intersection homology and the invariants of singular spaces it effectuates. The pace of the book is intentionally *not* uniform: I change gears by omitting some details precisely when there is a danger that a flood of technicalities may prevent the reader from seeing the forest for the trees.

Chapter 1 introduces sheaves, cohomology with coefficients in a sheaf (via injective resolutions and via the Čech resolution), complexes of sheaves (i.e. differential graded sheaves) and cohomology with coefficients in a complex of sheaves (the so-called hypercohomology). One way to construct the intersection homology groups is as the hypercohomology groups of a certain complex of sheaves, the Deligne–Goresky–MacPherson intersection chain complex. This chapter also contains the definition of truncation functors that are so important in the definition of the intersection chain complex.

Chapter 2 discusses concepts that are applicable to any abelian category, hence it does not depend on the previous chapter. We give a self-contained treatment of triangulated categories and localization of categories. In particular, localization with respect to quasi-isomorphisms of complexes will give rise to derived categories. Derived functors are constructed pragmatically.

Chapter 3 on Verdier duality introduces the functors $f_!$, $f^!$ operating on sheaf complexes, where f is a continuous map. The latter functor is well-defined only on the derived category, so that this chapter depends on Chap. 2. We then use $f^!$ to construct the Verdier dualizing complex of a space and the Verdier duality functor $\mathcal{D}$. Verdier self-dual sheaves, i.e. sheaves that are isomorphic to their own dual, play a central role in this book, since all invariants to be discussed are induced by self-duality isomorphisms.

Chapter 4 defines stratified pseudomanifolds and intersection homology on stratified pseudomanifolds. This is done in two different ways: In Sect. 4.1, intersection chains are defined for a piecewise linear pseudomanifold as a subcomplex of the complex of all piecewise linear chains, by placing certain conditions on how a chain is supposed to meet strata. This is the approach that was taken by Goresky and MacPherson in their first paper [GM80]. Readers who wish to get a geometrically intuitive first introduction to intersection homology without having to absorb all of the material from Chaps. 1 through 3 may wish to peruse Sect. 4.1, except Subsect. 4.1.4, before reading the rest of the book. In Sect. 4.2, all of Chaps. 1 through 3 are applied in constructing intersection homology sheaf-theoretically. This is the approach that was taken by Goresky and MacPherson in their second paper [GM83].

Chapter 5 reviews manifold theory. Apart from setting the stage for subsequent chapters, this is done in order to place the signature and L-classes into a larger context so that from this elevated vantage point, the less advanced reader may get a better impression of their importance and applications. We compute smooth oriented bordism groups rationally and give a rudimentary introduction to surgery theory. In order to understand the construction of L-classes for singular spaces, it is helpful to be familiar with Sect. 5.7, where Thom's method for obtaining L-classes from the signatures of submanifolds is explained.

The remaining chapters deal with the signature and L-class of singular spaces and are structured in a progression from more restricted stratifications towards less restricted ones.

Chapter 6 works on spaces that have restrictions along the strata of odd codimension. Section 6.1, for example, defines the signature, and Sect. 6.3 the L-class, of spaces that have no strata of odd codimension. This class of spaces includes all complex algebraic varieties.

The derived categories constructed in Chap. 2 are generally not abelian, but have the structure of a triangulated category. The distinguished triangles substitute for the exact sequences. Chapter 7 explains how one can obtain abelian subcategories of triangulated categories by endowing them with certain truncation-structures, or t-structures. Section 7.1 is purely algebraic and requires only Chap. 2 as prerequisite. The derived category of complexes of sheaves can be endowed with a particularly important t-structure, the perverse t-structure, which is discussed in Sects. 7.2 and 7.3.

Chapter 8 provides tools for the computation of the characteristic classes arising from self-dual sheaves. We are mainly interested in the behavior of these classes under stratified maps. We define when two self-dual complexes of sheaves on the same space are algebraically bordant. (This is not to be confused with geometric bordism of spaces covered with self-dual sheaves.) This chapter has two highlights: In Sect. 8.1, we give a fully detailed proof of the Cappell–Shaneson decomposition theorem for self-dual sheaves on a space with only even-codimensional strata. The theorem states that every self-dual sheaf is bordant to an orthogonal sum of intersection chain sheaves. All signature- and L-class-formulae are then implied by this. These formulae, however, contain twisted signatures and twisted L-classes. Section 8.3 shows how to calculate those when the underlying space is a Witt space.

All invariants previously constructed require the singular space to satisfy the Witt-condition. Generalized Poincaré–Verdier self-duality on singular spaces that do not necessarily satisfy the Witt condition ("non-Witt spaces") is discussed in Chap. 9.

We will see that self-duality on such completely general stratified spaces hinges on the existence of Lagrangian structures. If a space possesses no Lagrangian structures, then it possesses no self-dual homology groups compatible with intersection homology. The L-class of non-Witt spaces is constructed in Sect. 9.2. The behavior of these classes under stratified maps is investigated in Sect. 9.3, while Sect. 9.4 once more takes up twisted characteristic classes, this time on non-Witt spaces.

Chapter 10 gives a brief introduction to Cheeger's method of recovering Poincaré duality for singular Riemannian spaces using L^2 differential forms on the top stratum. The relation to the intersection homology of Goresky and MacPherson is discussed.

First and foremost, I would like to thank Peter Orlik. It was he who suggested in the spring of 2000 that I write up my seminar notes—this book is the result. Special thanks are due to Greg Friedman, whose careful reading of early drafts of the manuscript led to many improvements and elimination of errors. I would like to extend my gratitude to Aron Fischer, Filipp Levikov, Falk Loewner, Laurentiu George Maxim, Augusto Minatta and Jonathan Pakianathan for corrections and valuable suggestions. Thanks also to Carl McTague, who produced the images of the algebraic curves in Figs. 2–4 of Sect. 6.2. I am grateful to Sylvain Cappell for introducing me to stratified spaces and to the National Science Foundation for supporting the work on this book.

Markus Banagl
Universität Heidelberg
January 2006

Contents

1

Elementary Sheaf Theory

1.1 Sheaves

Very loosely speaking, a sheaf is a structure over a space which has horizontal topological structure and vertical algebraic structure. The horizontal topological structure mimics locally the topology of the space. Additional references on sheaf theory include [Bre97], [God58], [Ive86] as well as [KS90].

Definition 1.1.1 *Let X be a topological space. A* presheaf A (of abelian groups) on *X is an assignment $U \mapsto A(U)$ of an abelian group $A(U)$ to every open subset $U \subset X$, together with restriction homomorphisms $r_{U,V} : A(V) \to A(U)$ for every pair $U \subset V$ of open sets in X, satisfying*

1. *$A(\varnothing) = 0$, the trivial group,*
2. *$r_{U,U} = 1$ (identity map),*
3. *$r_{U,V} \circ r_{V,W} = r_{U,W}$ for open $U \subset V \subset W$.*

The group $A(U)$ is referred to as the "group of sections of A over U" (although at this stage they are not sections of anything). It is convenient to write $s|_U = r_{U,V}(s)$ for the image of a section s under restriction. Using the language of categories, a presheaf is thus a contravariant functor from the category of open subsets of X and inclusions to the category of abelian groups and homomorphisms. *Mutatis mutandis* one defines presheaves of rings, modules, groupoids, etc.

Examples 1.1.2 *Letting $A(U) = G$, $(U \neq \varnothing)$, with G a fixed group (and taking the restrictions to be the identity) one obtains the* constant presheaf. *Interesting examples arise from assigning to an open set the group of continuous real-valued functions defined on that set, the group of singular p-cochains of the set, on a differentiable manifold the group of differentiable functions or differentiable p-forms or vector fields over that set.*

Fix a point $x \in X$. Let U and V be open neighborhoods of x. Call two sections $s \in A(U)$ and $t \in A(V)$ equivalent if there exists an open neighborhood $W \subset U \cap V$

of x such that $s|_W = t|_W$. We denote the set of equivalence classes by $\mathbf{A}_x$, it is called the set of *germs of sections at* x. We write s_x for the equivalence class of the section s in $\mathbf{A}_x$; s_x is the *germ of* s *at* x. In other words, $\mathbf{A}_x$ is the direct limit

$$\mathbf{A}_x = \varinjlim A(U)$$

over the directed set of open neighborhoods of x. In particular, $\mathbf{A}_x$ inherits a group structure.

For x varying, we wish to collect the various $\mathbf{A}_x$ into a space. Form the disjoint union $\mathbf{A} = \coprod \mathbf{A}_x$ and make it into a topological space by declaring the collection of sets of the form $\{s_x \in \mathbf{A}_x | x \in U\}$, for $U \subset X$ open, $s \in A(U)$, to be a basis for its topology. The natural map $\pi : \mathbf{A} \to X$, sending $\mathbf{A}_x \to \{x\}$, is continuous and a local homeomorphism. Moreover, the group operations in the $\mathbf{A}_x$ are continuous for x varying. This means precisely that the map $\{(a, b) \in \mathbf{A} \times \mathbf{A} | \pi(a) = \pi(b)\} \to \mathbf{A}$, $(a, b) \mapsto a - b$, is continuous. The pair $(\mathbf{A}, \pi)$ is called the *sheaf generated by the presheaf* A. The fibers $\mathbf{A}_x$ of π are the *stalk of* $\mathbf{A}$ *at* x. The process described above is referred to as "sheafification" of a presheaf, and we will also write

$$\mathbf{A} = \mathbf{Sheaf}(A).$$

Remark 1.1.3 *In general, the topology of a sheaf space is non-Hausdorff: Let* $X = \mathbb{R}$ *and define a presheaf by* $A(U) = \mathbb{Z}/2$, *if* $0 \in U$, *and* $A(U) = 0$ *otherwise. The sheafification of* A *has stalk* $\mathbb{Z}/2$ *at* 0 *and stalk* 0 *for all* $x \neq 0$. *As the projection* π *to* $\mathbb{R}$ *is a local homeomorphism, the two points over* 0 *cannot be separated by open sets.*

Abstracting the above properties of a sheaf space, we define:

Definition 1.1.4 *A* sheaf (of abelian groups) on X *is a pair* $(\mathbf{A}, \pi)$ *with*

1. $\mathbf{A}$ *a topological space,*
2. $\pi : \mathbf{A} \to X$ *a local homeomorphism,*
3. *each* $\mathbf{A}_x := \pi^{-1}(x)$, $x \in X$, *an abelian group, the "stalk,"*
4. *group operations continuous.*

Definition 1.1.5 *A* homomorphism of sheaves $f : \mathbf{A} \to \mathbf{B}$ *is a continuous map preserving stalks,* $f(\mathbf{A}_x) \subset \mathbf{B}_x$, *all* x, *such that all restrictions* $f_x : \mathbf{A}_x \to \mathbf{B}_x$ *are homomorphisms.*

Thus sheaves on a space X form a category, $Sh(X)$. A homomorphism of sheaves f is a *monomorphism* (*epimorphism, isomorphism*), if all restrictions f_x are monomorphisms (epimorphisms, isomorphisms). The sheafification of the constant presheaf $A(U) = G$, G a fixed abelian group, is called the *constant sheaf with stalk* G. Its sheaf space is $X \times G$, where G has the discrete topology. Given a sheaf $(\mathbf{A}, \pi)$ and a subspace $Y \subset X$, $\pi^{-1}(Y)$ is a sheaf $\mathbf{A}|_Y$, the *restriction of* $\mathbf{A}$ *to* Y. A sheaf $\mathbf{A}$ on X is *locally constant*, if every point $x \in X$ has an open neighborhood U such that $\mathbf{A}|_U$ is isomorphic to a constant sheaf.

Examples 1.1.6

- *For an n-dimensional manifold M, we define its* orientation sheaf *by*

$$\mathcal{O}_M = \mathbf{Sheaf}(U \mapsto H_n^{sing}(M, M - U; \mathbb{Z})),$$

where H_*^{sing} denotes singular homology. If $\partial M = \varnothing$, then $\mathcal{O}_M$ is locally constant with stalks $\mathbb{Z}$. It is constant if M is orientable. If $\partial M \neq \varnothing$, then $\mathcal{O}_M|_{\partial M} = 0$.
- *Let X be a path connected space with fundamental group $G = \pi_1(X)$. Every G-module V determines a locally constant sheaf* **A** *by the diagonal action:*

$$\mathbf{A} = \widetilde{X} \times_G V^{dis},$$

where $\widetilde{X}$ is the universal cover of X with G acting as deck-transformations and V^{dis} specifies the discrete topology to be used on V. The sheaf **A** *has stalk V over each point. Conversely, every locally constant sheaf* **A** *on X is given (up to isomorphism) by this construction; the representation of G on the stalk of* **A** *over the base point is given by the "monodromy" (the linear operators arising from going around loops). Indeed one can show that this sets up an equivalence of categories between the category of locally constant sheaves on X and the category of G-modules. (For more information on this, see also Sect.* 8.2.4.)

The *group of sections over* $Y \subset X$ is $\Gamma(Y; \mathbf{A}) = \{s : Y \to \mathbf{A} \text{ continuous} | \pi s = 1\}$. The assignment $U \mapsto \Gamma(U; \mathbf{A})$ defines a presheaf (using the obvious restrictions), called the *presheaf of sections of* **A**. Given $a \in \mathbf{A}$, choose a local section s with $s(\pi(a)) = a$, and consider $s_{\pi(a)}$, the germ of s at $\pi(a)$, an element of the stalk at $\pi(a)$ of the sheafification of the presheaf of sections of **A**. Then $a \mapsto s_{\pi(a)}$ is a well-defined natural continuous map

$$\mathbf{A} \xrightarrow{\cong} \mathbf{Sheaf}(U \mapsto \Gamma(U; \mathbf{A}))$$

which is an isomorphism, because it is bijective on each stalk.

Conversely, one asks whether, starting with a presheaf A, one can canonically identify $\Gamma(U; \mathbf{Sheaf}(A)) \cong A(U)$. The answer, in general, is no.

Example 1.1.7 *To $U \subset \mathbb{R}$ open, assign the space $L^1(U, dx)$ of Lebesgue-integrable functions on U. Cover $\mathbb{R}$ with bounded open intervals U_α. Then the restriction of the constant function $1|_{U_\alpha}$ is an element in $L^1(U_\alpha, dx)$. It defines sections in $\Gamma(U_\alpha; \mathbf{Sheaf}(L^1))$ for each α. On overlaps $U_\alpha \cap U_\beta$, these sections agree, and so give rise to a global section $1 \in \Gamma(\mathbb{R}; \mathbf{Sheaf}(L^1))$, but $1 \notin L^1(\mathbb{R}, dx)$.*

What fails here is that for an arbitrary presheaf one may not be able to "glue" sections. This leads us to formulate the following *unique gluing* condition for a presheaf A:

(G) Given an open $U \subset X$, an open cover $\{U_\alpha\}$ of U, sections $s_\alpha \in A(U_\alpha)$ such that $s_\alpha|_{U_\alpha \cap U_\beta} = s_\beta|_{U_\alpha \cap U_\beta}$, all α, β, there exists a unique $s \in A(U)$ with $s|_{U_\alpha} = s_\alpha$, all α.

Note that we have a canonical map $\phi : A(U) \to \Gamma(U; \mathbf{Sheaf}(A))$ defined by sending $s \in A(U)$ to $(x \mapsto s_x)_{x \in U}$, i.e. the section of $\mathbf{Sheaf}(A)$ over U whose value at x is the germ of s at x.

Proposition 1.1.8 *For presheaves A that satisfy unique gluing* (G), *the canonical map*

$$\phi : A(U) \xrightarrow{\cong} \Gamma(U; \mathbf{Sheaf}(A))$$

is an isomorphism.

Proof. Put $\mathbf{A} = \mathbf{Sheaf}(A)$. Let $s, t \in A(U)$ such that $s_x = t_x \in \mathbf{A}_x$, all $x \in U$. Equality of the germs means that each x has an open neighborhood V_x with $s|_{V_x} = t|_{V_x}$. So s and t have the same restriction to each set in the cover $\{V_x\}_{x \in U}$ of U and $s = t$ by the uniqueness part of (G).

Given $s \in \Gamma(U; \mathbf{A})$ ($s(x)$ will denote the value of s at $x \in U$) there exist, for each x, a neighborhood V_x and an element $t^x \in A(V_x)$ whose germ at every point x' of V_x is the value of s at x': $(t^x)_{x'} = s(x')$ in $\mathbf{A}_{x'}$ for all $x' \in V_x$. The existence of such V_x and t^x uses that $\pi : \mathbf{A} \to X$ is a local homeomorphism. If $x, y \in U$, then $t^x|_{V_x \cap V_y} = t^y|_{V_x \cap V_y}$, by the above uniqueness argument. Hence (G) ensures the existence of $t \in A(U)$ such that $t|_{V_x} = t^x$ for all $x \in U$, and $t_x = (t|_{V_x})_x = (t^x)_x = s(x)$ means $\phi(t) = s$. □

Corollary 1.1.9 *Sheafification of a presheaf and taking the presheaf of sections of a sheaf set up a one-to-one correspondence between sheaves and presheaves satisfying* (G).

For this reason, it is customary not to distinguish between presheaves with unique gluing of sections (G) and sheaves.

A *subsheaf* $(\mathbf{A}, \pi_A)$ of sheaf $(\mathbf{B}, \pi_B)$ is an open subspace $\mathbf{A} \subset \mathbf{B}$ such that π_A is the restriction of π_B to $\mathbf{A}$ and $\pi_B^{-1}(x) \cap \mathbf{A}$ is a subgroup of $\mathbf{B}_x$ for all $x \in X$. Given a homomorphism $f : \mathbf{A} \to \mathbf{B}$, we define the kernel of f as $\mathbf{ker}(f) = \{a \in \mathbf{A} | f(a) = 0\}$, it is a subsheaf of $\mathbf{A}$. $\mathbf{im}(f)$ is the subsheaf of $\mathbf{B}$ whose sheaf space is the subset $\mathrm{Im}(f)$. The homomorphism f induces a map on sections $f_U : \Gamma(U; \mathbf{A}) \to \Gamma(U; \mathbf{B})$. Note that the presheaf

$$U \mapsto \ker(f_U)$$

satisfies condition (G) and so is a sheaf. In particular, we have the formula

$$\Gamma(U; \mathbf{ker}(f)) = \ker(f_U).$$

Such a formula however does not hold for $\mathbf{im}(f)$, but we have the description

$$\mathbf{im}(f) = \mathbf{Sheaf}(U \mapsto \mathrm{im}(f_U)).$$

With a notion of kernel and image sheaf in hand, we are now able to define a sequence of sheaf homomorphisms

$$\mathbf{A} \xrightarrow{f} \mathbf{B} \xrightarrow{g} \mathbf{C}$$

to be *exact* if $\mathbf{ker}(g) = \mathbf{im}(f)$. By definition, this is the case if and only if

$$\mathbf{A}_x \xrightarrow{f_x} \mathbf{B}_x \xrightarrow{g_x} \mathbf{C}_x$$

is exact for each $x \in X$. Thus exactness of a sequence of sheaf maps may be checked on stalks, which is a useful criterion in practice.

Example 1.1.10 *Let M be a smooth manifold and $\mathbf{C}_M$ be the sheaf of germs of smooth complex valued functions on M, i.e. the sheafification of the presheaf that assigns to an open set the additive group of smooth $\mathbb{C}$-valued functions on the open set. Denote by $\mathbf{C}_M^*$ the multiplicative sheaf of germs of smooth functions with values in the multiplicative group $\mathbb{C}^*$. The exponential map* $\exp : \mathbb{C} \to \mathbb{C}^*$ *induces a map of presheaves and, consequently, a map of sheaves*

$$\exp : \mathbf{C}_M \longrightarrow \mathbf{C}_M^*.$$

The kernel of $\exp : \mathbb{C} \to \mathbb{C}^*$ *is $2\pi\sqrt{-1}\mathbb{Z}$, and we consider the* exponential sequence

$$0 \longrightarrow 2\pi\sqrt{-1}\mathbb{Z}_M \longrightarrow \mathbf{C}_M \xrightarrow{\exp} \mathbf{C}_M^* \longrightarrow 0,$$

where $\mathbb{Z}_M$ denotes the constant sheaf with stalk $\mathbb{Z}$ on M. To show exactness, it remains to verify that exp *is onto. Given $x \in M$ and a germ $f_x \in (\mathbf{C}_M^*)_x$, there exists a contractible neighborhood U of x such that f_x is the germ of a smooth function $f : U \to \mathbb{C}^*$. Now every point $z_0 \in \mathbb{C}^*$ has a neighborhood in which a single branch* $\log z$ *can be chosen. Thus* $\log f : U \to \mathbb{C}$ *defines a germ in $\mathbf{C}_M$ mapping to f_x.*

Let $\mathbf{A} \subset \mathbf{B}$ be a subsheaf of sheaf $\mathbf{B}$. We define the quotient sheaf $\mathbf{B}/\mathbf{A}$ by

$$\mathbf{B}/\mathbf{A} = \mathbf{Sheaf}(U \mapsto \Gamma(U; \mathbf{B})/\Gamma(U; \mathbf{A})).$$

The exact sequences $0 \to \Gamma(U; \mathbf{A}) \to \Gamma(U; \mathbf{B}) \to \Gamma(U; \mathbf{B})/\Gamma(U; \mathbf{A}) \to 0$ induce an exact sequence

$$0 \longrightarrow \mathbf{A} \longrightarrow \mathbf{B} \longrightarrow \mathbf{B}/\mathbf{A} \longrightarrow 0$$

because the direct limit functor is exact. This implies, in particular, the identification

$$(\mathbf{B}/\mathbf{A})_x \cong \mathbf{B}_x/\mathbf{A}_x.$$

Looking at stalks, it is clear that the functor $\Gamma(Y; -)$, with $Y \subset X$, is left exact. It is however not exact in general: Consider the exponential sequence of Example 1.1.10, say for $M = \mathbb{R}^2$. Let U be an open annulus about the origin of $\mathbb{R}^2$ and $f(x, y) = x + \sqrt{-1}y$. Then f defines a section in $\Gamma(U; \mathbf{C}_M^*)$, but there is no section $g \in \Gamma(U; \mathbf{C}_M)$ such that $\exp(g) = f$. Therefore $\exp : \Gamma(U; \mathbf{C}_M) \to \Gamma(U; \mathbf{C}_M^*)$ is not surjective. It also follows that in general $\Gamma(U; \mathbf{B}/\mathbf{A}) \neq \Gamma(U; \mathbf{B})/\Gamma(U; \mathbf{A})$ for sheaves $\mathbf{A} \subset \mathbf{B}$.

Let X and Y be topological spaces, $f : X \to Y$ a continuous map.

- Given $\mathbf{A} \in Sh(X)$, consider the presheaf $U \mapsto \Gamma(f^{-1}(U); \mathbf{A})$ on Y. It satisfies condition (G) and hence is a sheaf $f_*\mathbf{A} \in Sh(Y)$, called the *direct image* or *push-forward* of $\mathbf{A}$. This construction is obviously functorial; f_* is left exact covariant (since $\Gamma(U; -)$ is left exact).
- Given $(\mathbf{B}, \pi) \in Sh(Y)$, we define the *inverse image* or *pullback* of $\mathbf{B}$: Its sheaf space is given by the usual pullback construction $f^*\mathbf{B} = \{(x, b) \in X \times \mathbf{B} | f(x) = \pi(b)\}$ endowed with the subspace topology. The projection $f^*\mathbf{B} \to X$ is $(x, b) \mapsto x$, a local homeomorphism. The group structure on $(f^*\mathbf{B})_x$ is declared by requiring the canonical bijection $\mathbf{B}_{f(x)} \xrightarrow{\cong} (f^*\mathbf{B})_x$, sending $b \mapsto (x, b)$, to be an isomorphism of groups. The covariant functor f^* is exact as a consequence of $\mathbf{B}_{f(x)} \cong (f^*\mathbf{B})_x$.

The next proposition shows that f^* and f_* are adjoint functors:

Proposition 1.1.11 *Let $f : X \to Y$ be a continuous map and $\mathbf{A} \in Sh(X), \mathbf{B} \in Sh(Y)$. Then*

$$\mathrm{Hom}_{Sh(X)}(f^*\mathbf{B}, \mathbf{A}) \cong \mathrm{Hom}_{Sh(Y)}(\mathbf{B}, f_*\mathbf{A}).$$

Proof. The map of presheaves

$$\begin{aligned} \Gamma(U; \mathbf{B}) &\to \Gamma(f^{-1}(U); f^*\mathbf{B}) \\ s &\mapsto (x \mapsto (x, s(f(x)))) \end{aligned}$$

induces a canonical homomorphism

$$\mathbf{B} \longrightarrow f_* f^*\mathbf{B}. \tag{1.1}$$

Identifying

$$(f^* f_*\mathbf{A})_x \cong (f_*\mathbf{A})_{f(x)} \cong \lim_{U \ni f(x)} \Gamma(U; f_*\mathbf{A}) = \lim_{U \ni f(x)} \Gamma(f^{-1}(U); \mathbf{A})$$

and using the canonical map between direct limits

$$\lim_{U \ni f(x)} \Gamma(f^{-1}(U); \mathbf{A}) \to \lim_{U \ni x} \Gamma(U; \mathbf{A}) \cong \mathbf{A}_x$$

we obtain a canonical homomorphism

$$f^* f_*\mathbf{A} \longrightarrow \mathbf{A}. \tag{1.2}$$

The morphisms (1.1) and (1.2) are of independent interest and referred to as the *canonical adjunction morphisms*. Apply (1.1) in defining the composition

$$\mathrm{Hom}(f^*\mathbf{B}, \mathbf{A}) \to \mathrm{Hom}(f_* f^*\mathbf{B}, f_*\mathbf{A}) \to \mathrm{Hom}(\mathbf{B}, f_*\mathbf{A})$$

and apply (1.2) in defining

$$\mathrm{Hom}(\mathbf{B}, f_*\mathbf{A}) \to \mathrm{Hom}(f^*\mathbf{B}, f^* f_*\mathbf{A}) \to \mathrm{Hom}(f^*\mathbf{B}, \mathbf{A}).$$

These two compositions are inverse to each other. □

To functors (of several variables) from the category of abelian groups to itself one can associate corresponding functors on sheaves. We carry this out for the bifunctors $\oplus$, $\otimes$ and $\mathrm{Hom}(-,-)$. Let $\mathbf{A}, \mathbf{B} \in Sh(X)$. The presheaf

$$U \mapsto \Gamma(U; \mathbf{A}) \oplus \Gamma(U; \mathbf{B})$$

satisfies unique gluing (G), and hence is a sheaf $\mathbf{A} \oplus \mathbf{B}$, the *direct sum of* $\mathbf{A}$ *and* $\mathbf{B}$. One may compute sections as

$$\Gamma(U; \mathbf{A} \oplus \mathbf{B}) = \Gamma(U; \mathbf{A}) \oplus \Gamma(U; \mathbf{B}),$$

and since direct sums commute with direct limits we have the identity

$$(\mathbf{A} \oplus \mathbf{B})_x \cong \mathbf{A}_x \oplus \mathbf{B}_x.$$

Define the *tensor product of* $\mathbf{A}$ *and* $\mathbf{B}$ by

$$\mathbf{A} \otimes \mathbf{B} = \mathbf{Sheaf}(U \mapsto \Gamma(U; \mathbf{A}) \otimes \Gamma(U; \mathbf{B})).$$

The following example shows that the defining presheaf does not satisfy (G), so that sheafification is necessary.

Example 1.1.12 *Think of the circle* S^1 *as* $[0,1]/(0 \sim 1)$. *Define an open cover consisting of two sets* $U_1 = (\frac{1}{10}, \frac{9}{10})$ *and* $U_2 = ([0, \frac{1}{5}) \cup (\frac{4}{5}, 1])/\sim$. *Let* $\mathbf{A}$ *be the sheaf* $[0,1] \times \mathbb{Z}/((0,n) \sim (1,-n))$ *over* S^1. *For all* $x \in U_1$, *define* $s_1(x) = 1$, $s_1'(x) = -1$. *For* $x \in [0, \frac{1}{5})$, *let* $s_2(x) = 1$ *and* $s_2'(x) = -1$, *and for* $x \in (\frac{4}{5}, 1]$, *set* $s_2(x) = -1$, $s_2'(x) = 1$. *We have defined sections* $s_1, s_1' \in \Gamma(U_1; \mathbf{A})$ *and* $s_2, s_2' \in \Gamma(U_2; \mathbf{A})$. *On the overlap* $(\frac{1}{10}, \frac{1}{5})$, *we have* $s_1 \otimes s_1' = 1 \otimes (-1) = s_2 \otimes s_2'$, *and on the overlap* $(\frac{4}{5}, \frac{9}{10})$, *we have* $s_1 \otimes s_1' = 1 \otimes (-1) = (-1) \otimes 1 = s_2 \otimes s_2'$, *so that the sections* $s_1 \otimes s_1'$ *and* $s_2 \otimes s_2'$ *agree on overlaps. However there exist no* $s, s' \in \Gamma(S^1; \mathbf{A})$ *with* $(s \otimes s')|_{U_1} = s_1 \otimes s_1'$, $(s \otimes s')|_{U_2} = s_2 \otimes s_2'$, *because* $\mathbf{A}$ *has* $\Gamma(S^1; \mathbf{A}) = 0$.

Since tensor products commute with direct limits, stalks can be calculated using

$$(\mathbf{A} \otimes \mathbf{B})_x \cong \mathbf{A}_x \otimes \mathbf{B}_x.$$

The presheaf

$$U \mapsto \mathrm{Hom}(\mathbf{A}|_U, \mathbf{B}|_U)$$

satisfies condition (G) and is thus a sheaf $\mathbf{Hom}(\mathbf{A}, \mathbf{B})$. Its group of sections is

$$\Gamma(U; \mathbf{Hom}(\mathbf{A}, \mathbf{B})) = \mathrm{Hom}(\mathbf{A}|_U, \mathbf{B}|_U),$$

but in general $\mathbf{Hom}(\mathbf{A}, \mathbf{B})_x \neq \mathrm{Hom}(\mathbf{A}_x, \mathbf{B}_x)$.

Example 1.1.13 *Over* $[0,1]$, *let* $\mathbf{A}$ *be the sheaf with stalks* $\mathbf{A}_0 = \mathbb{Z}$, $\mathbf{A}_t = 0$, *for* $t > 0$. *For* $\mathbf{B}$, *take the constant sheaf with stalks* $\mathbb{Z}$ *on* $[0,1]$. *If* $U = [0, \epsilon)$, *then* $\Gamma(U; \mathbf{Hom}(\mathbf{A}, \mathbf{B})) = \mathrm{Hom}(\mathbf{A}|_U, \mathbf{B}|_U) = 0$, *and therefore* $\mathbf{Hom}(\mathbf{A}, \mathbf{B})_0 = 0$, *but* $\mathbf{A}_0 = \mathbb{Z} = \mathbf{B}_0$ *and so* $\mathrm{Hom}(\mathbf{A}_0, \mathbf{B}_0) \cong \mathbb{Z}$.

The following theorem shows that the usual methods of homological algebra apply to the category of sheaves.

Theorem 1.1.14 *$Sh(X)$ is an abelian category.*

Proof. There exists a zero object $\mathbf{0} \in Sh(X)$ with $\mathrm{Hom}_{Sh(X)}(\mathbf{0}, \mathbf{0}) = 0$, and for any $\mathbf{A}, \mathbf{B} \in Sh(X)$, $\mathrm{Hom}_{Sh(X)}(\mathbf{A}, \mathbf{B})$ has the structure of an abelian group such that the composition law is bilinear. Direct sums $\mathbf{A} \oplus \mathbf{B}$ exist and hence $Sh(X)$ is an additive category.

For any morphism $f : \mathbf{A} \to \mathbf{B}$, $\mathbf{ker}\, f$ and $\mathbf{coker}\, f = \mathbf{B}/\mathbf{im} f$ exist. Let $\mathbf{coim}\, f = \mathbf{A}/\mathbf{ker}\, f$. To check that the canonical morphism $\mathbf{coim}\, f \to \mathbf{im} f$ is an isomorphism, look at stalks and use $(\mathbf{coim}\, f)_x = (\mathbf{A}/\mathbf{ker}\, f)_x \cong \mathbf{A}_x/(\mathbf{ker}\, f)_x$. □

1.2 Sheaf Cohomology

We shall discuss sheaf cohomology from two perspectives: The first method will be to employ injective resolutions. This has the advantage of fitting well into the general context of derived categories and derived functors to be developed in Sect. 2.4, but has the disadvantage of being almost useless as a means to carry out calculations in practice. To provide a geometrically more transparent way to construct cycles in sheaf cohomology, our second method will be to study the Čech complex.

1.2.1 Injective Resolutions

Definition 1.2.1 *A* resolution $\mathbf{K}^\bullet$ *of a sheaf* $\mathbf{A}$ *is a collection of sheaves* $(\mathbf{K}^i)_{i \geq 0}$ *fitting into an exact sequence* $0 \to \mathbf{A} \xrightarrow{d^{-1}} \mathbf{K}^0 \xrightarrow{d^0} \mathbf{K}^1 \xrightarrow{d^1} \cdots$. *Given a morphism of sheaves* $f : \mathbf{A} \to \mathbf{B}$ *and a resolution* $\mathbf{L}^\bullet$ *of* $\mathbf{B}$, *a* morphism of resolutions $\phi : \mathbf{K}^\bullet \to \mathbf{L}^\bullet$ *is a collection* $(\phi^i)_{i \geq 0}$ *of sheaf morphisms* $\phi^i : \mathbf{K}^i \to \mathbf{L}^i$ *such that the diagram*

$$\begin{array}{ccccccccc}
0 & \longrightarrow & \mathbf{A} & \longrightarrow & \mathbf{K}^0 & \longrightarrow & \mathbf{K}^1 & \longrightarrow & \cdots \\
 & & \downarrow{\scriptstyle f} & & \downarrow{\scriptstyle \phi^0} & & \downarrow{\scriptstyle \phi^1} & & \\
0 & \longrightarrow & \mathbf{B} & \longrightarrow & \mathbf{L}^0 & \longrightarrow & \mathbf{L}^1 & \longrightarrow & \cdots
\end{array}$$

commutes. If $\phi, \psi : \mathbf{K}^\bullet \to \mathbf{L}^\bullet$ *are two morphisms of resolutions, then a* homotopy s from ϕ to ψ *is a collection of sheaf maps* $(s^i)_{i \geq 0}$, $s^i : \mathbf{K}^i \to \mathbf{L}^{i-1}$ *such that* $d_L^{i-1} s^i + s^{i+1} d_K^i = \phi^i - \psi^i$ *(with* $\mathbf{L}^{-1} := \mathbf{B}$*).*

Example 1.2.2 *Let M be a smooth manifold. To an open set $U \subset M$, associate the group $\Omega^i(U)$ of differential i-forms on U. The presheaf $U \mapsto \Omega^i(U)$ satisfies unique gluing of sections* (G) *and so is a sheaf* $\boldsymbol{\Omega}^i$ *on M that satisfies $\Gamma(U; \boldsymbol{\Omega}^i) = \Omega^i(U)$. Exterior derivation gives a map $d : \Omega^i(U) \to \Omega^{i+1}(U)$, which induces sheaf maps*

$$d : \boldsymbol{\Omega}^i \to \boldsymbol{\Omega}^{i+1},$$

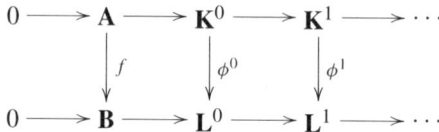

and we can form the de Rham complex

$$\boldsymbol{\Omega}^\bullet = \boldsymbol{\Omega}^0 \xrightarrow{d} \boldsymbol{\Omega}^1 \xrightarrow{d} \boldsymbol{\Omega}^2 \longrightarrow \cdots .$$

Let $\mathbb{R}_M$ be the constant sheaf with stalk $\mathbb{R}$ on M. Assigning to a real number $r \in (\mathbb{R}_M)_x$ the germ of the constant function r in $(\boldsymbol{\Omega}^0)_x$ defines a monomorphism $\mathbb{R}_M \hookrightarrow \boldsymbol{\Omega}^0$. By the Poincaré lemma ("every closed form on Euclidean space is exact"), the sequence

$$0 \longrightarrow \mathbb{R}_M \longrightarrow \boldsymbol{\Omega}^0 \xrightarrow{d} \boldsymbol{\Omega}^1 \xrightarrow{d} \boldsymbol{\Omega}^2 \longrightarrow \cdots$$

is exact. Thus the de Rham complex $\boldsymbol{\Omega}^\bullet$ is a resolution of the constant sheaf $\mathbb{R}_M$.

Definition 1.2.3 $\mathbf{I} \in Sh(X)$ *is* injective *if, whenever we are given a monomorphism $\mathbf{A} \hookrightarrow \mathbf{B}$ and a sheaf map $\mathbf{A} \to \mathbf{I}$, there exists an extension $\mathbf{B} \to \mathbf{I}$:*

$$\begin{array}{ccccc} 0 & \longrightarrow & \mathbf{A} & \longrightarrow & \mathbf{B} \\ & & \downarrow & \swarrow & \\ & & \mathbf{I} & & \end{array}$$

In other words: $\mathbf{I}$ is injective if and only if $\mathrm{Hom}(-, \mathbf{I})$ is exact.

Lemma 1.2.4 *Every sheaf $\mathbf{A} \in Sh(X)$ is a subsheaf of some injective sheaf.*

Proof. For each $x \in X$, let $\mathbf{A}_x \hookrightarrow I_x$ be the canonical embedding of the stalk of $\mathbf{A}$ into an injective abelian group I_x as described for instance in [God58] (if $\mathbf{A}$ were a sheaf of vector spaces, $\mathbf{A}_x$ would already be injective, so we could take $I_x = \mathbf{A}_x$). Regarding I_x as a sheaf on the one-point space $\{x\}$, let $\mathbf{I}(x) = j_{x*} I_x$, where j_x is the inclusion $j_x : \{x\} \hookrightarrow X$. If $\mathbf{B}$ is any sheaf, then restriction to stalks is an isomorphism $\mathrm{Hom}(\mathbf{B}, \mathbf{I}(x)) \xrightarrow{\cong} \mathrm{Hom}(\mathbf{B}_x, I_x)$, whence $\mathbf{I}(x)$ is an injective sheaf and we have a canonical morphism $\mathbf{A} \to \mathbf{I}(x)$. Let $\mathbf{I} = \prod_{x \in X} \mathbf{I}(x)$. The product of injective sheaves is injective, so the canonical map $\mathbf{A} \to \mathbf{I}$ is a monomorphism into an injective sheaf. □

One says that the category of sheaves $Sh(X)$ has "enough injectives." An *injective resolution* $\mathbf{I}^\bullet$ is a resolution such that each $\mathbf{I}^i$ is injective.

Proposition 1.2.5 *Every sheaf $\mathbf{A} \in Sh(X)$ has a canonical injective resolution.*

Proof. By induction: Construct $\mathbf{I}^0$ and $d^{-1} : \mathbf{A} \to \mathbf{I}^0$ invoking Lemma 1.2.4. Assuming $\mathbf{I}^i$ has been constructed, embed the cokernel $\mathbf{I}^i/\mathbf{im}(d^{i-1})$ into an injective $\mathbf{I}^{i+1}$ using Lemma 1.2.4, and let d^i be the composition $\mathbf{I}^i \twoheadrightarrow \mathbf{I}^i/\mathbf{im}(d^{i-1}) \hookrightarrow \mathbf{I}^{i+1}$. □

Proposition 1.2.6 *Let $f : \mathbf{A} \to \mathbf{B}$ be a morphism, $\mathbf{A} \to \mathbf{K}^\bullet$ a resolution and $\mathbf{B} \to \mathbf{I}^\bullet$ an injective resolution of $\mathbf{B}$. Then there exists a morphism of resolutions $\phi : \mathbf{K}^\bullet \to \mathbf{I}^\bullet$.*

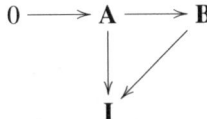

Proof. Use the injective property of $\mathbf{I}^0$ to find ϕ^0:

$$\begin{array}{ccccc} 0 & \longrightarrow & \mathbf{A} & \xrightarrow{d_A^{-1}} & \mathbf{K}^0 \\ & & & \searrow^{d_I^{-1} f} & \downarrow \phi^0 \\ & & & & \mathbf{I}^0 \end{array}$$

Assume ϕ^i has been constructed. Use the injective property of $\mathbf{I}^i$ to find ϕ^{i+1}:

$$\begin{array}{ccccc} 0 & \longrightarrow & \mathbf{K}^i/\mathbf{im}(d_K^{i-1}) = \mathbf{K}^i/\mathbf{ker}(d_K^i) & \xrightarrow{d_K^i} & \mathbf{K}^{i+1} \\ & & & \searrow^{d_I^i \phi^i} & \downarrow \phi^{i+1} \\ & & & & \mathbf{I}^{i+1} \end{array}$$

□

In fact, ϕ is unique up to homotopy by similar extension arguments.

Definition 1.2.7 *Let* **A** *be a sheaf of abelian groups on the topological space X. The* i-th sheaf cohomology group of X with coefficients in **A** *is defined to be*

$$H^i(X; \mathbf{A}) = H^i\Gamma(X; \mathbf{I}^\bullet)$$

where $\mathbf{A} \to \mathbf{I}^\bullet$ *is the canonical injective resolution of* **A** (*note that* $\mathbf{I}^\bullet$ *does* not *contain* **A** *itself*), *and* $\Gamma(X; \mathbf{I}^\bullet)$ *denotes the complex of global section groups induced by* $\mathbf{I}^\bullet$.

Since $\Gamma(X; -)$ is left exact, we have an exact sequence

$$0 \to \Gamma(X; \mathbf{A}) \to \Gamma(X; \mathbf{I}^0) \to \Gamma(X; \mathbf{I}^1).$$

Thus

$$H^0(X; \mathbf{A}) = \Gamma(X; \mathbf{A}).$$

A morphism $f : \mathbf{A} \to \mathbf{B}$ induces a map on cohomology as follows: Let $\mathbf{A} \to \mathbf{I}^\bullet$ be the canonical injective resolution of **A** and $\mathbf{B} \to \mathbf{J}^\bullet$ be the canonical injective resolution of **B**. By Proposition 1.2.6, there exists a morphism of resolutions $\phi : \mathbf{I}^\bullet \to \mathbf{J}^\bullet$, which induces maps $\Gamma(X; \mathbf{I}^\bullet) \to \Gamma(X; \mathbf{J}^\bullet)$ on the global section groups and so homomorphisms

$$H^i(X; \mathbf{A}) \to H^i(X; \mathbf{B}).$$

These homomorphisms are well-defined, since ϕ is unique up to homotopy.

If **A** is any sheaf, $\mathbf{I}^\bullet$ its canonical and $\mathbf{J}^\bullet$ any injective resolution, then the identity $\mathbf{A} \to \mathbf{A}$ induces a morphism of resolutions $\phi : \mathbf{I}^\bullet \to \mathbf{J}^\bullet$ as well as $\psi : \mathbf{J}^\bullet \to \mathbf{I}^\bullet$. By uniqueness up to homotopy, $\psi\phi$ is homotopic to the identity on $\mathbf{I}^\bullet$ and $\phi\psi$ is homotopic to the identity on $\mathbf{J}^\bullet$. Thus ϕ is a homotopy equivalence and the induced maps $H^i\Gamma(X; \mathbf{I}^\bullet) \overset{\cong}{\to} H^i\Gamma(X; \mathbf{J}^\bullet)$ are isomorphisms. Thus to calculate $H^i(X; \mathbf{A})$, we are free to choose any injective resolution of **A**.

Every short exact sequence of sheaves induces a long exact sequence on sheaf cohomology. We outline this process: Given the short exact sequence $0 \to \mathbf{A} \to \mathbf{B} \to \mathbf{C} \to 0$, choose injective resolutions $\mathbf{A} \to \mathbf{I}^\bullet$, $\mathbf{C} \to \mathbf{K}^\bullet$ (for example the canonical ones). It is not difficult to construct inductively an injective resolution $\mathbf{B} \to \mathbf{J}^\bullet$ (take $\mathbf{J}^i = \mathbf{I}^i \oplus \mathbf{K}^i$; the differential is of course not in general the direct sum) and morphisms $\phi : \mathbf{I}^\bullet \to \mathbf{J}^\bullet$, $\psi : \mathbf{J}^\bullet \to \mathbf{K}^\bullet$ such that

$$\begin{array}{ccccccccc} 0 & \longrightarrow & \mathbf{A} & \longrightarrow & \mathbf{B} & \longrightarrow & \mathbf{C} & \longrightarrow & 0 \\ & & \downarrow & & \downarrow & & \downarrow & & \\ 0 & \longrightarrow & \mathbf{I}^\bullet & \xrightarrow{\phi} & \mathbf{J}^\bullet & \xrightarrow{\psi} & \mathbf{K}^\bullet & \longrightarrow & 0 \end{array}$$

commutes and the bottom row is an exact sequence of resolutions, i.e. each $0 \to \mathbf{I}^i \stackrel{\phi^i}{\to} \mathbf{J}^i \stackrel{\psi^i}{\to} \mathbf{K}^i \to 0$ is short exact. Note that any short exact sequence $0 \to \mathbf{I} \to \mathbf{X} \to \mathbf{Y} \to 0$, with $\mathbf{I}$ injective, splits. Furthermore, any additive functor between abelian categories takes split exact sequences to split exact sequences. Now $\Gamma(X; -)$ is such an additive functor, hence $0 \to \Gamma(X; \mathbf{I}^i) \to \Gamma(X; \mathbf{J}^i) \to \Gamma(X; \mathbf{K}^i) \to 0$ is exact for all i. Then

$$0 \to \Gamma(X; \mathbf{I}^\bullet) \to \Gamma(X; \mathbf{J}^\bullet) \to \Gamma(X; \mathbf{K}^\bullet) \to 0$$

is a short exact sequence of complexes of abelian groups, and hence there exist connecting morphisms $\delta^i : H^i\Gamma(X; \mathbf{K}^\bullet) \to H^{i+1}\Gamma(X; \mathbf{I}^\bullet)$ such that the sequence

$$\cdots \to H^i\Gamma(X; \mathbf{I}^\bullet) \to H^i\Gamma(X; \mathbf{J}^\bullet) \to H^i\Gamma(X; \mathbf{K}^\bullet) \stackrel{\delta^i}{\to} H^{i+1}\Gamma(X; \mathbf{I}^\bullet) \to \cdots$$

is exact. This is the long exact sequence on sheaf cohomology:

$$\cdots \to H^i(X; \mathbf{A}) \to H^i(X; \mathbf{B}) \to H^i(X; \mathbf{C}) \stackrel{\delta^i}{\to} H^{i+1}(X; \mathbf{A}) \to \cdots .$$

Examples 1.2.8 *We give some applications of the long exact cohomology sequence to complex analysis. Let $X \subset \mathbb{C}$ be an open subset of the complex plane; we have $H^2(X; \mathbb{Z}) = 0$, as X is a noncompact two-dimensional manifold. Denote by $\mathcal{O}_X$ the sheaf of germs of holomorphic functions on X and by $\mathcal{O}_X^*$ the multiplicative sheaf of germs of nowhere vanishing holomorphic functions on X. Let us, for the moment, accept the fact that $H^i(X; \mathcal{O}_X) = 0$ for all $i > 0$. This fact will be explained in Example* 1.2.15 *below. We have a short exact exponential sequence*

$$0 \longrightarrow 2\pi\sqrt{-1}\mathbb{Z}_X \longrightarrow \mathcal{O}_X \stackrel{\exp}{\longrightarrow} \mathcal{O}_X^* \longrightarrow 0, \tag{1.3}$$

analogous to the one discussed earlier in Example 1.1.10. *Applying the induced long exact sequence, we can now prove the following statement: Every zero-free holomorphic function on X has a holomorphic logarithm if and only if $H^1(X; \mathbb{Z}) = 0$. Indeed,* (1.3) *induces*

$$H^0(X; \mathcal{O}_X) \to H^0(X; \mathcal{O}_X^*) \to H^1(X; \mathbb{Z}) \to H^1(X; \mathcal{O}_X) = 0,$$

so that $H^1(X; \mathbb{Z}) = 0$ is equivalent to $\exp : \Gamma(X; \mathcal{O}_X) \to \Gamma(X; \mathcal{O}_X^*)$ *being onto.*

Now let $\mathcal{M}_X^*$ denote the multiplicative sheaf of germs of not identically vanishing meromorphic functions on X; there is a natural embedding $\mathcal{O}_X^* \hookrightarrow \mathcal{M}_X^*$. To an open subset $U \subset X$, assign the additive group $\mathrm{Div}(U)$ *of functions $\nu : U \to \mathbb{Z}$ such that $\nu(x) \neq 0$ only on a discrete subset of U. An element $\nu \in \mathrm{Div}(U)$ is called a* divisor on U. *The presheaf*

$$U \mapsto \mathrm{Div}(U)$$

is a sheaf $\mathbf{Div}_X$ on X, $\Gamma(U; \mathbf{Div}_X) = \mathrm{Div}(U)$. If $f \in \Gamma(U; \mathcal{M}_X^*)$ is a meromorphic function, let $\nu_x(f)$ be its order at $x \in U$. Then

$$\begin{array}{ccc} \Gamma(U; \mathcal{M}_X^*) & \to & \Gamma(U; \mathbf{Div}_X) \\ f & \mapsto & \nu(f) \end{array}$$

defines a sheaf morphism

$$\mathcal{M}_X^* \longrightarrow \mathbf{Div}_X,$$

since $\nu_x(fg) = \nu_x(f) + \nu_x(g)$ and the order of a meromorphic function is nonzero only on a discrete set of points. As the divisor $\nu(f)$ of a meromorphic function f is zero precisely when f is nowhere vanishing holomorphic, we have an exact sequence

$$0 \longrightarrow \mathcal{O}_X^* \longrightarrow \mathcal{M}_X^* \longrightarrow \mathbf{Div}_X \longrightarrow 0 \tag{1.4}$$

(observe that locally we can achieve any divisor by writing down a Laurent series with the prescribed order). Using the exponential sequence (1.3), *we get an exact sequence*

$$H^1(X; \mathcal{O}_X) \longrightarrow H^1(X; \mathcal{O}_X^*) \longrightarrow H^2(X; \mathbb{Z}),$$

which shows that $H^1(X; \mathcal{O}_X^*) = 0$. The cohomology sequence associated to (1.4) *is*

$$H^0(X; \mathcal{M}_X^*) \longrightarrow H^0(X; \mathbf{Div}_X) \longrightarrow H^1(X; \mathcal{O}_X^*) = 0.$$

Thus we have proved Weierstrass' theorem: Given any divisor ν on X, there exists a meromorphic function on X with divisor ν.

Definition 1.2.9 *A sheaf $\mathbf{A} \in Sh(X)$ is* acyclic *if $H^i(X; \mathbf{A}) = 0$, all $i > 0$.*

Proposition 1.2.10 *Injective sheaves are acyclic.*

Proof. If $\mathbf{I}$ is an injective sheaf, then

$$0 \to \mathbf{I} \xrightarrow{1} \mathbf{I} \to 0 \to \cdots$$

is an injective resolution, so $H^i(X; \mathbf{I}) = 0$, $i > 0$. □

Proposition 1.2.11 *Let $\mathbf{K}^\bullet$ be a resolution of a sheaf $\mathbf{A}$ by acyclic sheaves. Then there is a natural isomorphism*

$$H^i\Gamma(X; \mathbf{K}^\bullet) \xrightarrow{\cong} H^i(X; \mathbf{A}).$$

Proof. Put $\mathbf{Z}^i = \mathbf{ker}(\mathbf{K}^i \to \mathbf{K}^{i+1})$. Note that we have $\Gamma(X;\mathbf{Z}^i) = \ker(\Gamma(X;\mathbf{K}^i) \to \Gamma(X;\mathbf{K}^{i+1}))$, and $\operatorname{im}(\Gamma(X;\mathbf{K}^{i-1}) \to \Gamma(X;\mathbf{K}^i)) = \operatorname{im}(\Gamma(X;\mathbf{K}^{i-1}) \to \Gamma(X;\mathbf{Z}^i))$. Hence

$$H^i\Gamma(X;\mathbf{K}^\bullet) = \Gamma(X;\mathbf{Z}^i)/\operatorname{im}(\Gamma(X;\mathbf{K}^{i-1}) \to \Gamma(X;\mathbf{Z}^i)).$$

For $i > 0$, the short exact sequence $0 \to \mathbf{Z}^{i-1} \to \mathbf{K}^{i-1} \to \mathbf{Z}^i \to 0$ induces the long exact cohomology sequence

$$\begin{aligned}0 \to \Gamma(X;\mathbf{Z}^{i-1}) \to \Gamma(X;\mathbf{K}^{i-1}) \to \Gamma(X;\mathbf{Z}^i) \xrightarrow{\delta} H^1(X;\mathbf{Z}^{i-1})\\ \to H^1(X;\mathbf{K}^{i-1}) = 0.\end{aligned}$$

The connecting homomorphism δ induces

$$\Gamma(X;\mathbf{Z}^i)/\operatorname{im}(\Gamma(X;\mathbf{K}^{i-1}) \to \Gamma(X;\mathbf{Z}^i)) \xrightarrow{\cong} H^1(X;\mathbf{Z}^{i-1}).$$

Summarizing, we have constructed an isomorphism

$$H^i\Gamma(X;\mathbf{K}^\bullet) \xrightarrow{\cong} H^1(X;\mathbf{Z}^{i-1}).$$

The short exact sequence $0 \to \mathbf{Z}^{i-2} \to \mathbf{K}^{i-2} \to \mathbf{Z}^{i-1} \to 0$ induces a long exact sequence on cohomology with

$$\delta : H^1(X;\mathbf{Z}^{i-1}) \xrightarrow{\cong} H^2(X;\mathbf{Z}^{i-2})$$

an isomorphism, since the neighboring terms vanish by acyclicity of $\mathbf{K}^\bullet$. Continuing, we obtain a sequence of isomorphisms

$$H^i\Gamma(X;\mathbf{K}^\bullet) \cong H^1(X;\mathbf{Z}^{i-1}) \cong \cdots \cong H^i(X;\mathbf{Z}^0),$$

but $\mathbf{Z}^0 = \mathbf{ker}(\mathbf{K}^0 \to \mathbf{K}^1) = \mathbf{im}(\mathbf{A} \to \mathbf{K}^0) \cong \mathbf{A}$. □

The point of Proposition 1.2.11 is that one can identify several classes of sheaves such as the so-called fine, soft or flabby sheaves which on the one hand can be proven to be acyclic (assume X is paracompact), and on the other hand occur more naturally than injective sheaves. Then the proposition tells us that to compute cohomology, we may as well use resolutions consisting of sheaves in that class instead of having to construct injective resolutions.

We shall in particular have occasion to consider the class of soft sheaves, so we define this concept here:

Definition 1.2.12 *Let X be a paracompact space. A sheaf $\mathbf{A} \in Sh(X)$ is* soft, *if the restriction map $\Gamma(X;\mathbf{A}) \to \Gamma(K;\mathbf{A})$ is surjective for all closed subsets $K \subset X$.*

The assumption of paracompactness is necessary to obtain the following

Proposition 1.2.13 *Soft sheaves are acyclic.*

For a proof consult e.g. [Bre97, Theorem II.9.11], or [Bry93, Theorem 1.4.6]. This result means that in order to calculate sheaf cohomology, we are allowed to resolve by soft sheaves. Two examples will illustrate this technique:

Example 1.2.14 *Let M be a (paracompact) smooth manifold and $\mathbb{R}_M \to \boldsymbol{\Omega}^\bullet$ the de Rham resolution introduced in Example 1.2.2. Every $\boldsymbol{\Omega}^i$ is a soft sheaf.*[1]

Thus to determine $H^i(M;\mathbb{R}_M)$, the sheaf cohomology of M with coefficients in the constant sheaf $\mathbb{R}_M$, we may use $\boldsymbol{\Omega}^\bullet$:

$$H^i(M;\mathbb{R}_M) \cong H^i\Gamma(M;\boldsymbol{\Omega}^\bullet) = H^i(\Omega^\bullet(M)) = H^i_{DR}(M;\mathbb{R}),$$

the de Rham cohomology groups of M.

Example 1.2.15 *Let $X \subset \mathbb{C}$ be a connected open subset of the complex plane. Our goal is to show $H^i(X;\mathcal{O}_X) = 0$ for $i > 0$, an assertion that we used in Example 1.2.8. Consider the linear partial differential operator*

$$\frac{\partial}{\partial \bar{z}} = \frac{1}{2}\left(\frac{\partial}{\partial x} + \sqrt{-1}\frac{\partial}{\partial y}\right).$$

The map $f \mapsto \partial f/\partial \bar{z}$ induces a morphism of sheaves $\bar{\partial} : \mathbf{C}_X \to \mathbf{C}_X$ (recall $\mathbf{C}_X$ is the sheaf of germs of smooth complex valued functions on X). Now f is holomorphic if and only if $\partial f/\partial \bar{z} = 0$ (Cauchy–Riemann equations). Thus we have an exact sequence

$$0 \longrightarrow \mathcal{O}_X \longrightarrow \mathbf{C}_X \xrightarrow{\bar{\partial}} \mathbf{C}_X.$$

We shall employ the following fact from analytic function theory [Gun66, §3, Theorem 4]:

(∗) *Let U be a connected, open subset of the complex plane. Given a smooth function $g : U \to \mathbb{C}$, there exists a smooth $f : U \to \mathbb{C}$ such that $\partial f/\partial \bar{z} = g$ on U.*

Applying (∗) to subsets $U \subset X$, we obtain the short exact sequence

$$0 \longrightarrow \mathcal{O}_X \longrightarrow \mathbf{C}_X \xrightarrow{\bar{\partial}} \mathbf{C}_X \longrightarrow 0. \tag{1.5}$$

Applying (∗) to $U = X$, we see that

$$\bar{\partial} : \Gamma(X;\mathbf{C}_X) \longrightarrow \Gamma(X;\mathbf{C}_X)$$

is onto. The sheaf $\mathbf{C}_X$ is soft, so (1.5) is a soft resolution of $\mathcal{O}_X$. We use it to calculate $H^i(X;\mathcal{O}_X)$:

$$H^1(X;\mathcal{O}_X) = \Gamma(X;\mathbf{C}_X)/\bar{\partial}\Gamma(X;\mathbf{C}_X) = 0,$$

and trivially $H^i(X;\mathcal{O}_X) = 0, i \geq 2$.

[1] One proves first that the sheaf $\boldsymbol{\Omega}^0$ of real valued C^∞-functions on M is soft. Therefore, any sheaf of $\boldsymbol{\Omega}^0$-modules on M is soft, in particular $\boldsymbol{\Omega}^i$.

1.2.2 Čech Cohomology

To provide an alternative view of sheaf cohomology, we shall define the Čech complex associated to an open cover. The advantage is that many geometric and analytic objects naturally give rise to a Čech cocycle. Čech cohomology groups of a space are obtained as the direct limit over all open covers of the groups associated to the covers. There is a canonical map from Čech cohomology to sheaf cohomology; it is an isomorphism for paracompact spaces.

Let $\mathcal{U} = \{U_i\}_{i\in I}$ be an open cover of the space X and let A be a presheaf on X. With $U_{i_0,\dots,i_p} = U_{i_0} \cap \cdots \cap U_{i_p}$ for $i_0, \dots, i_p \in I$, define $C^p(\mathcal{U}; A)$ to be the group of all functions s on I^{p+1} with values $s(i_0, \dots, i_p) \in A(U_{i_0,\dots,i_p})$. We define a coboundary homomorphism

$$d : C^p(\mathcal{U}; A) \to C^{p+1}(\mathcal{U}; A)$$

by

$$(ds)(i_0, \dots, i_{p+1}) = \sum_{j=0}^{p+1} (-1)^j s(i_0, \dots, \hat{\imath}_j, \dots, i_{p+1})|_{U_{i_0,\dots,i_{p+1}}}.$$

In applying d twice, each term appears twice and with opposite sign. Thus $d^2 = 0$. An element $s \in C^p(\mathcal{U}; A)$ is called a *Čech p-cochain* and is a *Čech p-cocycle* if $d(s) = 0$. The complex

$$\cdots \xrightarrow{d} C^p(\mathcal{U}; A) \xrightarrow{d} C^{p+1}(\mathcal{U}; A) \xrightarrow{d} \cdots$$

is the *Čech complex* $C^\bullet(\mathcal{U}; A)$ *of* $\mathcal{U}$ *with coefficients in* A.

Definition 1.2.16 *The* degree p Čech cohomology group *of the cover* $\mathcal{U}$ *with coefficients in the presheaf* A *is*

$$\check{H}^p(\mathcal{U}; A) = H^p C^\bullet(\mathcal{U}; A),$$

the degree p cohomology of the Čech complex of $\mathcal{U}$.

Proposition 1.2.17 *If the presheaf* A *satisfies unique gluing* (G), *then*

$$\check{H}^0(\mathcal{U}; A) = A(X).$$

Proof. An element $s \in \check{H}^0(\mathcal{U}; A) = \ker(d : C^0(\mathcal{U}; A) \to C^1(\mathcal{U}; A))$ is a family of sections

$$s(i) \in A(U_i), \quad i \in I,$$

such that

$$(ds)(i_0, i_1) = s(i_1)|_{U_{i_0,i_1}} - s(i_0)|_{U_{i_0,i_1}} = 0.$$

Thus the sections $s(i)$ agree on overlaps and condition (G) ensures the existence of a unique $\bar{s} \in A(X)$ with $\bar{s}|_{U_i} = s(i)$. □

If $\mathbf{A} \in Sh(X)$ is a sheaf, we use its canonical presheaf to define Čech cohomology:

$$\check{H}^p(\mathcal{U}; \mathbf{A}) = \check{H}^p(\mathcal{U}; (U \mapsto \Gamma(U; \mathbf{A}))).$$

Example 1.2.18 *Let Σ be a Riemann surface and $x_1, \ldots, x_n$ points on Σ together with given polar parts*

$$p_k(z) = \frac{a_{-r_k}^{(k)}}{(z - x_k)^{r_k}} + \cdots + \frac{a_{-2}^{(k)}}{(z - x_k)^2} + \frac{a_{-1}^{(k)}}{z - x_k}$$

at $x_k, k = 1, \ldots, n$, z a local coordinate. Given this data, Cousin's problem (in dimension 1) *asks: Does there exist a meromorphic function f on Σ having at most $x_1, \ldots, x_n$ as poles and with the prescribed polar parts p_k at x_k?*

Locally, of course, the problem can be solved: There exists a cover $\mathcal{U} = \{U_i\}_{i \in I}$ of Σ and meromorphic f_i on U_i such that each f_i has the required polar parts at $x_1, \ldots, x_n$. Then $s(i, j) = f_i|_{U_{i,j}} - f_j|_{U_{i,j}}$ is holomorphic on $U_{i,j}$, since the polar parts at all x_k cancel. Therefore s is a Čech 1*-cochain with coefficients in $\mathcal{O}_\Sigma$, the sheaf of germs of holomorphic functions on Σ. In fact, s is a cocycle, as*

$$\begin{aligned}(ds)(i_0, i_1, i_2) &= s(i_1, i_2) - s(i_0, i_2) + s(i_0, i_1) \\ &= f_{i_1} - f_{i_2} - (f_{i_0} - f_{i_2}) + f_{i_0} - f_{i_1} \\ &= 0.\end{aligned}$$

We shall show that Cousin's problem has a solution if and only if the cohomology class of s is zero in $\check{H}^1(\mathcal{U}; \mathcal{O}_\Sigma)$. Suppose the class of s is zero. There is a cochain $t \in C^0(\mathcal{U}; \mathcal{O}_\Sigma)$ with $dt = s$, i.e. we have holomorphic $t_i : U_i \to \mathbb{C}$ such that $t_i - t_j = s(i, j) = f_i - f_j$, hence $\{f_i - t_i\}_{i \in I}$ is a collection of meromorphic functions with correct polar parts which agree on overlaps and so define a global meromorphic function having polar part p_k at $x_k, k = 1, \ldots, n$.

Now assume that f is a solution to Cousin's problem. Then $t \in C^0(\mathcal{U}; \mathcal{O}_\Sigma)$ with $t(i) = f_i - f$ is a Čech 0*-cochain such that*

$$(dt)(i, j) = t(j) - t(i) = f_j - f - (f_i - f) = s(i, j).$$

Remark 1.2.19 *If Σ is an open Riemann surface, then $H^1(\Sigma; \mathcal{O}_\Sigma) = 0$.*

To define Čech cohomology of the space X, we put an order on the set of open covers $\mathcal{U}$ of X : $\mathcal{V} = \{V_j\}_{j \in J}$ is *finer* than $\mathcal{U} = \{U_i\}_{i \in I}$ if every V_j is contained in some U_i, that is, if there exists a function $h : J \to I$ with $V_j \subset U_{h(j)}$ for all $j \in J$. Such a function h induces a morphism of complexes

$$h_* : C^\bullet(\mathcal{U}; A) \to C^\bullet(\mathcal{V}; A)$$

by setting

$$(h_* s)(j_0, \ldots, j_p) = s(h(j_0), \ldots, h(j_p))|_{V_{j_0, \ldots, j_p}}.$$

If $h' : J \to I$ is another function with $V_j \subset U_{h'(j)}$, then h'_* and h_* are homotopic and so the induced map $h_* : \check{H}^p(\mathcal{U}; A) \to \check{H}^p(\mathcal{V}; A)$ is independent of h. Thus $\{\check{H}^p(\mathcal{U}; A)\}_{\mathcal{U}}$ is a direct system.

Definition 1.2.20 *The* Čech cohomology *of X with coefficients in a presheaf A is the direct limit*

$$\check{H}^p(X; A) = \varinjlim \check{H}^p(\mathcal{U}; A),$$

taken over the directed set of open covers $\mathcal{U}$ of X.

Again we set $\check{H}^p(X; \mathbf{A}) = \check{H}^p(X; (U \mapsto \Gamma(U; \mathbf{A})))$ for a sheaf $\mathbf{A}$ on X.

What is the relation between Čech cohomology $\check{H}(X; \mathbf{A})$ and sheaf cohomology $H(X; \mathbf{A})$? We shall first construct a natural map

$$\check{H}^p(X; \mathbf{A}) \longrightarrow H^p(X; \mathbf{A}).$$

Let $\mathcal{U} = \{U_i\}_{i \in I}$ be an open cover of X, and $V \subset X$ be an open set. Then $\mathcal{U}|_V := \{U_i \cap V\}_{i \in I}$ is an open cover of V and we can consider the group of Čech p-cochains over V, $C^p(\mathcal{U}|_V; \mathbf{A})$. The assignment

$$V \mapsto C^p(\mathcal{U}|_V; \mathbf{A})$$

defines a presheaf (with the natural restriction maps), which satisfies condition (G), and so is a sheaf $\mathbf{C}^p(\mathcal{U}; \mathbf{A})$. The homomorphisms

$$\begin{array}{ccc} \Gamma(V; \mathbf{A}) & \to & \Gamma(V; \mathbf{C}^0(\mathcal{U}; \mathbf{A})) = C^0(\mathcal{U}|_V; \mathbf{A}) \\ s & \mapsto & (i \mapsto s|_{U_i \cap V}) \end{array}$$

induce a monomorphism of sheaves $\mathbf{A} \hookrightarrow \mathbf{C}^0(\mathcal{U}; \mathbf{A})$. In fact, one can show (see e.g. [God58, Théorème 5.2.1]) that the sequence

$$0 \to \mathbf{A} \to \mathbf{C}^0(\mathcal{U}; \mathbf{A}) \xrightarrow{d} \mathbf{C}^1(\mathcal{U}; \mathbf{A}) \xrightarrow{d} \cdots$$

is exact. Thus $\mathbf{A} \to \mathbf{C}^\bullet(\mathcal{U}; \mathbf{A})$ is a resolution of the sheaf $\mathbf{A}$. Let $\mathbf{A} \to \mathbf{I}^\bullet$ be the canonical injective resolution of $\mathbf{A}$. By Proposition 1.2.6, there exists a morphism of resolutions

$$\phi : \mathbf{C}^\bullet(\mathcal{U}; \mathbf{A}) \longrightarrow \mathbf{I}^\bullet.$$

If $\mathcal{V} = \{V_j\}_{j \in J}$ is finer than $\mathcal{U}$, witnessed by $h : J \to I$, $V_j \subset U_{h(j)}$ all j, then for each $p \geq 0$ there is a morphism of sheaves

$$h_* : \mathbf{C}^p(\mathcal{U}; \mathbf{A}) \to \mathbf{C}^p(\mathcal{V}; \mathbf{A}),$$

induced by the maps

$$h_* : C^p(\mathcal{U}|_V; \mathbf{A}) \to C^p(\mathcal{V}|_V; \mathbf{A}),$$

for all $V \subset X$ open. That way we obtain a morphism of resolutions

$$h_* : \mathbf{C}^\bullet(\mathcal{U}; \mathbf{A}) \to \mathbf{C}^\bullet(\mathcal{V}; \mathbf{A}).$$

Using Proposition 1.2.6 once more provides a morphism of resolutions

$$\psi : \mathbf{C}^\bullet(\mathcal{V}; \mathbf{A}) \to \mathbf{I}^\bullet.$$

Now both ϕ and ψh_* are morphisms of resolutions from $\mathbf{C}^\bullet(\mathcal{U}; \mathbf{A})$ to $\mathbf{I}^\bullet$, so they are homotopic. Thus on cohomology we have a commutative diagram

$$\begin{array}{ccc} \check{H}^p(\mathcal{U}; \mathbf{A}) & \longrightarrow & \check{H}^p(\mathcal{V}; \mathbf{A}) \\ & \searrow \quad \swarrow & \\ & H^p(X; \mathbf{A}) & \end{array}$$

which defines a morphism from the direct system $\{\check{H}^p(\mathcal{U}; \mathbf{A})\}_{\mathcal{U}}$ to (the constant direct system) $H^p(X; \mathbf{A})$. This morphism induces a map on the direct limit

$$\check{H}^p(X; \mathbf{A}) = \varinjlim \check{H}^p(\mathcal{U}; \mathbf{A}) \longrightarrow H^p(X; \mathbf{A}),$$

which is the desired canonical map from Čech cohomology to sheaf cohomology. The main result is the following:

Theorem 1.2.21 *If X is paracompact, then the canonical map*

$$\check{H}^p(X; \mathbf{A}) \longrightarrow H^p(X; \mathbf{A})$$

is an isomorphism.

For a proof, see [God58, Théorème 5.10.1]. Roughly, one shows that if X is paracompact and A is a presheaf on X such that $\mathbf{Sheaf}(A) = 0$, then $\check{H}^p(X; A) = 0$ for all $p \geq 0$. Knowing this, one applies the "Čech cohomology spectral sequence" whose E_2-term is $E_2^{p,q} = \check{H}^p(X; (U \mapsto H^q(U; \mathbf{A})))$ and whose E_∞-term is the associated bigraded group of a suitable filtration of $H^\bullet(X; \mathbf{A})$, $\mathbf{A} \in Sh(X)$.

Example 1.2.22 (*Classification of complex line bundles*) *Let M be a smooth (paracompact) manifold. First, we describe how to interpret a smooth complex line bundle ξ over M, with total space $E(\xi)$ and projection $\pi : E(\xi) \to M$, as a cohomology class. Given such a bundle ξ, we can choose an open cover $\mathcal{U} = \{U_i\}_{i \in I}$ of M together with smooth trivializations $\phi_i : U_i \times \mathbb{C} \xrightarrow{\cong} \pi^{-1}(U_i)$. Over $U_{i,j} = U_i \cap U_j$, $\phi_j(x, \lambda) = \phi_i(x, g_{ij}(x)\lambda)$, for a unique smooth function $g_{ij} : U_{i,j} \to \mathbb{C}^*$. Over $U_{i,j,k}$, we obtain by uniqueness the relation $g_{ik} = g_{ij} g_{jk}$. The collection $g = \{g_{ij}\}_{i,j \in I}$ is called a* system of transition functions *for ξ.*

Call two systems of transition functions (over the same cover) $\{g_{ij}\}$ and $\{g'_{ij}\}$ equivalent *if there exist maps $f_i : U_i \to \mathbb{C}^*$, all $i \in I$, satisfying $g'_{ij}(x) = f_i(x)^{-1} g_{ij}(x) f_j(x)$. The key fact is: Suppose ξ, ξ' are two bundles trivialized over the same cover with systems of transition functions $\{g_{ij}\}$, $\{g'_{ij}\}$ respectively. Then ξ and ξ' are isomorphic if and only if $\{g_{ij}\}$ and $\{g'_{ij}\}$ are equivalent.*

We may regard g_{ij} as a section $g_{ij} \in \Gamma(U_{i,j}; \mathbf{C}^*_M)$ so that $g = \{g_{ij}\}$ is a Čech 1-cochain $g \in C^1(\mathcal{U}; \mathbf{C}^*_M)$. Moreover,

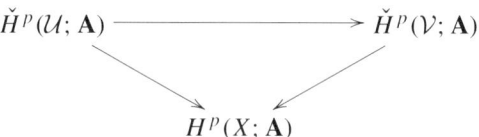

$$(dg)_{ijk} = g_{jk}g_{ik}^{-1}g_{ij} = 1$$

shows that g is a Čech cocycle and thus defines an element in $\check{H}^1(\mathcal{U}; \mathbf{C}^*_M)$. *We claim that the class of g in* $\check{H}^1(M; \mathbf{C}^*_M)$ *depends only on the isomorphism class of* ξ. *To see this, let* ξ' *be a line bundle isomorphic to* ξ, $\mathcal{V} = \{V_j\}_{j\in J}$ *an open cover together with trivializations of* ξ' *whose transition functions are* $g' = \{g'_{ij}\}_{i,j\in J}$. *Consider the common refinement* $\mathcal{U} \cap \mathcal{V} = \{U_i \cap V_j\}_{(i,j)\in I\times J}$ *with natural refining functions* $h : I \times J \to I$, $h' : I \times J \to J$. *Then* h_*g *is a system of transition functions for* ξ *over* $\mathcal{U} \cap \mathcal{V}$, *and* h'_*g' *is a system of transition functions for the isomorphic bundle* ξ' *over* $\mathcal{U} \cap \mathcal{V}$. *Hence* h_*g *and* h'_*g' *are equivalent and there exist functions* $f_{(i,j)} : U_i \cap V_j \to \mathbb{C}^*$ *with*

$$(h'_*g')_{kl} = f_k^{-1}(h_*g)_{kl}f_l,$$

all $k, l \in K := I \times J$. *This means that we have a Čech* 0*-cochain* $f = \{f_k\}_{k\in K} \in C^0(\mathcal{U} \cap \mathcal{V}; \mathbf{C}^*_M)$ *whose coboundary*

$$(df)_{kl} = f_l f_k^{-1} = (h'_*g')_{kl}(h_*g)_{kl}^{-1}$$

is the difference between h'_*g' *and* h_*g. *Thus* h'_*g' *and* h_*g *define the same class in* $\check{H}^1(\mathcal{U} \cap \mathcal{V}; \mathbf{C}^*_M)$, *which proves the claim.*

Now let $Pic^\infty(M)$ *be the set of isomorphism classes of smooth complex line bundles over* M. *Given* $\xi \in Pic^\infty(M)$, *write* $c_1(\xi)$ *for the class in* $\check{H}^1(M; \mathbf{C}^*_M)$ *of a system of transition functions for* ξ, *it is called the* first Chern class of ξ.

Then c_1 *is a well-defined function*

$$c_1 : Pic^\infty(M) \longrightarrow \check{H}^1(M; \mathbf{C}^*_M).$$

$Pic^\infty(M)$ *acquires the structure of an abelian group via the tensor product* $\xi \otimes \xi'$ *of two line bundles* ξ *and* ξ'. $Pic^\infty(M)$ *is the* smooth Picard group of M, *an analogue of the classical Picard group* $Pic(M)$ *of holomorphic line bundles over a complex manifold. If* $\{g_{ij}\}$, $\{g'_{ij}\}$ *are systems of transition functions over some cover* $\mathcal{U}$ *for* ξ, ξ' *respectively, then* $\{g_{ij}g'_{ij}\}$ *is a system of transition functions for* $\xi \otimes \xi'$. *Thus* c_1 *is a group homomorphism. We prove next that* c_1 *is indeed an isomorphism:*

Given $c_1(\xi) = 1 \in \check{H}^1(M; \mathbf{C}^*_M)$ *and representing* $c_1(\xi)$ *by* $g = \{g_{ij}\}$ *over* $\mathcal{U} = \{U_i\}$, *there exist* $f_i : U_i \to \mathbb{C}^*$ *such that* $df = g$, *i.e.* $g_{ij} = f_j f_i^{-1}$. *Now the trivial bundle has transition functions* $g'_{ij} = 1$ *for all* i, j. *Therefore,*

$$g_{ij} = f_j g'_{ij} f_i^{-1}$$

and ξ *is isomorphic to the trivial bundle. This proves* c_1 *to be injective.*

Given a class $g \in \check{H}^1(M; \mathbf{C}^*_M)$, *we represent it by* $\{g_{ij}\}$ *over* $\mathcal{U} = \{U_i\}_{i\in I}$ *and put*

$$E(\xi) = \left(\coprod_{i\in I} U_i \times \mathbb{C}\right) \Big/ \sim,$$

where $\sim$ *is the equivalence relation defined by* $(i, x, \lambda) \sim (j, y, \mu)$ *if* $x = y$ *and* $\mu = g_{ij}\lambda$. *Then* $E(\xi)$ *is the total space of a line bundle* ξ *with* $c_1(\xi) = g$, *verifying the surjectivity of* c_1.

Summarizing, the first Chern class is a group isomorphism

$$c_1 : Pic^{\infty}(M) \xrightarrow{\cong} \check{H}^1(M; \mathbf{C}^*_M) \cong H^1(M; \mathbf{C}^*_M)$$

(using that Čech and sheaf cohomology are isomorphic on paracompact spaces, Theorem 1.2.21*). The exponential sequence*

$$0 \to 2\pi\sqrt{-1}\mathbb{Z}_M \to \mathbf{C}_M \overset{\exp}{\to} \mathbf{C}^*_M \to 0$$

induces the exact sequence

$$H^1(M; \mathbf{C}_M) \to H^1(M; \mathbf{C}^*_M) \to H^2(M; \mathbb{Z}) \to H^2(M; \mathbf{C}_M).$$

Since $\mathbf{C}_M$ *is soft and soft sheaves are acyclic (Proposition* 1.2.13*), we have* $H^i(M; \mathbf{C}_M) = 0$, $i > 0$, *and thus*

$$H^1(M; \mathbf{C}^*_M) \xrightarrow{\cong} H^2(M; \mathbb{Z}).$$

Under this identification, we may regard the first Chern class as an isomorphism

$$c_1 : Pic^{\infty}(M) \xrightarrow{\cong} H^2(M; \mathbb{Z}).$$

Remark 1.2.23 *Analogous classifications exist for line bundles with more structure. For example, the Picard group* $Pic(M)$ *of holomorphic line bundles on a complex manifold can be identified with the sheaf cohomology group* $H^1(M; \mathcal{O}^*_M)$, *and the group of isomorphism classes of flat line bundles (*g_{ij} *locally constant functions) can be identified with* $H^1(M; \mathbb{C}^*_M)$, *with* $\mathbb{C}^*_M$ *the constant sheaf with stalk* $\mathbb{C}^*$ *on* M.

1.3 Complexes of Sheaves

Traditionally, complexes of abelian groups have been used in topology to collect algebraically encoded geometric information in various dimensions ("chain complex") and relate this information via maps ("boundary maps"). Sheaves on the other hand provide a systematic way to organize local data on a space into a single object, which is a particularly attractive theory on singular spaces, where the nonuniformity of the topology on the space prompts changing local data as well. To take into account coefficients that change over the space as well as multi-dimensional phenomena, it is thus natural to introduce complexes of sheaves.

Definition 1.3.1 *A* complex of sheaves (*or* differential graded sheaf) $\mathbf{A}^\bullet$ *on* X *is a collection of sheaves* $\mathbf{A}^i$, $i \in \mathbb{Z}$, *and morphisms* $d^i : \mathbf{A}^i \to \mathbf{A}^{i+1}$ *such that* $d^{i+1}d^i = 0$. *The* i-th cohomology sheaf (*or* derived sheaf) *of* $\mathbf{A}^\bullet$ *is*

$$\mathbf{H}^i(\mathbf{A}^\bullet) = \frac{\mathbf{ker}(d^i)}{\mathbf{im}(d^{i-1})} = \mathbf{Sheaf}(U \mapsto H^i\Gamma(U; \mathbf{A}^\bullet)).$$

Note that as restriction to stalks is an exact functor, we have $(x \in X)$

$$\mathbf{H}^i(\mathbf{A}^\bullet)_x \cong H^i(\mathbf{A}^\bullet_x).$$

A morphism of complexes $f : \mathbf{A}^\bullet \to \mathbf{B}^\bullet$ *is a collection of sheaf homomorphisms* $f^i : \mathbf{A}^i \to \mathbf{B}^i$, $i \in \mathbb{Z}$, *such that each square*

$$\begin{array}{ccc} \mathbf{A}^{i+1} & \xrightarrow{f^{i+1}} & \mathbf{B}^{i+1} \\ {\scriptstyle d_A^i}\uparrow & & \uparrow{\scriptstyle d_B^i} \\ \mathbf{A}^i & \xrightarrow{f^i} & \mathbf{B}^i \end{array}$$

commutes. A morphism f *of complexes induces sheaf maps* $\mathbf{H}^i(f) : \mathbf{H}^i(\mathbf{A}^\bullet) \to \mathbf{H}^i(\mathbf{B}^\bullet)$. *We call* f *a* quasi-isomorphism *if* $\mathbf{H}^i(f)$ *is an isomorphism for all* i.

If $\mathbf{A}$ is a sheaf, we may regard it as a complex

$$\cdots \to 0 \to 0 \overset{d^{-1}}{\to} \mathbf{A} \overset{d^0}{\to} 0 \to 0 \to \cdots$$

(put $\mathbf{A}$ in degree 0). A resolution $\mathbf{A} \to \mathbf{K}^\bullet$ of $\mathbf{A}$ may then be viewed as a quasi-isomorphism of complexes. Thus one generalizes the notion of a resolution as follows: A *resolution of a complex of sheaves* $\mathbf{A}^\bullet$ is a quasi-isomorphism $f : \mathbf{A}^\bullet \to \mathbf{K}^\bullet$. Sheaf complexes with the above notion of morphisms form a category $C(Sh(X))$ (we shall usually briefly write $C(X)$). With the obvious definition of kernel- and cokernel-complexes, $C(X)$ is easily seen to be an abelian category.

Our next objective will be to associate global cohomology groups to a complex of sheaves that agree with the sheaf cohomology as defined previously for a single sheaf when that sheaf is regarded as a complex concentrated in one dimension. To accomplish this, we first need a canonical injective resolution for a complex of sheaves. Such a resolution can be obtained as the simple complex associated to a suitable double complex, we review the relevant definitions:

Definition 1.3.2 *A* double complex of sheaves $\mathbf{A}^{\bullet\bullet}$ *is a collection of sheaves* $\mathbf{A}^{p,q}$, $p, q \in \mathbb{Z}$, *together with horizontal differentials* $d' : \mathbf{A}^{p,q} \to \mathbf{A}^{p+1,q}$ *and vertical differentials* $d'' : \mathbf{A}^{p,q} \to \mathbf{A}^{p,q+1}$ *such that* $d'^2 = 0$, $d''^2 = 0$ *and* $d'd'' + d''d' = 0$.

The associated simple complex (*or* total complex) $\mathbf{A}^\bullet$ *is given by*

$$\mathbf{A}^i = \bigoplus_{p+q=i} \mathbf{A}^{p,q}$$

and differential $d = d' + d''$.

Let us call a complex $\mathbf{A}^\bullet$ *bounded below*, if there is an n such that $\mathbf{A}^i = 0$ for $i < n$.

Proposition 1.3.3 *Every bounded below complex* $\mathbf{A}^\bullet$ *has a canonical injective resolution* $\mathbf{A}^\bullet \to \mathbf{I}^\bullet$, *where* $\mathbf{I}^\bullet$ *is the simple complex of a double complex* $\mathbf{I}^{\bullet\bullet}$ *such that for each* q, $\mathbf{I}^{\bullet,q}$ *is an injective resolution of* $\mathbf{A}^q$, $d''(\mathbf{I}^{\bullet,q-1})$ *is an injective resolution of* $d_A(\mathbf{A}^{q-1})$, $\mathbf{ker}(d'' : \mathbf{I}^{\bullet,q} \to \mathbf{I}^{\bullet,q+1})$ *is an injective resolution of* $\mathbf{ker}(d_A : \mathbf{A}^q \to \mathbf{A}^{q+1})$ *and* $\mathbf{H}^{\bullet,q}_{d''}(\mathbf{I}^{\bullet\bullet})$ *(the vertical cohomology) is an injective resolution of* $\mathbf{H}^q(\mathbf{A}^\bullet)$.

The properties of $\mathbf{I}^{\bullet\bullet}$ in the above proposition are carefully chosen so that one can conveniently use $\mathbf{I}^{\bullet\bullet}$ to obtain the so-called "hypercohomology spectral sequence," Proposition 1.3.5. For a proof of Proposition 1.3.3, see e.g. [Bry93, Proposition 1.2.6].

Definition 1.3.4 *The* hypercohomology groups $\mathcal{H}^i(X; \mathbf{A}^\bullet)$ of a space X with coefficients in a bounded below complex of sheaves $\mathbf{A}^\bullet$ *are*

$$\mathcal{H}^i(X; \mathbf{A}^\bullet) = H^i\Gamma(X; \mathbf{I}^\bullet),$$

where $\mathbf{A}^\bullet \to \mathbf{I}^\bullet$ *is the canonical injective resolution of Proposition* 1.3.3.

To handle hypercohomology, the following *hypercohomology spectral sequence* is frequently helpful:

Proposition 1.3.5 *Let* $\mathbf{A}^\bullet \in C(X)$ *be a bounded below complex of sheaves on* X. *There exists a convergent spectral sequence with*

$$E_2^{p,q} \cong H^p(X; \mathbf{H}^q(\mathbf{A}^\bullet))$$

and

$$E_\infty^{p,q} \cong G_p\mathcal{H}^{p+q}(X; \mathbf{A}^\bullet),$$

the graded quotients associated to a suitable filtration of the hypercohomology $\mathcal{H}(X; \mathbf{A}^\bullet)$.

Proof. Let $\mathbf{I}^{\bullet\bullet}$ be the injective double complex given by Proposition 1.3.3, whose simple complex $\mathbf{I}^\bullet$ is an injective resolution of $\mathbf{A}^\bullet$. Denote the horizontal differential by $d' : \mathbf{I}^{p,q} \to \mathbf{I}^{p+1,q}$ and the vertical differential by $d'' : \mathbf{I}^{p,q} \to \mathbf{I}^{p,q+1}$. Let $I^{\bullet\bullet} = \Gamma(X; \mathbf{I}^{\bullet\bullet})$ and $I^\bullet$ be the simple complex of $I^{\bullet\bullet}$. By definition, $\mathcal{H}^p(X; \mathbf{A}^\bullet) = H^p(I^\bullet)$. The "first filtration" F on $I^\bullet$ is $F_iI^n = \bigoplus_{p\geq i} I^{p,n-p}$. We consider the spectral sequence of the filtered complex $I^\bullet$ (i.e. the "first spectral sequence" of the double complex $I^{\bullet\bullet}$). As $\mathbf{I}^{\bullet\bullet}$ satisfies $\mathbf{I}^{p,q} = 0$ for all $p < 0$ and there exists an integer l such that $\mathbf{I}^{p,q} = 0$ whenever $q < l$ (recall $\mathbf{A}^\bullet$ is bounded below), this spectral sequence is convergent (in the strong sense), i.e. for each p, q there exists an r such that

$$E_r^{p,q} \cong E_{r+1}^{p,q} \cong \cdots \cong E_\infty^{p,q}.$$

We calculate the E_1-term:

$$E_1^{p,q} = H^q(F_pI^\bullet/F_{p+1}I^\bullet) = H^q\Big(\bigoplus_{i\geq p} I^{i,\bullet} \Big/ \bigoplus_{i\geq p+1} I^{i,\bullet}\Big) = H^q_{d''}(I^{p,\bullet}).$$

Let $\mathbf{Z}^{\bullet,q} = \mathbf{ker}(d'' : \mathbf{I}^{\bullet,q} \to \mathbf{I}^{\bullet,q+1})$, $\mathbf{B}^{\bullet,q+1} = \mathbf{I}^{\bullet,q}/\mathbf{Z}^{\bullet,q}$, $\mathbf{H}^{\bullet,q} = \mathbf{Z}^{\bullet,q}/\mathbf{B}^{\bullet,q}$. By the properties of $\mathbf{I}^{\bullet\bullet}$, $\mathbf{Z}^{\bullet,q}$ is an injective resolution of $\mathbf{ker}(d_A : \mathbf{A}^q \to \mathbf{A}^{q+1})$, $\mathbf{B}^{\bullet,q}$ is an injective resolution of $\mathbf{im}(d_A : \mathbf{A}^{q-1} \to \mathbf{A}^q)$, and $\mathbf{H}^{\bullet,q}$ is an injective resolution of $\mathbf{H}^q(\mathbf{A}^\bullet)$.

Now

$$\Gamma(X; \mathbf{H}^{p,q}) = \Gamma(X; \mathbf{Z}^{p,q}/\mathbf{B}^{p,q}) = \Gamma(X; \mathbf{Z}^{p,q})/\Gamma(X; \mathbf{B}^{p,q}),$$

since the involved sheaves are injective. As for the sections of the kernel-sheaf,

$$\Gamma(X; \mathbf{Z}^{p,q}) = \ker(d'' : \Gamma(X; \mathbf{I}^{p,q}) \to \Gamma(X; \mathbf{I}^{p,q+1})) = \ker(d'' : I^{p,q} \to I^{p,q+1}).$$

The sections of the image sheaves are

$$\begin{aligned} \Gamma(X; \mathbf{B}^{p,q+1}) &= \Gamma(X; \mathbf{I}^{p,q}/\mathbf{Z}^{p,q}) = \Gamma(X; \mathbf{I}^{p,q})/\Gamma(X; \mathbf{Z}^{p,q}) \\ &= I^{p,q}/\ker(d'' : I^{p,q} \to I^{p,q+1}) = d''(I^{p,q}). \end{aligned}$$

Hence,

$$E_1^{p,q} = H^q_{d''}(I^{p,\bullet}) = \ker(d'' : I^{p,q} \to I^{p,q+1})/d''(I^{p,q-1}) = \Gamma(X; \mathbf{H}^{p,q}).$$

Let us move on to the E_2-term. The differential $d_1 : E_1^{p,q} \to E_1^{p+1,q}$ is the map on vertical cohomology induced by the horizontal differential d'. Under the above identification, $d_1 : \Gamma(X; \mathbf{H}^{p,q}) \to \Gamma(X; \mathbf{H}^{p+1,q})$ is thus induced by $d' : \mathbf{H}^{p,q} \to \mathbf{H}^{p+1,q}$. But the complex $\mathbf{H}^{\bullet,q}$ is an injective resolution of $\mathbf{H}^q(\mathbf{A}^\bullet)$. Therefore,

$$E_2^{p,q} = H_{d_1}(E_1^{\bullet\bullet}) = H^p_{d'}\Gamma(X; \mathbf{H}^{\bullet,q}) = H^p(X; \mathbf{H}^q(\mathbf{A}^\bullet)).$$

As $\mathbf{I}^\bullet$ is regularly filtered, $E_\infty^{p,q} \cong G_pH^{p+q}(I^\bullet)$, the graded module $GH(I^\bullet)$ associated to the filtration $F_pH^n(I^\bullet) = \mathrm{im}(H^n(F_pI^\bullet) \to H^n(I^\bullet))$. That is (using $H^n(I^\bullet) = \mathcal{H}^n(X; \mathbf{A}^\bullet)$),

$$E_\infty^{p,q} \cong G_p\mathcal{H}^{p+q}(X; \mathbf{A}^\bullet) = \frac{F_p\mathcal{H}^{p+q}(X; \mathbf{A}^\bullet)}{F_{p+1}\mathcal{H}^{p+q}(X; \mathbf{A}^\bullet)}. \qquad \square$$

As an application of the spectral sequence, it follows that a quasi-isomorphism $f : \mathbf{A}^\bullet \to \mathbf{B}^\bullet$ induces isomorphisms $\mathcal{H}^i(X; \mathbf{A}^\bullet) \cong \mathcal{H}^i(X; \mathbf{B}^\bullet)$ on hypercohomology.

Let us introduce some further operations on complexes. Let $n \in \mathbb{Z}$. The *shift functor* $[n] : C(X) \to C(X)$ assigns to a complex $\mathbf{A}^\bullet \in C(X)$ the complex $\mathbf{A}^\bullet[n]$ defined by

$$\begin{aligned} (\mathbf{A}^\bullet[n])^i &= \mathbf{A}^{i+n} \\ d^i_{\mathbf{A}[n]} &= (-1)^n d^{i+n}_{\mathbf{A}}. \end{aligned}$$

If $f : \mathbf{A}^\bullet \to \mathbf{B}^\bullet$ is a morphism of complexes, then $f[n] : \mathbf{A}^\bullet[n] \to \mathbf{B}^\bullet[n]$ is $f[n]^i = f^{n+i}$. The *truncation functors* $\tau_{\le n}, \tau_{\ge n} : C(X) \to C(X)$ are given by

$$\tau_{\le n}\mathbf{A}^\bullet = \cdots \to \mathbf{A}^{n-2} \to \mathbf{A}^{n-1} \to \mathbf{ker}\, d^n \to 0 \to 0 \to \cdots$$

and

$$\tau_{\geq n}\mathbf{A}^\bullet = \cdots \to 0 \to 0 \to \mathbf{coker}\, d^{n-1} \to \mathbf{A}^{n+1} \to \mathbf{A}^{n+2} \to \cdots .$$

Induced morphisms $\tau_{\leq n} f$, $\tau_{\geq n} f$ are obtained in the obvious way. These truncations are designed to satisfy

$$\mathbf{H}^i(\tau_{\leq n}\mathbf{A}^\bullet) = \begin{cases} \mathbf{H}^i(\mathbf{A}^\bullet), & \text{for } i \leq n, \\ 0, & \text{for } i > n \end{cases}$$

and

$$\mathbf{H}^i(\tau_{\geq n}\mathbf{A}^\bullet) = \begin{cases} 0, & \text{for } i < n, \\ \mathbf{H}^i(\mathbf{A}^\bullet), & \text{for } i \geq n. \end{cases}$$

They will play a crucial role in constructing the intersection chain complex, see Sect. 4.1. Note that

$$\tau_{\leq n}\tau_{\leq m}\mathbf{A}^\bullet = \tau_{\leq \min(n,m)}\mathbf{A}^\bullet$$

and

$$\tau_{\leq n}(\mathbf{A}^\bullet[m]) = (\tau_{\leq n+m}\mathbf{A}^\bullet)[m].$$

Moreover, if $f : \mathbf{A}^\bullet \to \mathbf{B}^\bullet$ is a morphism of complexes such that $\mathbf{H}^i(f) : \mathbf{H}^i(\mathbf{A}^\bullet) \to \mathbf{H}^i(\mathbf{B}^\bullet)$ is an isomorphism for $i \leq n$, then $\tau_{\leq n} f : \tau_{\leq n}\mathbf{A}^\bullet \to \tau_{\leq n}\mathbf{B}^\bullet$ is a quasi-isomorphism.

The *direct sum* $\mathbf{A}^\bullet \oplus \mathbf{B}^\bullet$ of complexes $\mathbf{A}^\bullet$ and $\mathbf{B}^\bullet$ is defined by taking direct sums in each degree and taking the differential to be the sum of the differentials in $\mathbf{A}^\bullet$ and $\mathbf{B}^\bullet$. For $\mathbf{A}^\bullet, \mathbf{B}^\bullet \in C(X)$, the collection $\{\mathbf{A}^p \otimes \mathbf{B}^q\}_{p,q\in\mathbb{Z}}$ becomes a double complex if we define horizontal differentials

$$d' = d_{\mathbf{A}} \otimes 1 : \mathbf{A}^p \otimes \mathbf{B}^q \longrightarrow \mathbf{A}^{p+1} \otimes \mathbf{B}^q$$

and vertical differentials

$$d'' = (-1)^p 1 \otimes d_{\mathbf{B}} : \mathbf{A}^p \otimes \mathbf{B}^q \longrightarrow \mathbf{A}^p \otimes \mathbf{B}^{q+1}.$$

(The sign in d'' guarantees $d''d' + d'd'' = 0$.) Then the *tensor product* $\mathbf{A}^\bullet \otimes \mathbf{B}^\bullet \in C(X)$ is by definition the simple complex associated to the double complex $(\{\mathbf{A}^p \otimes \mathbf{B}^q\}_{p,q}, d', d'')$.

The *complex of homomorphisms*, $\mathbf{Hom}^\bullet(\mathbf{A}^\bullet, \mathbf{B}^\bullet) \in C(X)$, is constructed as follows: In degree n, take

$$\mathbf{Hom}^n(\mathbf{A}^\bullet, \mathbf{B}^\bullet) = \prod_{p\in\mathbb{Z}} \mathbf{Hom}(\mathbf{A}^p, \mathbf{B}^{p+n}).$$

We shall describe the differentials $d^n : \mathbf{Hom}^n(\mathbf{A}^\bullet, \mathbf{B}^\bullet) \to \mathbf{Hom}^{n+1}(\mathbf{A}^\bullet, \mathbf{B}^\bullet)$ by their operation on sections

$$\Gamma(U; \mathbf{Hom}^n(\mathbf{A}^\bullet, \mathbf{B}^\bullet)) \xrightarrow{d^n} \Gamma(U; \mathbf{Hom}^{n+1}(\mathbf{A}^\bullet, \mathbf{B}^\bullet))$$

($U \subset X$ open).

Note that

$$\begin{aligned}\Gamma(U;\mathbf{Hom}^n(\mathbf{A}^\bullet,\mathbf{B}^\bullet)) &= \Gamma\Big(U;\prod_p \mathbf{Hom}(\mathbf{A}^p,\mathbf{B}^{p+n})\Big)\\ &\cong \prod_p \Gamma(U;\mathbf{Hom}(\mathbf{A}^p,\mathbf{B}^{p+n}))\\ &\cong \prod_p \mathrm{Hom}(\mathbf{A}^p|_U,\mathbf{B}^{p+n}|_U),\end{aligned}$$

so we need a map

$$\prod_p \mathrm{Hom}(\mathbf{A}^p|_U,\mathbf{B}^{p+n}|_U) \xrightarrow{d^n} \prod_p \mathrm{Hom}(\mathbf{A}^p|_U,\mathbf{B}^{p+n+1}|_U).$$

If $\{f^p\}_{p\in\mathbb{Z}} \in \prod_p \mathrm{Hom}(\mathbf{A}^p|_U,\mathbf{B}^{p+n}|_U)$, let

$$d^n(f^p) = d_{\mathbf{B}}^{p+n} \circ f^p + (-1)^{n+1} f^{p+1} \circ d_{\mathbf{A}}^p. \tag{1.6}$$

Remark 1.3.6 $\mathbf{Hom}^\bullet(\mathbf{A}^\bullet,\mathbf{B}^\bullet)$ *can be defined from a double complex* $\mathbf{Hom}^{\bullet\bullet}$ *with* $\mathbf{Hom}^{p,q} = \mathbf{Hom}(\mathbf{A}^{-p},\mathbf{B}^q)$ *and appropriate differentials, but in taking the simple complex one must use direct products instead of direct sums (the "second total complex"). It is necessary to use direct products as then the basic adjointness relation*

$$\mathrm{Hom}(\mathbf{A}^\bullet \otimes \mathbf{B}^\bullet, \mathbf{C}^\bullet) \cong \mathrm{Hom}(\mathbf{A}^\bullet, \mathbf{Hom}^\bullet(\mathbf{B}^\bullet,\mathbf{C}^\bullet))$$

holds.

Remark 1.3.7 *By* (1.6), *the n-cycles of* $\Gamma(X;\mathbf{Hom}^\bullet(\mathbf{A}^\bullet,\mathbf{B}^\bullet))$ *are in one-to-one correspondence with morphisms of complexes* $\mathbf{A}^\bullet \to \mathbf{B}^\bullet[n]$, *and the n-boundaries correspond to morphisms which are homotopic to zero. Thus,* $H^n\Gamma(X;\mathbf{Hom}^\bullet(\mathbf{A}^\bullet,\mathbf{B}^\bullet))$ *can be identified with the group of homotopy classes of morphisms from* $\mathbf{A}^\bullet$ *to* $\mathbf{B}^\bullet[n]$.

If X and Y are topological spaces, $f : X \to Y$ a continuous map and $\mathbf{A}^\bullet \in C(X)$, $\mathbf{B}^\bullet \in C(Y)$ are complexes of sheaves, then we define the *direct image complex* $f_*\mathbf{A}^\bullet$ by

$$\begin{aligned}(f_*\mathbf{A}^\bullet)^n &= f_*(\mathbf{A}^n)\\ d^n_{f_*\mathbf{A}^\bullet} &= f_*(d^n_{\mathbf{A}^\bullet})\end{aligned}$$

and the *pullback complex* $f^*\mathbf{B}^\bullet$ by

$$\begin{aligned}(f^*\mathbf{B}^\bullet)^n &= f^*(\mathbf{B}^n)\\ d^n_{f^*\mathbf{B}^\bullet} &= f^*(d^n_{\mathbf{B}^\bullet}).\end{aligned}$$

As f^* is an exact functor, we have

$$\mathbf{H}^n(f^*\mathbf{B}^\bullet) \cong f^*\mathbf{H}^n(\mathbf{B}^\bullet).$$

Furthermore, pullback commutes with truncation:

$$\tau_{\le n} f^*\mathbf{B}^\bullet \cong f^*\tau_{\le n}\mathbf{B}^\bullet.$$

2

Homological Algebra

2.1 The Homotopy Category of Complexes

Frequently, diagrams involving morphisms of complexes commute only up to homotopy. It is then convenient to devise a category in which such diagrams give rise to commutative diagrams and where homotopy equivalences become isomorphisms. The construction is applicable to complexes of objects in any abelian category A, and we will describe it in this generality first. Later on, we will be mostly concerned with the case $A = Sh(X)$.

Let A be an abelian category and $C(A)$ its category of differential complexes. Recall the following

Definition 2.1.1 *Two morphisms $f, g : X^\bullet \to Y^\bullet$, $X^\bullet, Y^\bullet \in C(A)$, are* homotopic *if there exists a collection $\{s^i\}_{i\in\mathbb{Z}}$ of maps $s^i : X^i \to Y^{i-1}$* (a homotopy), *such that $d_Y^{i-1}s^i + s^{i+1}d_X^i = f^i - g^i$ for all $i \in \mathbb{Z}$.*

Let us write $[f]$ for the homotopy class of f.

Definition 2.1.2 *The* homotopy category of A, $K(A)$, *is defined by letting*

$$ObK(A) = ObC(A),$$
$$\mathrm{Hom}_{K(A)}(X^\bullet, Y^\bullet) = \{[f] : f \in \mathrm{Hom}_{C(A)}(X^\bullet, Y^\bullet)\}.$$

Note that the composition of morphisms $[f] \circ [g] = [f \circ g]$ is well-defined as compositions of homotopic maps are homotopic. Moreover, the sum $[f] + [g] = [f + g]$ of morphisms is well-defined and $K(A)$ is an additive category.

If f is a homotopy equivalence, then $[f]$ is invertible in $K(A)$, with inverse $[g]$, where g is a homotopy inverse of f. This advantage comes at a cost: $K(A)$ is not an abelian category anymore!

Example 2.1.3 *Take A to be the category of abelian groups and let $F : C(A) \longrightarrow K(A)$, $F(X^\bullet) = X^\bullet$, $F(f) = [f]$, be the quotient functor. We will show that $K(A)$ is not an abelian category such that F is an exact functor. Consider the following complexes:*

$$X^\bullet = \cdots \to 0 \to 0 \to \mathbb{Z} \to 0 \to \cdots$$

$$Y^\bullet = \cdots \to 0 \to \mathbb{Z} \xrightarrow{1} \mathbb{Z} \to 0 \to \cdots$$

$$Z^\bullet = \cdots \to 0 \to \mathbb{Z} \to 0 \to 0 \to \cdots$$

We have a short exact sequence

$$\begin{array}{ccccccc}
0 & & & & 0 & & 0 \\
\downarrow & & & & \downarrow & & \downarrow \\
X^\bullet = & \cdots & \longrightarrow & & 0 & \longrightarrow & \mathbb{Z} & \longrightarrow \cdots \\
\downarrow{\scriptstyle i} & & & & \downarrow & & \downarrow{\scriptstyle 1} \\
Y^\bullet = & \cdots & \longrightarrow & & \mathbb{Z} & \xrightarrow{1} & \mathbb{Z} & \longrightarrow \cdots \\
\downarrow{\scriptstyle j} & & & & \downarrow{\scriptstyle 1} & & \downarrow \\
Z^\bullet = & \cdots & \longrightarrow & & \mathbb{Z} & \longrightarrow & 0 & \longrightarrow \cdots \\
\downarrow & & & & \downarrow & & \downarrow \\
0 & & & & 0 & & 0
\end{array}$$

Now the zero map $Y^\bullet \to 0^\bullet$ from $Y^\bullet$ to the zero complex $0^\bullet$ is a homotopy equivalence since the zero map $Y^\bullet \overset{0}{\to} Y^\bullet$ is homotopic to the identity:

$$\begin{array}{ccccccccccc}
Y^\bullet & \longrightarrow & 0 & \longrightarrow & \mathbb{Z} & \xrightarrow{1} & \mathbb{Z} & \longrightarrow & 0 & \longrightarrow \\
{\scriptstyle 1}\downdownarrows{\scriptstyle 0} & & {\scriptstyle 1}\downdownarrows{\scriptstyle 0} & \swarrow{\scriptstyle 0} & {\scriptstyle 1}\downdownarrows{\scriptstyle 0} & \swarrow{\scriptstyle 1} & {\scriptstyle 1}\downdownarrows{\scriptstyle 0} & \swarrow{\scriptstyle 0} & {\scriptstyle 1}\downdownarrows{\scriptstyle 0} & \\
Y^\bullet & \longrightarrow & 0 & \longrightarrow & \mathbb{Z} & \xrightarrow{1} & \mathbb{Z} & \longrightarrow & 0 & \longrightarrow
\end{array}$$

is a homotopy. Thus $F(Y^\bullet) \cong F(0^\bullet) = 0^\bullet$. If $K(A)$ were abelian and F exact, then the short exact sequence

$$0 \to F(X^\bullet) \xrightarrow{F(i)} F(Y^\bullet) \xrightarrow{F(j)} F(Z^\bullet) \to 0$$

would imply

$$F(X^\bullet) \cong \operatorname{im} F(i) \cong \ker F(j) \cong 0^\bullet.$$

However, cohomology is a homotopy invariant and $H^\bullet(F(X^\bullet)) = H^\bullet(X^\bullet) \neq 0$, a contradiction.

Consequently, we have to look for an effective substitute for short exact sequences. Such a substitute should still have the property that it induces long exact sequences on cohomology. The concept of a triangulated category with its "distinguished triangles" provides a solution.

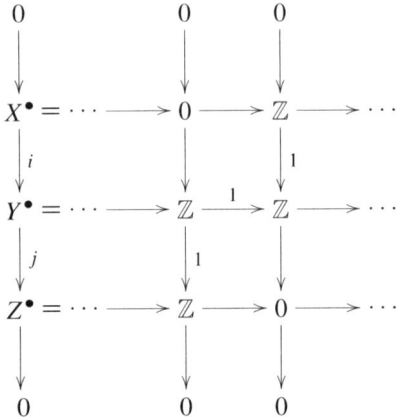

2.2 Triangulated Categories

Let K be an additive category with an automorphism[1] $T : K \to K$. A *triangle* is a sextuple (X, Y, Z, u, v, w), where $X, Y, Z \in ObK$ and $u : X \to Y$, $v : Y \to Z$, $w : Z \to T(X)$ are morphisms. For a triangulated category, one singles out collections of triangles that satisfy certain axioms. The members of such a collection are then to be called *distinguished triangles*. A *morphism of triangles* $(X, Y, Z, u, v, w) \to (X', Y', Z', u', v', w')$ is a commutative diagram

$$\begin{array}{ccccccc} X & \xrightarrow{u} & Y & \xrightarrow{v} & Z & \xrightarrow{w} & T(X) \\ \downarrow & & \downarrow & & \downarrow & & \downarrow \\ X' & \xrightarrow{u'} & Y' & \xrightarrow{v'} & Z' & \xrightarrow{w'} & T(X') \end{array}$$

Definition 2.2.1 *A* triangulated category *is an additive category K, together with*

1. *an automorphism $T : K \to K$, called the* translation functor, *and*
2. *a collection of triangles $\{(X, Y, Z, u, v, w)\}$, called the* distinguished triangles, *satisfying the following axioms*:
   (TR0): *Any triangle (X, Y, Z, u, v, w) isomorphic to a distinguished triangle is distinguished.*
   (TR1): *The triangle $(X, X, 0, 1_X, 0, 0)$ is distinguished for all $X \in ObK$.*
   (TR2): *Every morphism $u : X \to Y$ can be embedded in a distinguished triangle (X, Y, Z, u, v, w).*
   (TR3): Turning Triangles: *(X, Y, Z, u, v, w) is distinguished if and only if $(Y, Z, T(X), v, w, -T(u))$ is distinguished.*
   (TR4): *Given two distinguished triangles (X, Y, Z, u, v, w) and (X', Y', Z', u', v', w'), a commutative square*

$$\begin{array}{ccc} X & \xrightarrow{u} & Y \\ \downarrow & & \downarrow \\ X' & \xrightarrow{u'} & Y' \end{array}$$

   can be embedded in a morphism of triangles ($Z \to Z'$ need not be unique in general, however).
   (TR5): *The* octahedral axiom, *to be explained below.*

We shall also denote triangles (X, Y, Z, u, v, w) as

$$X \overset{u}{\to} Y \overset{v}{\to} Z \overset{w}{\to} T(X), \qquad X \to Y \to Z \overset{[1]}{\to}$$

[1] Our prime example will be $K = K(A)$, the homotopy category of complexes, and $T = [1]$, the shift on complexes.

or

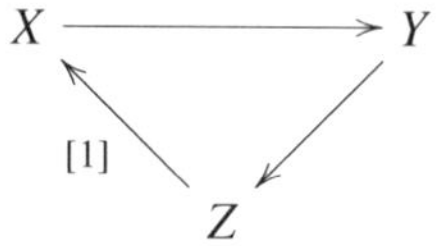

where the [1] indicates the morphism having as target the translated object ($T(X)$ in this case).

The *octahedral axiom* (TR5) requires that for three given distinguished triangles

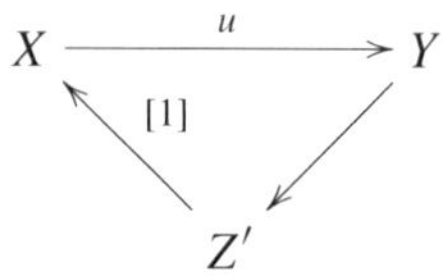

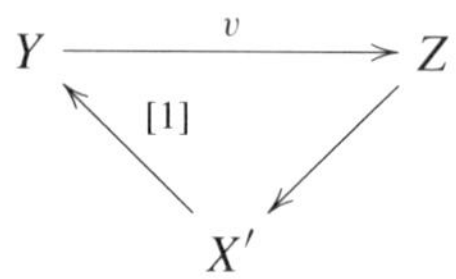

and

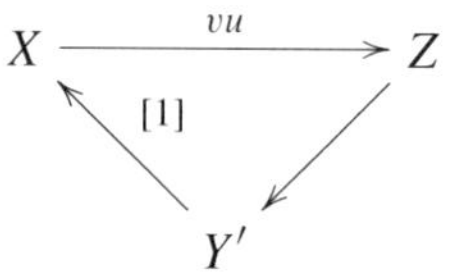

there exists a distinguished triangle

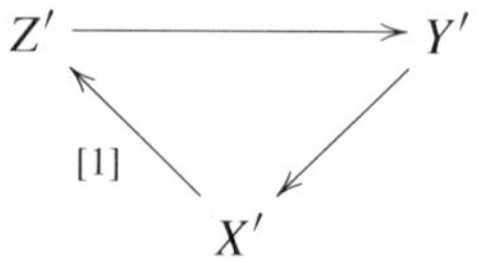

which completes the *octahedral diagram*

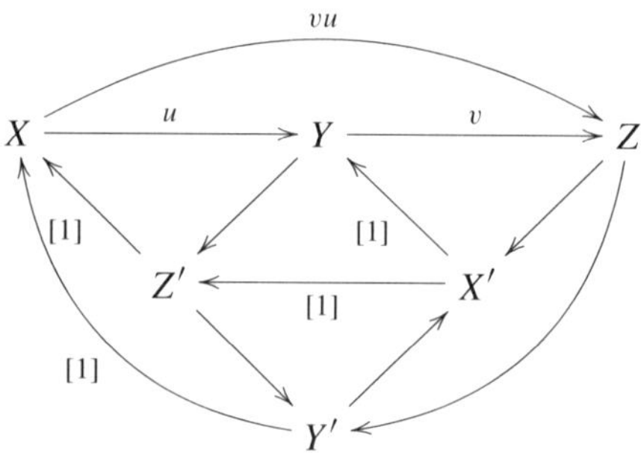

such that the following commute:

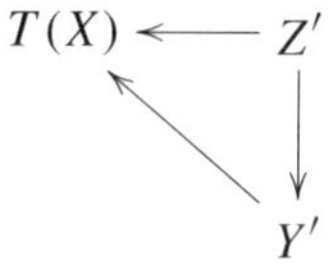

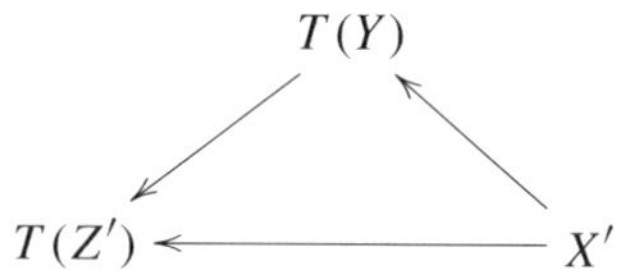

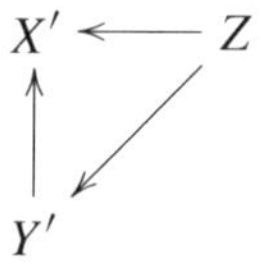

as well as the two squares

$$\begin{array}{ccc} T(X) & \xrightarrow{T(u)} & T(Y) \\ \uparrow & & \uparrow \\ Y' & \longrightarrow & X' \end{array}$$

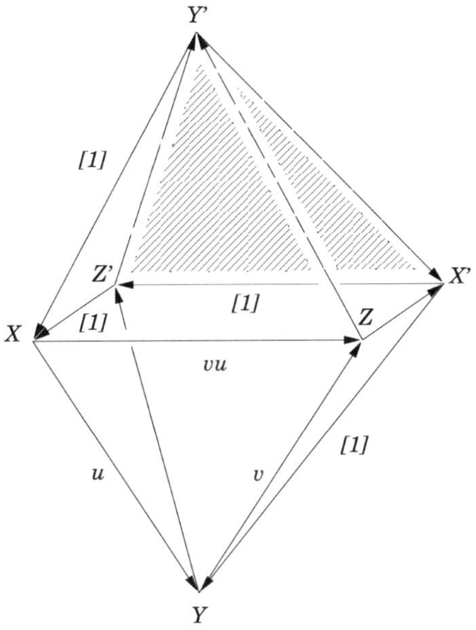

Fig. 2.1. The octahedral axiom

and

$$\begin{array}{ccc} Y & \xrightarrow{v} & Z \\ \downarrow & & \downarrow \\ Z' & \longrightarrow & Y' \end{array}$$

The name of the axiom derives from the fact that the graph of the above diagram is the graph of an octahedron, see Fig. 2.1.

Proposition 2.2.2 *If* $X \xrightarrow{u} Y \xrightarrow{v} Z \xrightarrow{[1]}$ *is a distinguished triangle, then* $vu = 0$.

Proof. The triangle $X \xrightarrow{1} X \longrightarrow 0 \xrightarrow{[1]}$ is distinguished according to (TR1). Using (TR4), there is a morphism $0 \to Z$ such that the diagram

$$\begin{array}{ccccccc} X & \xrightarrow{1} & X & \longrightarrow & 0 & \xrightarrow{[1]} & \\ {\scriptstyle 1}\downarrow & & {\scriptstyle u}\downarrow & & \downarrow & & \\ X & \xrightarrow{u} & Y & \xrightarrow{v} & Z & \xrightarrow{[1]} & \end{array}$$

commutes. □

Definition 2.2.3 *Let* K *and* K' *be triangulated categories. An additive functor* $F : K \to K'$ *is called a* functor of triangulated categories *if it commutes with the translation automorphisms and takes distinguished triangles in* K *to distinguished triangles in* K'.

Definition 2.2.4 *Let K be a triangulated category and A an abelian category. An additive functor $F : K \to A$ is called a* cohomological functor *if applying F to a distinguished triangle $X \to Y \to Z \to T(X)$ induces an exact sequence $F(X) \to F(Y) \to F(Z)$.*

Writing $F^i = F \circ T^i$ and using (TR3) repeatedly, note that a cohomological functor induces in fact a long exact sequence

$$\cdots \to F^{i-1}(Z) \to F^i(X) \to F^i(Y) \to F^i(Z) \to F^{i+1}(X) \to \cdots .$$

Example 2.2.5 *Applying axioms* (TR1), (TR3) *and* (TR4), *one sees that for any $X \in ObK$, $\mathrm{Hom}_K(X, -)$ and $\mathrm{Hom}_K(-, X)$ are cohomological functors. This in turn has the following useful corollary: If*

$$\begin{array}{ccccccc} X & \longrightarrow & Y & \longrightarrow & Z & \longrightarrow & T(X) \\ \downarrow & & \downarrow & & \downarrow & & \downarrow \\ X' & \longrightarrow & Y' & \longrightarrow & Z' & \longrightarrow & T(X') \end{array}$$

is a morphism of distinguished triangles with $X \to X'$ and $Y \to Y'$ being isomorphisms, then $Z \to Z'$ is an isomorphism as well: For any $P \in ObK$, apply $\mathrm{Hom}_K(P, -)$ to the diagram and use the five-lemma.

The following lemma is another straightforward consequence of the axioms for a triangulated category (see for example [KS90]). We will use it for instance in building the theory of t-structures, Sect. 7.1.

Lemma 2.2.6 *Let K be a triangulated category and*

$$X \xrightarrow{f} Y \xrightarrow{g} Z \overset{h_1}{\underset{h_2}{\rightrightarrows}} X[1]$$

be two distinguished triangles. If $\mathrm{Hom}_K(X[1], Z) = 0$, then $h_1 = h_2$.

2.3 The Triangulation of the Homotopy Category

Let us return to the homotopy category $K(A)$ of an abelian category A. As we have seen in Sect. 2.1, $K(A)$ is not an abelian category; but it turns out that it can be endowed with the structure of a triangulated category. To do this, we need to specify a translation automorphism T and a collection of distinguished triangles. Let

$$T = [1],$$

the shift on complexes.

Given $X^\bullet, Y^\bullet \in ObC(A)$ and a morphism of complexes $u : X^\bullet \to Y^\bullet$, define the *mapping cone of u*, $C^\bullet(u) \in ObC(A)$, by

$$C^n(u) = X^{n+1} \oplus Y^n,$$
$$d^n_{C(u)} = \begin{pmatrix} d^n_{X[1]} & 0 \\ u^{n+1} & d^n_Y \end{pmatrix}.$$

We have the natural inclusion $i : Y^\bullet \to C^\bullet(u)$ and projection $p : C^\bullet(u) \to X^\bullet[1]$. The triangle

$$X^\bullet \xrightarrow{u} Y^\bullet \xrightarrow{i} C^\bullet(u) \xrightarrow{p} X^\bullet[1]$$

is called a *standard triangle*. Let Δ be the collection of all triangles in $K(A)$ isomorphic to standard triangles.

Theorem 2.3.1 *The homotopy category $K(A)$ together with the shift functor $[1] : K(A) \to K(A)$ and Δ as the collection of distinguished triangles forms a triangulated category.*

Proof. The closure property (TR0) is clear from the definition of Δ. Embedding a given $u : X^\bullet \to Y^\bullet$ in the standard triangle $X^\bullet \xrightarrow{u} Y^\bullet \longrightarrow C^\bullet(u) \longrightarrow X^\bullet[1]$ proves (TR2).

To prove (TR3), we may assume that the given triangle is standard $X^\bullet \xrightarrow{u} Y^\bullet \xrightarrow{i} C^\bullet(u) \xrightarrow{p} X^\bullet[1]$. We have to see that the turned triangle $Y^\bullet \xrightarrow{i} C^\bullet(u) \xrightarrow{p} X^\bullet[1] \xrightarrow{-u[1]} Y^\bullet[1]$ is isomorphic to a standard one in $K(A)$. Consider the mapping cone $C^\bullet(i)$ of i with the natural maps $C^\bullet(u) \xrightarrow{i'} C^\bullet(i) \xrightarrow{p'} Y^\bullet[1]$. Define a morphism

$$f^n : X^\bullet[1]^n = X^{n+1} \longrightarrow C^n(i) = Y^{n+1} \oplus C^n(u) = Y^{n+1} \oplus X^{n+1} \oplus Y^n$$

by

$$f^n = \begin{pmatrix} -u^{n+1} \\ 1_{X^{n+1}} \\ 0 \end{pmatrix}.$$

Then $f = \{f^n\} : X^\bullet[1] \to C^\bullet(i)$ is a morphism of complexes and the diagram

$$\begin{array}{ccccccc} Y^\bullet & \xrightarrow{i} & C^\bullet(u) & \xrightarrow{p} & X^\bullet[1] & \xrightarrow{-u[1]} & Y^\bullet[1] \\ \| & & \| & & \downarrow f & & \| \\ Y^\bullet & \xrightarrow{i} & C^\bullet(u) & \xrightarrow{i'} & C^\bullet(i) & \xrightarrow{p'} & Y^\bullet[1] \end{array}$$

commutes. The morphisms

$$g^n : C^n(i) \longrightarrow X^\bullet[1]^n$$

defined by

$$g^n = (0, 1_{X^{n+1}}, 0)$$

form a morphism of complexes $g = \{g^n\} : C^\bullet(i) \longrightarrow X^\bullet[1]$. Now

$$g^n \circ f^n = (0, 1_{X^{n+1}}, 0) \begin{pmatrix} -u^{n+1} \\ 1_{X^{n+1}} \\ 0 \end{pmatrix} = 1_{X^{n+1}}$$

and so $gf = 1_{X^\bullet[1]}$. Defining

$$s^n : C^n(i) = Y^{n+1} \oplus X^{n+1} \oplus Y^n \longrightarrow Y^n \oplus X^n \oplus Y^{n-1} = C^{n-1}(i),$$

$$s^n = \begin{pmatrix} 0 & 0 & 1_{Y^n} \\ 0 & 0 & 0 \\ 0 & 0 & 0 \end{pmatrix}$$

we calculate

$$\begin{aligned}
&s^{n+1} \circ d^n_{C(i)} + d^{n-1}_{C(i)} \circ s^n \\
&= \begin{pmatrix} 0 & 0 & 1_{Y^{n+1}} \\ 0 & 0 & 0 \\ 0 & 0 & 0 \end{pmatrix} \begin{pmatrix} d^n_{Y[1]} & 0 & 0 \\ 0 & d^n_{X[1]} & 0 \\ 1_{Y^{n+1}} & u^{n+1} & d^n_Y \end{pmatrix} \\
&\quad + \begin{pmatrix} d^{n-1}_{Y[1]} & 0 & 0 \\ 0 & d^{n-1}_{X[1]} & 0 \\ 1_{Y^n} & u^n & d^{n-1}_Y \end{pmatrix} \begin{pmatrix} 0 & 0 & 1_{Y^n} \\ 0 & 0 & 0 \\ 0 & 0 & 0 \end{pmatrix} \\
&= \begin{pmatrix} 1_{Y^{n+1}} & u^{n+1} & d^n_Y \\ 0 & 0 & 0 \\ 0 & 0 & 0 \end{pmatrix} + \begin{pmatrix} 0 & 0 & d^{n-1}_{Y[1]} \\ 0 & 0 & 0 \\ 0 & 0 & 1_{Y^n} \end{pmatrix} \\
&= \begin{pmatrix} 1_{Y^{n+1}} & u^{n+1} & 0 \\ 0 & 0 & 0 \\ 0 & 0 & 1_{Y^n} \end{pmatrix} \\
&= \begin{pmatrix} 1_{Y^{n+1}} & 0 & 0 \\ 0 & 1_{X^{n+1}} & 0 \\ 0 & 0 & 1_{Y^n} \end{pmatrix} - \begin{pmatrix} -u^{n+1} \\ 1_{X^{n+1}} \\ 0 \end{pmatrix} (0, 1_{X^{n+1}}, 0) \\
&= 1_{C(i)^n} - f^n \circ g^n.
\end{aligned}$$

Thus fg is homotopic to $1_{C(i)}$ and f is an isomorphism in $K(A)$. (Note that this would not hold in $C(A)$.) Axiom (TR1) follows readily: As the mapping cone of $0^\bullet \to X^\bullet$ is $X^\bullet$, the triangle $0^\bullet \longrightarrow X^\bullet \xrightarrow{1_X} X^\bullet \longrightarrow 0^\bullet[1]$ is distinguished. Use (TR3) to see that $X^\bullet \xrightarrow{1_X} X^\bullet \longrightarrow 0^\bullet \xrightarrow{[1]}$ is distinguished (this shows in particular that the mapping cone of the identity map is homotopy equivalent to the zero complex).

We prove (TR4): We may assume that the given triangles are standard, say

$$X^\bullet \xrightarrow{u} Y^\bullet \xrightarrow{i} C^\bullet(u) \xrightarrow{p} X^\bullet[1]$$

and

$$W^\bullet \xrightarrow{v} Z^\bullet \xrightarrow{i'} C^\bullet(v) \xrightarrow{p'} W^\bullet[1].$$

For a commutative square in $K(A)$,

$$\begin{array}{ccc} X^\bullet & \xrightarrow{u} & Y^\bullet \\ f\downarrow & & \downarrow g \\ W^\bullet & \xrightarrow{v} & Z^\bullet \end{array}$$

there exists a homotopy $\{s^n\}$, $s^n : X^n \to Z^{n-1}$, from vf to gu, i.e. $g^n \circ u^n - v^n \circ f^n = s^{n+1} \circ d_X^n + d_Z^{n-1} \circ s^n$. Define

$$h^n : C^n(u) = X^{n+1} \oplus Y^n \longrightarrow C^n(v) = W^{n+1} \oplus Z^n$$

by

$$h^n = \begin{pmatrix} f^{n+1} & 0 \\ s^{n+1} & g^n \end{pmatrix}.$$

Then $h = \{h^n\} : C^\bullet(u) \to C^\bullet(v)$ is a morphism of complexes and the squares

$$\begin{array}{ccc} Y^\bullet & \xrightarrow{i} & C^\bullet(u) \\ g\downarrow & & \downarrow h \\ Z^\bullet & \xrightarrow{i'} & C^\bullet(v) \end{array}$$

and

$$\begin{array}{ccc} C^\bullet(u) & \xrightarrow{p} & X^\bullet[1] \\ h\downarrow & & \downarrow f[1] \\ C^\bullet(v) & \xrightarrow{p'} & W^\bullet[1] \end{array}$$

commute (already in $C(A)$).

It remains to verify the octahedral axiom (TR5). We assume that all triangles are standard, that is, $Z'^\bullet = C^\bullet(u)$, $X'^\bullet = C^\bullet(v)$, $Y'^\bullet = C^\bullet(vu)$. We define the triangle

$$Z'^\bullet \xrightarrow{f} Y'^\bullet \xrightarrow{g} X'^\bullet \xrightarrow{h} Z'^\bullet[1] \tag{2.1}$$

as follows: Let $f^n : C^n(u) = X^{n+1} \oplus Y^n \to C^n(vu) = X^{n+1} \oplus Z^n$ be given by

$$f^n = \begin{pmatrix} 1_{X^{n+1}} & 0 \\ 0 & v^n \end{pmatrix},$$

let $g^n : C^n(vu) = X^{n+1} \oplus Z^n \to C^n(v) = Y^{n+1} \oplus Z^n$ be given by

$$g^n = \begin{pmatrix} u^{n+1} & 0 \\ 0 & 1_{Z^n} \end{pmatrix},$$

and let $h : C^\bullet(v) \to C^\bullet(u)[1]$ be the composition

$$C^\bullet(v) \xrightarrow{p} Y^\bullet[1] \xrightarrow{i} C^\bullet(u)[1].$$

Then all the commutativity relations in the octahedron are satisfied. Proving triangle (2.1) to be distinguished will complete the proof: We shall exhibit a homotopy equivalence $k : C^\bullet(f) \to X'^\bullet$ (an isomorphism in $K(A)$) such that the diagram

$$\begin{array}{ccccccc} Z'^\bullet & \xrightarrow{f} & Y'^\bullet & \xrightarrow{i'} & C^\bullet(f) & \xrightarrow{p'} & Z'^\bullet[1] \\ \| & & \| & & \downarrow k & & \| \\ Z'^\bullet & \xrightarrow{f} & Y'^\bullet & \xrightarrow{g} & X'^\bullet & \xrightarrow{h} & Z'^\bullet[1] \end{array}$$

commutes. Define

$$\begin{array}{ll} C^n(f) & = C^{n+1}(u) \oplus C^n(vu) = X^{n+2} \oplus Y^{n+1} \oplus X^{n+1} \oplus Z^n \\ {\scriptstyle k^n}\downarrow & \\ X'^n & = C^n(v) = Y^{n+1} \oplus Z^n \end{array}$$

by

$$k^n = \begin{pmatrix} 0 & 1_{Y^{n+1}} & u^{n+1} & 0 \\ 0 & 0 & 0 & 1_{Z^n} \end{pmatrix}.$$

Then $ki' = g$. The morphism $l : X'^\bullet \to C^\bullet(f)$,

$$l^n = \begin{pmatrix} 0 & 0 \\ 1_{Y^{n+1}} & 0 \\ 0 & 0 \\ 0 & 1_{X^{n+1}} \end{pmatrix}$$

is a homotopy inverse for k with $p'l = h$. □

On the category $C(A)$, we can consider the functor

$$\begin{aligned} H^0(-) : C(A) &\to A \\ X^\bullet &\mapsto \ker(d_X^0)/\operatorname{im}(d_X^{-1}). \end{aligned}$$

We may indeed regard $H^0(-)$ as a well-defined functor on $K(A)$, because homotopic maps of complexes induce the same map on cohomology.

Proposition 2.3.2 *$H^0(-) : K(A) \longrightarrow A$ is a cohomological functor.*

Proof. We have to show that given a standard triangle

$$X^\bullet \xrightarrow{u} Y^\bullet \longrightarrow C^\bullet(u) \longrightarrow X^\bullet[1],$$

applying $H^0(-)$ induces an exact sequence

$$H^0(Y^\bullet) \longrightarrow H^0(C^\bullet(u)) \longrightarrow H^0(X^\bullet[1]).$$

Now in $C(A)$ the sequence

$$0 \longrightarrow Y^\bullet \longrightarrow C^\bullet(u) \longrightarrow X^\bullet[1] \longrightarrow 0$$

is short exact, so that the result follows from standard homological algebra. □

Remark 2.3.3 *If we write $H^n = H^0 \circ [n]$, we obtain as in Sect. 2.2 a long exact sequence*

$$\cdots \to H^{n-1}(C^\bullet(u)) \to H^n(X^\bullet) \to H^n(Y^\bullet) \to H^n(C^\bullet(u)) \\ \to H^{n+1}(X^\bullet) \to \cdots .$$

Note however that to obtain $\delta^n : H^n(C^\bullet(u)) \to H^{n+1}(X^\bullet)$, no connecting morphism has to be constructed as $\delta^n = H^n(p)$ is directly induced by the morphism $p : C^\bullet(u) \to X^\bullet[1]$ in $C(A)$. This is an advantage of distinguished triangles over exact sequences.

If we specialize to $A = Sh(X)$, the category of sheaves on the topological space X, then Proposition 2.3.2 asserts that given a distinguished triangle of complexes of sheaves

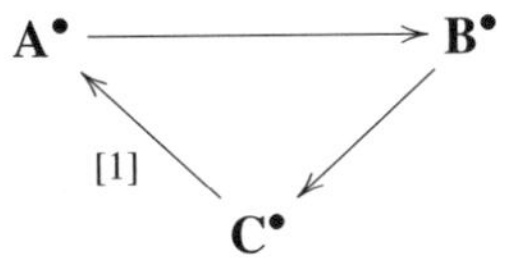

there is an induced long exact sequence of derived sheaves

$$\cdots \to \mathbf{H}^n(\mathbf{A}^\bullet) \to \mathbf{H}^n(\mathbf{B}^\bullet) \to \mathbf{H}^n(\mathbf{C}^\bullet) \to \mathbf{H}^{n+1}(\mathbf{A}^\bullet) \to \cdots .$$

Denote the homotopy category of complexes of sheaves on X by $K(X) = K(Sh(X))$. Given bounded below $\mathbf{A}^\bullet, \mathbf{B}^\bullet \in C(X)$ and two homotopic morphisms in $\mathrm{Hom}_{C(X)}(\mathbf{A}^\bullet, \mathbf{B}^\bullet)$, the hypercohomology spectral sequence (Proposition 1.3.5) implies that the induced maps $\mathcal{H}(X; \mathbf{A}^\bullet) \to \mathcal{H}(X; \mathbf{B}^\bullet)$ are equal. Thus hypercohomology $\mathcal{H}(X; -)$ is a well-defined functor on $K^+(X)$, where $K^+(X)$ denotes the homotopy category of bounded below complexes.

Proposition 2.3.4 *Hypercohomology $\mathcal{H}(X; -)$ is a cohomological functor on $K^+(X)$.*

Proof. We have to see that applying $\mathcal{H}(X; -)$ to a standard triangle

$$\mathbf{X}^\bullet \xrightarrow{u} \mathbf{Y}^\bullet \longrightarrow \mathbf{C}^\bullet(u) \to \mathbf{X}^\bullet[1]$$

yields an exact sequence

$$\mathcal{H}^0(X;\mathbf{Y}^\bullet) \longrightarrow \mathcal{H}^0(X;\mathbf{C}^\bullet(u)) \longrightarrow \mathcal{H}^0(X;\mathbf{X}^\bullet[1]).$$

Choose injective resolutions $\mathbf{X}^\bullet \to \mathbf{I}^\bullet$ and $\mathbf{Y}^\bullet \to \mathbf{J}^\bullet$. Let $u' : \mathbf{I}^\bullet \to \mathbf{J}^\bullet$ be a morphism completing the commutative square

$$\begin{array}{ccc} \mathbf{X}^\bullet & \longrightarrow & \mathbf{I}^\bullet \\ {\scriptstyle u}\downarrow & & \downarrow{\scriptstyle u'} \\ \mathbf{Y}^\bullet & \longrightarrow & \mathbf{J}^\bullet \end{array}$$

and consider the short exact sequence in $C(X)$:

$$0 \longrightarrow \mathbf{J}^\bullet \longrightarrow \mathbf{C}^\bullet(u') \longrightarrow \mathbf{I}^\bullet[1] \longrightarrow 0. \tag{2.2}$$

Note that $\mathbf{C}^\bullet(u')$ is an injective resolution of $\mathbf{C}^\bullet(u)$. As $\Gamma(X;-)$ is exact on injective sheaves, taking global sections yields a short exact sequence

$$0 \longrightarrow \Gamma(X;\mathbf{J}^\bullet) \longrightarrow \Gamma(X;\mathbf{C}^\bullet(u')) \longrightarrow \Gamma(X;\mathbf{I}^\bullet[1]) \longrightarrow 0.$$

By standard homological algebra,

$$H^0\Gamma(X;\mathbf{J}^\bullet) \longrightarrow H^0\Gamma(X;\mathbf{C}^\bullet(u')) \longrightarrow H^0\Gamma(X;\mathbf{I}^\bullet[1])$$

is exact. □

We return to the general situation, with A any abelian category. We ask the following question: To what extent does a short exact sequence in $C(A)$ give rise to an associated distinguished triangle in $K(A)$? Given the exact sequence in $C(A)$

$$0 \longrightarrow X^\bullet \xrightarrow{u} Y^\bullet \xrightarrow{v} Z^\bullet \longrightarrow 0, \tag{2.3}$$

we can always form the standard triangle

$$X^\bullet \xrightarrow{u} Y^\bullet \longrightarrow C^\bullet(u) \longrightarrow X^\bullet[1]. \tag{2.4}$$

The question is: What is the relation between $C^\bullet(u)$ and $Z^\bullet$? There is an obvious morphism $f : C^\bullet(u) \to Z^\bullet$ given by

$$\begin{aligned} f^n : C^n(u) = X^{n+1} \oplus Y^n &\longrightarrow Z^n, \\ f^n &= (0, v^n). \end{aligned}$$

Looking at the long exact sequences on cohomology induced by (2.3) and (2.4) and using the five-lemma, we see that f is a quasi-isomorphism. Moreover, it can be shown that if (2.3) splits, then f is a homotopy equivalence.

Thus in the case where (2.3) is split, we can replace $C^\bullet(u)$ in (2.4) by the isomorphic (in $K(A)$) $Z^\bullet$ to get a distinguished triangle

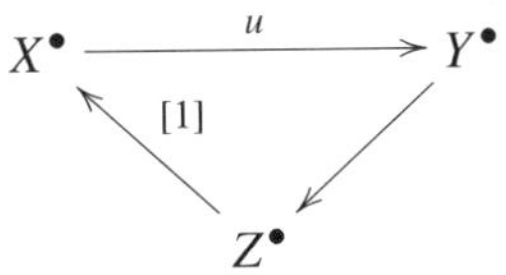

Once we have introduced the derived category $D(A)$ (see Sect. 2.4.2), it will follow that in $D(A)$ we can always (even when (2.3) is not split) replace $C^\bullet(u)$ by $Z^\bullet$ in (2.4) to get an associated distinguished triangle.

Remark 2.3.5 *Let A and B be abelian categories. An additive functor $F : A \to B$ induces in an obvious manner a functor $F : C(A) \to C(B)$ between the associated categories of complexes. If $f \in \mathrm{Hom}_{C(A)}(X^\bullet, Y^\bullet)$ is null-homotopic with homotopy $\{s^i\}$, then $\{F(s^i)\}$ is a null-homotopy for $F(f)$. Thus F induces a functor $F : K(A) \to K(B)$ between the homotopy categories.*

As an example, let X and Y be topological spaces and $f : X \to Y$ a continuous map, and take $A = Sh(X)$, $B = Sh(Y)$. Then direct image $f_* : K(X) \to K(Y)$ and inverse image $f^* : K(Y) \to K(X)$ are well-defined functors on the homotopy category of complexes of sheaves.

2.4 Derived Categories

Let A be an abelian category. In Sect. 2.1, we formed the homotopy category of complexes $K(A)$ which has the property that homotopy equivalences between complexes in $C(A)$ become isomorphisms in $K(A)$. It turns out that requiring isomorphic complexes to be homotopy equivalent is still too restrictive for many purposes. For instance, we saw in Sect. 2.3 that in comparing a short exact sequence $0 \to X^\bullet \xrightarrow{u} Y^\bullet \to Z^\bullet \to 0$ to the distinguished triangle $X^\bullet \xrightarrow{u} Y^\bullet \to C^\bullet(u) \xrightarrow{[1]}$, we may identify $C^\bullet(u)$ with $Z^\bullet$ only up to quasi-isomorphism. Thus if we had a natural triangulated category $D(A)$ where quasi-isomorphic complexes are *isomorphic* objects, we could speak of the distinguished triangle $X^\bullet \to Y^\bullet \to Z^\bullet \xrightarrow{[1]}$ associated to a short exact sequence $0 \to X^\bullet \to Y^\bullet \to Z^\bullet \to 0$ in $C(A)$. The axiomatic characterization of the intersection chain sheaves (see Sect. 4.1.4) as well as Verdier duality (see Sect. 3) rely heavily on regarding quasi-isomorphic complexes as isomorphic objects. Now such a category $D(A)$ exists; it is called the derived category of A. Its construction is described in the present section.

Our goal is the following: We want morphisms in $K(A)$ which induce isomorphisms on cohomology to be invertible in the category to be constructed. When we formed $K(A)$, a homotopy equivalence already had its natural candidate for an inverse morphism, namely the homotopy inverse. If we want a quasi-isomorphism to become an isomorphism, it has to have an inverse. Unfortunately, there is no candidate for an inverse around. So to define a derived category, we have to use a more elaborate process called localization of categories.

2.4.1 Localization of Categories

Definition 2.4.1 *Let K be a category. A collection S of morphisms of K is a* multiplicative system, *if it satisfies the following axioms*:

(S0): *The identity $1_X \in S$ for all $X \in ObK$.*
(S1): *S is closed under composition of morphisms.*
(S2): *Any diagram*

$$\begin{array}{ccc} & & Z \\ & & \big\downarrow{\scriptstyle s} \\ X & \xrightarrow{u} & Y \end{array}$$

with $s \in S$ can be completed to a commutative square

$$\begin{array}{ccc} W & \longrightarrow & Z \\ {\scriptstyle t}\big\downarrow & & \big\downarrow{\scriptstyle s} \\ X & \xrightarrow{u} & Y \end{array}$$

with $t \in S$. The analogous statement with all arrows reversed is required as well.
(S3): *Given two morphisms $u, v : X \to Y$, there exists an $s : Y \to Y'$ in S such that $su = sv$ if and only if there exists a $t : X' \to X$ in S such that $ut = vt$.*

Given a multiplicative system S, we shall construct a new category K_S. Take

$$ObK_S = ObK.$$

Morphisms in $\mathrm{Hom}_{K_S}(X, Y)$ will be represented by diagrams of the form (so-called "fractions," also referred to as "roofs"):

$$\begin{array}{ccccc} & & X' & & \\ & {\scriptstyle s}\swarrow & & \searrow{\scriptstyle u} & \\ X & & & & Y \end{array} \tag{2.5}$$

with $s \in S$. It is convenient to label arrows in S by the symbol Ⓢ. Diagram (2.5) for instance will be denoted

$$\begin{array}{ccccc} & & X' & & \\ & \swarrow\text{Ⓢ} & & \searrow & \\ X & & & & Y \end{array}$$

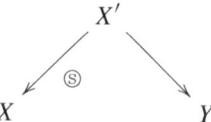

We define an equivalence relation on the collection of all fractions: Two fractions

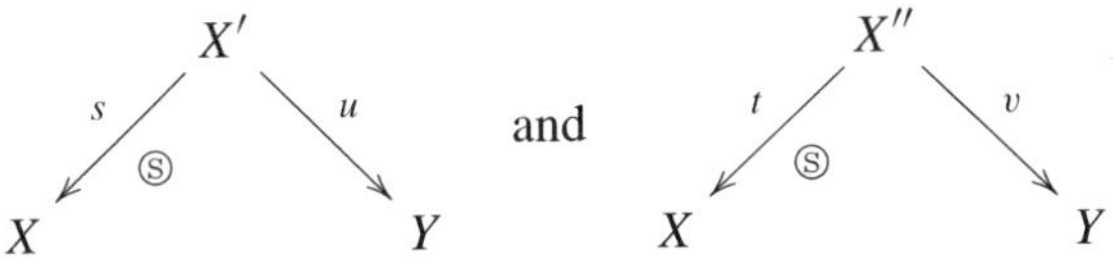

are equivalent, if there exists a commutative diagram

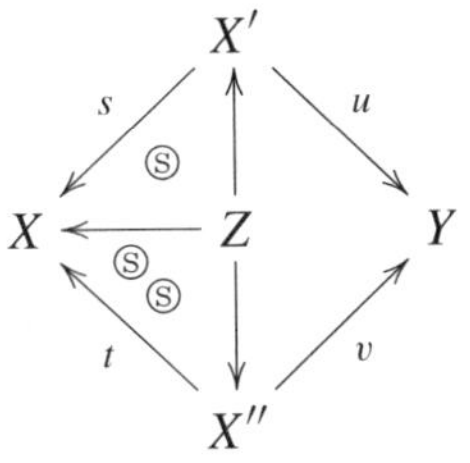

i.e. if there exists a fraction $X \xleftarrow{(S)} Z \to Y$ that lies over both $X \xleftarrow{(S)} X' \to Y$ and $X \xleftarrow{(S)} X'' \to Y$. We check transitivity for this relation, reflexivity and symmetry being trivial:

Given

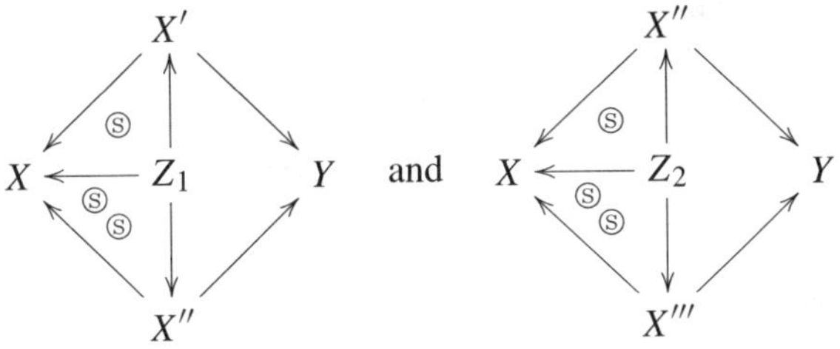

use (S2) to obtain the diagram

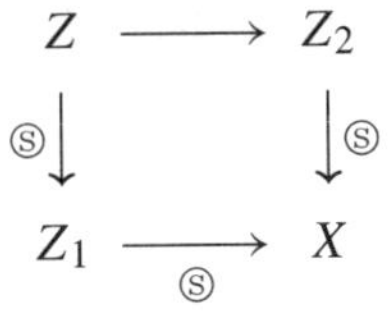

$$Z \longrightarrow Z_2$$
$$\downarrow{\scriptstyle (S)} \qquad \downarrow{\scriptstyle (S)}$$
$$Z_1 \xrightarrow{(S)} X$$

Let $Z \xrightarrow{(S)} X$ be the composition $Z \xrightarrow{(S)} Z_1 \xrightarrow{(S)} X$ (using (S1)), let $Z \to X'$ be $Z \xrightarrow{(S)} Z_1 \to X'$, and let $Z \to X'''$ be $Z \to Z_2 \to X'''$. Then we have a commutative diagram

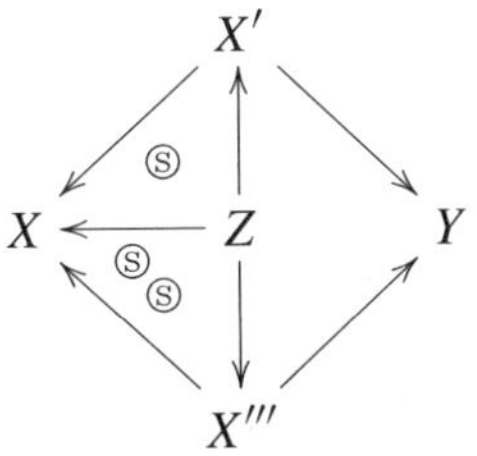

By definition, $\mathrm{Hom}_{K_S}(X, Y)$ is the set of equivalence classes of fractions (2.5). The composition of two fractions represented by

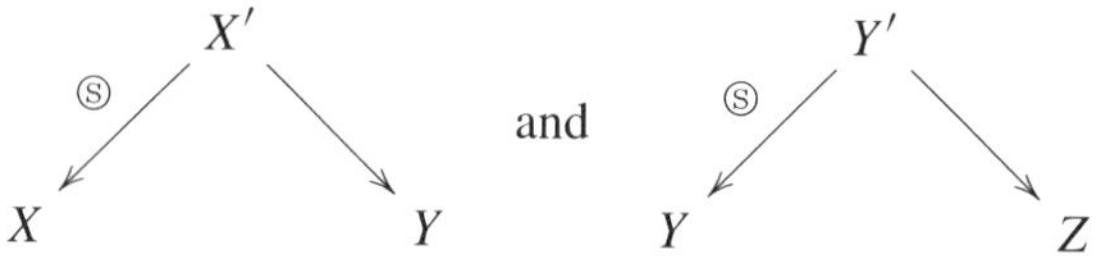

is defined to be the class of $X \overset{\textcircled{s}}{\leftarrow} X' \overset{\textcircled{s}}{\leftarrow} X'' \to Y' \to Z$ obtained by applying (S2) to complete the diagram

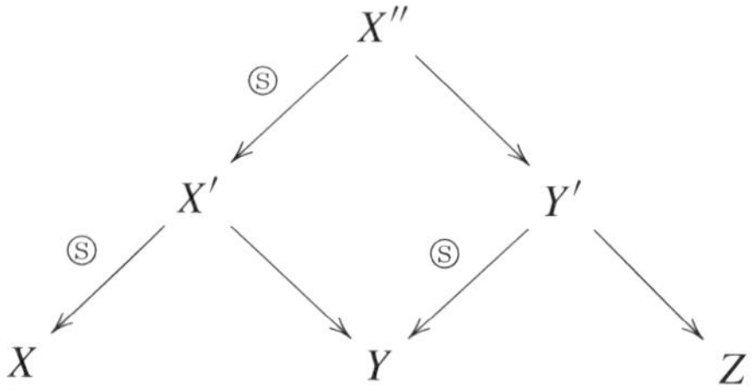

It is easily checked that the class of this composition is independent of the choice of representatives and commutative diagram.

This completes the construction of the category K_S.

Definition 2.4.2 *K_S is called the* localization of the category K *with respect to the multiplicative system S.*

There is a natural functor $Q : K \to K_S$: If $X \in ObK$, let

$$Q(X) = X,$$

and for $u \in \mathrm{Hom}_K(X, Y)$ define $Q(u)$ to be the class of the fraction

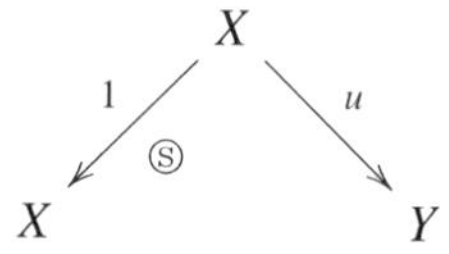

The point is that in the localization all arrows of the multiplicative system become isomorphisms. Precisely, we have the following statement: If $s \in S$, then $Q(s)$ is an isomorphism. For if $s : X \to Y$ is in S, then

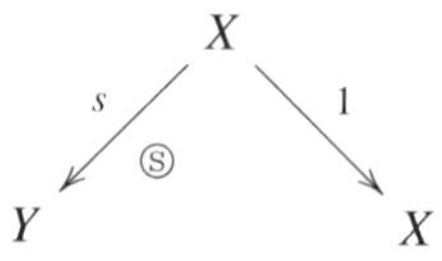

is an inverse for $Q(s)$ as represented by

$$\begin{array}{ccc} & X & \\ {}^{1}\swarrow & \circledS & \searrow^{s} \\ X & & Y \end{array}$$

(observe here that $Y \overset{s}{\leftarrow} X \overset{s}{\rightarrow} Y$ is equivalent to $Y \overset{1}{\leftarrow} Y \overset{1}{\rightarrow} Y$). Furthermore, the localization is the smallest construction with that property:

Proposition 2.4.3 *Let K and L be categories and S a multiplicative system in K. Any functor $F : K \to L$ such that $F(s)$ is an isomorphism for all $s \in S$ factors uniquely through $Q : K \to K_S$.*

Note that if K is an additive category, then K_S is additive as well (for details consult e.g. [GM99, Chap. 4, 2.8]).

Definition 2.4.4 *Let K be a triangulated category and S a multiplicative system in K. We call S* compatible with the triangulation *if the following hold*:

(S4): *$s \in S$ if and only if $T(s) \in S$ (where T is the translation functor).*

(S5): *Given two distinguished triangles (X, Y, Z, u, v, w) and (X', Y', Z', u', v', w'), a commutative square*

$$\begin{array}{ccc} X & \xrightarrow{u} & Y \\ \circledS\big\downarrow & & \big\downarrow\circledS \\ X' & \xrightarrow{u'} & Y' \end{array}$$

can be embedded in a morphism of triangles with $Z \to Z'$ in S (compare this to (TR4)).

Proposition 2.4.5 *Let K be a triangulated category and S a multiplicative system compatible with the triangulation. Then*:

1. *K_S has a unique triangulation such that $Q : K \to K_S$ is a functor of triangulated categories.*
2. *Any functor of triangulated categories $F : K \to L$ such that $F(s)$ is an isomorphism for all $s \in S$ factors uniquely as $F = F'Q$ with $F' : K_S \to L$ a functor of triangulated categories.*

The collection of distinguished triangles for K_S is given by all triangles isomorphic in K_S to triangles of the form

$$Q(X) \overset{Q(u)}{\longrightarrow} Q(Y) \overset{Q(v)}{\longrightarrow} Q(Z) \overset{Q(w)}{\longrightarrow} Q(T(X))$$

where

$$X \overset{u}{\longrightarrow} Y \overset{v}{\longrightarrow} Z \overset{w}{\longrightarrow} T(X)$$

is a distinguished triangle in K.

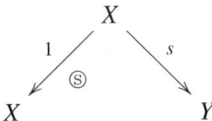

Remark 2.4.6 *Let K be any category and S a multiplicative system. Let K'_S denote the "opposite" localization, i.e. morphisms are represented by "opposite" fractions*

$$\begin{array}{ccccc} & & Y' & & \\ & \nearrow & & \nwarrow \circledS & \\ X & & & & Y \end{array} \tag{2.6}$$

subject to an analogous equivalence relation and with an analogous composition law. Define a functor

$$F : K_S \longrightarrow K'_S$$

as follows: Put $F(X) = X$ for $X \in ObK_S$. If $f \in \mathrm{Hom}_{K_S}(X, Y)$ is represented by the fraction $X \underset{\circledS}{\overset{s}{\longleftarrow}} X' \overset{u}{\longrightarrow} Y$, use (S2) *to find a commutative square*

$$\begin{array}{ccc} Y' & \overset{v}{\longleftarrow} & X \\ {\scriptstyle t}\uparrow{\scriptstyle \circledS} & & {\scriptstyle \circledS}\uparrow{\scriptstyle s} \\ Y & \overset{u}{\longleftarrow} & X' \end{array}$$

and let $F(f)$ be the equivalence class of $X \overset{v}{\longrightarrow} Y' \underset{\circledS}{\overset{t}{\longleftarrow}} Y$. If $X \overset{v'}{\longrightarrow} Y'' \underset{\circledS}{\overset{t'}{\longleftarrow}} Y$ is a different completion of the square, again use (S2) *to find*

$$\begin{array}{ccc} V & \overset{t'_0}{\longleftarrow} & Y' \\ {\scriptstyle t_0}\uparrow{\scriptstyle \circledS} & & {\scriptstyle \circledS}\uparrow{\scriptstyle t} \\ Y'' & \underset{\circledS}{\overset{t'}{\longleftarrow}} & Y \end{array}$$

and note that $t'_0 vs = t_0 v's : X' \to V$. By (S3)*, there exists $V \underset{\circledS}{\overset{w}{\longrightarrow}} W$ such that $wt'_0 v = wt_0 v' : X \to W$. Then the commutative diagram*

$$\begin{array}{ccccc} & & Y' & & \\ & {\scriptstyle v}\nearrow & \downarrow{\scriptstyle wt'_0\ \circledS} & \nwarrow{\scriptstyle t} & \\ X & & W & \overset{\circledS}{\longleftarrow} & Y \\ & {\scriptstyle v'}\searrow & {\scriptstyle \circledS}\uparrow{\scriptstyle wt_0} & \swarrow{\scriptstyle t'\ \circledS} & \\ & & Y'' & & \end{array}$$

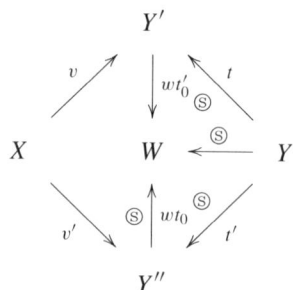

shows that $X \overset{v}{\longrightarrow} Y' \underset{\circledS}{\overset{t}{\longleftarrow}} Y$ and $X \overset{v'}{\longrightarrow} Y'' \underset{\circledS}{\overset{t'}{\longleftarrow}} Y$ are equivalent fractions. Similarly, $F(f)$ does not depend on the chosen representative of f. Thus $F(f)$ is *well-defined. Clearly F takes identity morphisms to identity morphisms. The following diagram shows that F is indeed a functor:*

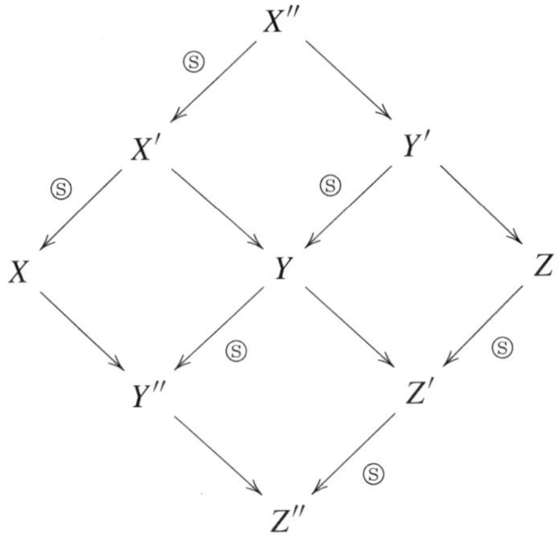

where X'', Y'', Z' and Z'' are obtained by applications of axiom (S2). *Now it is straightforward to check that F sets up an equivalence of categories*

$$K_S \xrightarrow{\simeq} K'_S.$$

The upshot is that we may as well work with fractions of the form (2.6).

2.4.2 Localization With Respect to Quasi-Isomorphisms

Let A be an abelian category and $K(A)$ its homotopy category of complexes. We denote by Qis the collection of all quasi-isomorphisms in $K(A)$.

Proposition 2.4.7 *Qis is a multiplicative system in $K(A)$, compatible with the triangulation.*

Proof. (S0) and (S1) are clearly satisfied. We prove (S2): Given the diagram

$$\begin{array}{ccc} & & Z^\bullet \\ & & \big\downarrow{\scriptstyle s} \\ X^\bullet & \xrightarrow{u} & Y^\bullet \end{array}$$

with $s \in Qis$, embed s in a triangle

$$Z^\bullet \xrightarrow{s} Y^\bullet \xrightarrow{v} V^\bullet \longrightarrow Z^\bullet[1] \tag{2.7}$$

and embed vu in a triangle

$$W^\bullet \xrightarrow{t} X^\bullet \xrightarrow{vu} V^\bullet \longrightarrow W^\bullet[1]. \tag{2.8}$$

By (TR4), the commutative square

$$\begin{array}{ccc} Y^\bullet & \xrightarrow{v} & V^\bullet \\ {\scriptstyle u}\uparrow & & \| \\ X^\bullet & \xrightarrow{vu} & V^\bullet \end{array}$$

can be embedded into a morphism of triangles

$$\begin{array}{ccccccc} Z^\bullet & \xrightarrow{s} & Y^\bullet & \xrightarrow{v} & V^\bullet & \longrightarrow & Z^\bullet[1] \\ \uparrow & & {\scriptstyle u}\uparrow & & \| & & \uparrow \\ W^\bullet & \xrightarrow{t} & X^\bullet & \xrightarrow{vu} & V^\bullet & \longrightarrow & W^\bullet[1] \end{array}$$

As s is a quasi-isomorphism, the long exact sequence on cohomology for triangle (2.7) (Proposition 2.3.2) implies that $H^i(V^\bullet) = 0$ for all i. Using this on the long exact cohomology sequence for triangle (2.8), we have that $t \in Qis$. Hence the desired square is

$$\begin{array}{ccc} W^\bullet & \longrightarrow & Z^\bullet \\ {\scriptstyle t}\downarrow & & \downarrow{\scriptstyle s} \\ X^\bullet & \xrightarrow{u} & Y^\bullet \end{array}$$

The opposite statement in (S2) is handled similarly.

Let us proceed to show (S3). Assume we are given two morphisms $u, v : X^\bullet \to Y^\bullet$ such that $su = sv$ for some quasi-isomorphism $s : Y^\bullet \to Y'^\bullet$. Let $f = u - v$ so that $sf = 0$. Embed s in a distinguished triangle

$$Z^\bullet \xrightarrow{h} Y^\bullet \xrightarrow{s} Y'^\bullet \longrightarrow Z^\bullet[1]. \tag{2.9}$$

By Example 2.2.5, $\mathrm{Hom}(X^\bullet, -)$ is a cohomological functor, which applied to the above triangle gives the exact sequence

$$\mathrm{Hom}(X^\bullet, Z^\bullet) \xrightarrow{h\circ} \mathrm{Hom}(X^\bullet, Y^\bullet) \xrightarrow{s\circ} \mathrm{Hom}(X^\bullet, Y'^\bullet).$$

Thus, as $sf = 0$, there exists a $g \in \mathrm{Hom}(X^\bullet, Z^\bullet)$ with $hg = f$. Now embed g in a distinguished triangle

$$X'^\bullet \xrightarrow{t} X^\bullet \xrightarrow{g} Z^\bullet \longrightarrow X'^\bullet[1]. \tag{2.10}$$

Using Proposition 2.2.2, we have $ft = (hg)t = h(gt) = h \circ 0 = 0$, i.e. $ut = vt$. Since $s \in Qis$, the long exact cohomology sequence for (2.9) shows $H^i(Z^\bullet) = 0$ for all i. Considering the long exact sequence for (2.10), this means that $t \in Qis$. The converse direction is proven by a similar argument.

Axiom (S4) is clear, and (S5) follows from applying the five-lemma to the diagram of associated long exact cohomology sequences. □

Remark 2.4.8 *The above proof only uses the axioms of a triangulated category, it does not depend on the specific triangulation of $K(A)$ given in Theorem 2.3.1. Therefore, Proposition 2.4.7 is valid for any triangulated category K, cohomological functor $H : K \to A$, A an abelian category, and S the collection of morphisms s in K with $H(T^i(s))$ an isomorphism for all i.*

Remark 2.4.9 *In general, the collection of all quasi-isomorphisms in $C(A)$ does not form a multiplicative system. Thus we cannot localize $C(A)$ with respect to Qis—another reason to construct the homotopy category $K(A)$ first.*

Definition 2.4.10 *The derived category $D(A)$ of A is the localization of $K(A)$ with respect to the multiplicative system Qis:*

$$D(A) = K(A)_{Qis}.$$

As pointed out above, the localization of an additive category is additive. Since $K(A)$ is additive it follows then that $D(A)$ is additive as well. Furthermore, according to Proposition 2.4.7, the multiplicative system Qis is compatible with the triangulation of $K(A)$, allowing us to invoke Proposition 2.4.5. We conclude that the derived category $D(A)$ is triangulated and the canonical $Q : K(A) \to D(A)$ is a functor of triangulated categories.

Example 2.4.11 *Let $f : X^\bullet \to Y^\bullet$ be a morphism of complexes in $C(A)$. It is worthwhile to compare and analyze the statements $f = 0$ (in $C(A)$), f is homotopic to 0, $Q(f) = 0$ in $D(A)$ and $H^\bullet(f) = 0$. The results are summarized in Fig. 2.2.*

First, let us clarify what $Q(f) = 0$ means. The morphism $Q(f)$ is represented by the fraction

$$\begin{array}{ccc} & X^\bullet & \\ {}^{1}\swarrow & & \searrow^{f} \\ X^\bullet & & Y^\bullet \end{array}$$

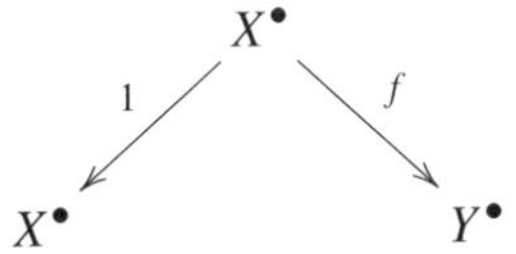

If $Q(f) = 0$, then there exists a diagram

$$\begin{array}{c} f = 0 \end{array} \underset{\Leftarrow}{\Rightarrow} \begin{array}{c} f \overset{h}{\sim} 0 \end{array} \underset{\Leftarrow}{\Rightarrow} \begin{array}{c} Q(f) = 0 \text{ in } D(A) \end{array} \underset{\Leftarrow}{\Rightarrow} \begin{array}{c} H^\bullet(f) = 0 \end{array}$$

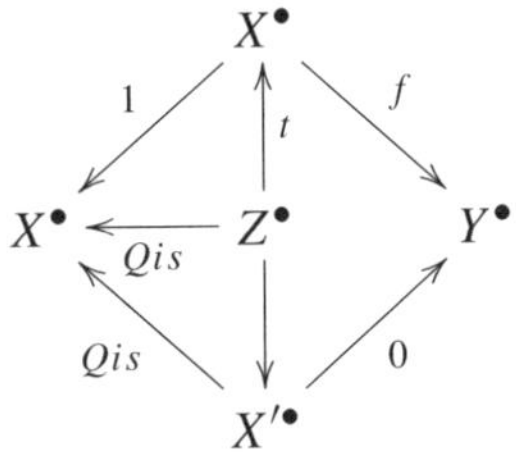

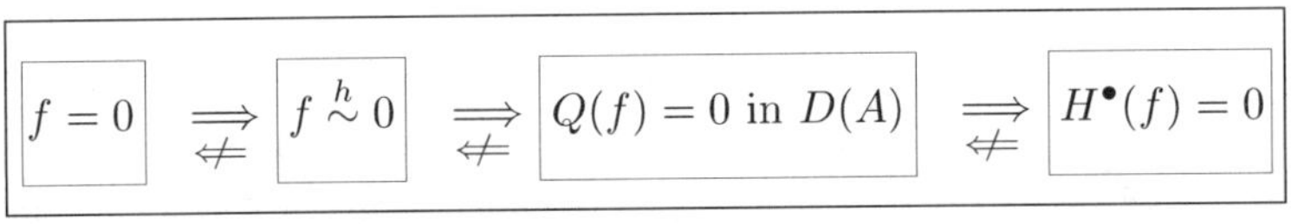

Fig. 2.2. Trivial morphisms in $C(A)$, $K(A)$ and $D(A)$.

Therefore,

$$Q(f) = 0 \quad \Longleftrightarrow \quad \exists t : Z^\bullet \to X^\bullet \text{ in } Qis \text{ such that } ft \text{ is homotopic to } 0.$$

(*Recall that by axiom* (S3) (*Definition* 2.4.1), *the above condition is equivalent to the existence of a quasi-isomorphism* $s : Y^\bullet \to W^\bullet$ *with* $sf = 0$.)

1. $f \overset{h}{\sim} 0$ *versus* $Q(f) = 0$: *If* f *is null-homotopic, then* $f = 0$ *in* $K(A)$ *and thus* $Q(f) = 0$ *as* $Q : K(A) \to D(A)$ *is an additive functor. The following example shows that the converse is false in general: Let A be the category of abelian groups,* $X^\bullet$ *the acyclic complex*

$$X^\bullet = \cdots \to 0 \to \mathbb{Z} \overset{2}{\to} \mathbb{Z} \to \mathbb{Z}/2 \to 0 \to \cdots$$

and take $f = 1_X : X^\bullet \to X^\bullet$. *The zero morphism* $s : X^\bullet \overset{0}{\to} 0^\bullet$ *is a quasi-isomorphism with* $sf = 0$, *whence* $Q(f) = 0$. *On the other hand,* f *is not null-homotopic, as for the existence of a homotopy*

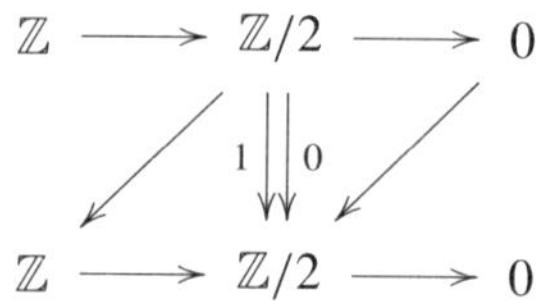

we would need a nonzero homomorphism $\mathbb{Z}/2 \to \mathbb{Z}$, *which does not exist.*

2. $Q(f) = 0$ *versus* $H^\bullet(f) = 0$: *Suppose* $t : Z^\bullet \to X^\bullet$ *is a quasi-isomorphism such that* ft *is null-homotopic. Then* $H^\bullet(f)H^\bullet(t) = 0$, $H^\bullet(t)$ *being an isomorphism, implies* $H^\bullet(f) = 0$. *Again, the converse is generally false: Let* f *be the morphism*

$$\begin{array}{ccccccc} X^\bullet & = \cdots 0 & \longrightarrow & \mathbb{Z} & \overset{2}{\longrightarrow} & \mathbb{Z} & \longrightarrow 0 \cdots \\ {\scriptstyle f}\downarrow & & & \downarrow{\scriptstyle 1} & & \downarrow{\scriptstyle 2} & \\ Y^\bullet & = \cdots 0 & \longrightarrow & \mathbb{Z} & \overset{1}{\longrightarrow} & \mathbb{Z}/3 & \longrightarrow 0 \cdots \end{array}$$

where $1 : \mathbb{Z} \to \mathbb{Z}$ *is in degree* 0, *say. We have* $H^0(X^\bullet) = 0$, $H^1(X^\bullet) = \mathbb{Z}/2$, $H^0(Y^\bullet) = 3\mathbb{Z}$ *and* $H^1(Y^\bullet) = 0$, *so* $H^\bullet(f) = 0$. *We claim there is no quasi-isomorphism* $t : Z^\bullet \to X^\bullet$ *with* ft *null-homotopic. Suppose we had such a* $t = \{t^n\}_{n \in \mathbb{Z}}$. *Pick a cycle* $z \in Z^1$ *such that the class of* $t^1(z)$ *is the generator of* $H^1(X^\bullet) = \mathbb{Z}/2$. *This implies that* $t^1(z)$ *is an odd integer. Now* $t^1(2z) = 2t^1(z)$ *is even, so* 0 *in* $H^1(X^\bullet)$. *As* t *is in* Qis, *it follows that the class of* $2z$ *is* 0 *in* $H^1(Z^\bullet)$ *and* $2z = d_Z(z')$ *for some* $z' \in Z^0$. *Let* $s = \{s^n\}_{n \in \mathbb{Z}}$ *be a null-homotopy for* ft:

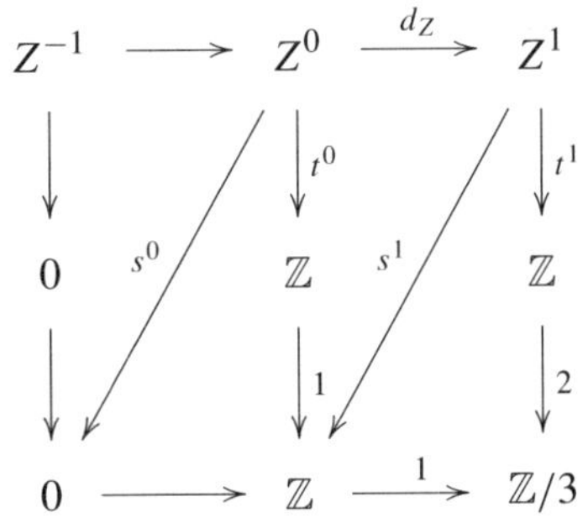

We have

$$2t^1(z) = t^1(d_Z(z')) = 2t^0(z') = 2s^1(d_Z(z'))$$

and hence

$$t^1(z) = s^1(d_Z(z')) = s^1(2z) = 2s^1(z),$$

contradicting $t^1(z)$ odd.

Remark 2.4.12 *Two complexes $X^\bullet, Y^\bullet \in ObK(A)$ are called* quasi-isomorphic *if there exists a fraction*

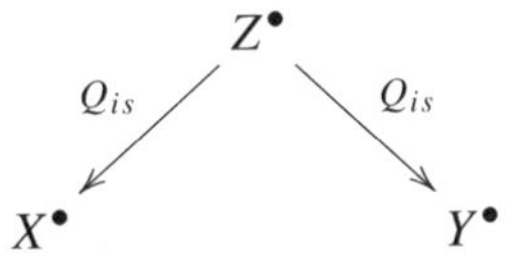

Then the following statement holds: $Q(X^\bullet) \cong 0^\bullet$ in $D(A)$ if and only if $X^\bullet$ is quasi-isomorphic to $0^\bullet$ in $K(A)$. Proof: Supposing $Q(X^\bullet) \cong 0^\bullet$, let

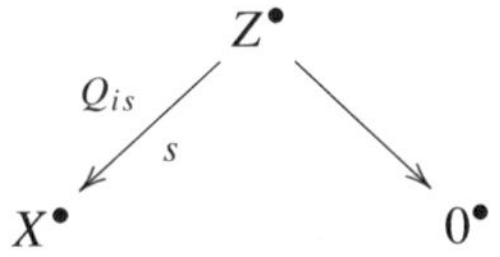

represent an isomorphism and let

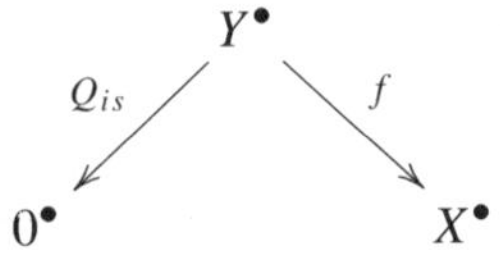

represent its inverse, so that the composition

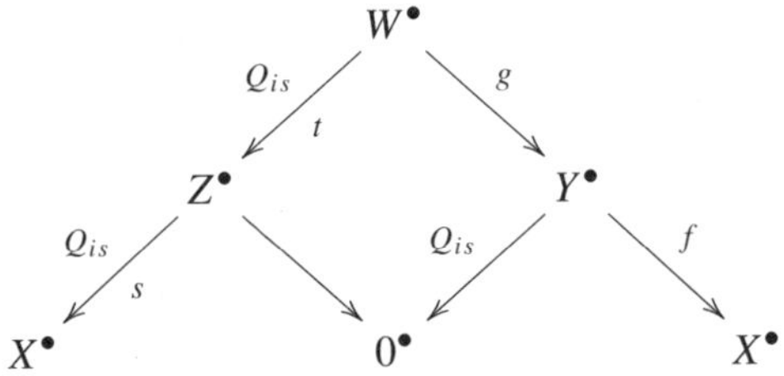

is equivalent to the identity:

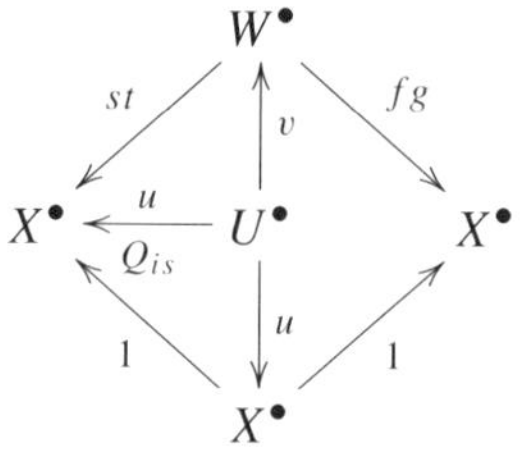

It follows that $fgv = u \in Qis$, *and consequently that* $fg \in Qis$, *as* $v \in Qis$. *Since* $Y^\bullet \to 0^\bullet$ *is in* Qis, *we have* $H^\bullet(Y^\bullet) = 0$ *so that* $H^\bullet(fg) = 0$. *Therefore,* $H^\bullet(X^\bullet) \cong \operatorname{im} H^\bullet(fg) \cong 0^\bullet$ *and* $H^\bullet(Z^\bullet) = 0^\bullet$ *as well, using* $s \in Qis$. *We conclude that* $Z^\bullet \to 0^\bullet$ *is a quasi-isomorphism so that* $X^\bullet$ *and* $0^\bullet$ *are quasi-isomorphic.*

Conversely, if we are presented with a fraction

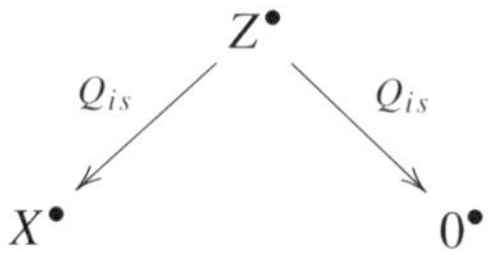

(2.11)

then

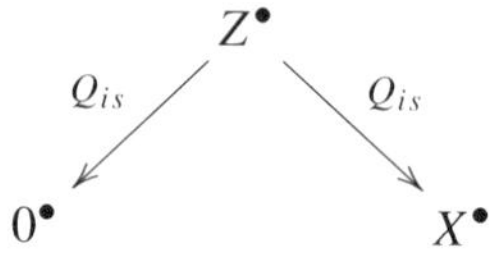

represents an inverse for (2.11) *and* $Q(X^\bullet) \cong 0^\bullet$ *in* $D(A)$.

Let X be an object in A. We may regard X as a complex concentrated in degree zero, thus obtaining a functor $A \to D(A)$ (to a morphism $f : X \to Y$ assign $Q(f)$, interpreting f as a morphism of complexes; note also that the homotopy class of f contains only f). This functor gives an equivalence of A with the full subcategory of $D(A)$ of complexes $X^\bullet$ whose cohomology is concentrated in degree 0.

Remark 2.4.13 *In Sect.* 1.3, *we have defined the truncations* $\tau_{\leq n}, \tau_{\geq n}$ *for complexes in* $C(X) = C(Sh(X))$; *clearly the definition applies to complexes in* $C(A)$, A *any abelian category. Since* $\tau_{\leq n}, \tau_{\geq n} : C(A) \to C(A)$ *take null-homotopic morphisms to null-homotopic morphisms, truncation is well-defined on the homotopy category:*

$$\tau_{\leq n}, \tau_{\geq n} : K(A) \longrightarrow K(A).$$

Moreover, as $\tau_{\leq n}, \tau_{\geq n}$ *take quasi-isomorphisms to quasi-isomorphisms, truncation is well-defined on the derived category:*

$$\tau_{\leq n}, \tau_{\geq n} : D(A) \longrightarrow D(A).$$

Remark 2.4.14 *For an arbitrary functor $F : C(A) \to C(B)$ (A, B abelian categories) it is by no means clear whether it induces a functor $D(A) \to D(B)$. An analysis of when and how this is possible leads to the notion of "derived functors," see Sect.* 2.4.3.

In Sect. 2.3, we have discussed the problem of associating a distinguished triangle to a short exact sequence in $C(A)$. Now that we have access to the derived category, the matter can be settled in a beautiful way: Every short exact sequence in $C(A)$ induces a distinguished triangle in $D(A)$. Concretely, let

$$0 \longrightarrow X^\bullet \xrightarrow{u} Y^\bullet \xrightarrow{v} Z^\bullet \longrightarrow 0 \tag{2.12}$$

be short exact in $C(A)$. Then

$$\begin{array}{ccc} X^\bullet & \xrightarrow{\quad Q([u]) \quad} & Y^\bullet \\ & {\scriptstyle [1]}\nwarrow \qquad \swarrow & \\ & C^\bullet(u) & \end{array}$$

is a distinguished triangle in $D(A)$. We have seen that $f = (0, v) : C^\bullet(u) \to Z^\bullet$ is a quasi-isomorphism, so $C^\bullet(u) \cong Z^\bullet$ in $D(A)$. Then the isomorphism of triangles

$$\begin{array}{ccccccc} X^\bullet & \xrightarrow{Q([u])} & Y^\bullet & \xrightarrow{Q([v])} & Z^\bullet & \longrightarrow & X^\bullet[1] \\ \| & & \| & & \simeq \uparrow Q([f]) & & \| \\ X^\bullet & \xrightarrow{Q([u])} & Y^\bullet & \longrightarrow & C^\bullet(u) & \longrightarrow & X^\bullet[1] \end{array}$$

shows that

$$\begin{array}{ccc} X^\bullet & \xrightarrow{\quad Q([u]) \quad} & Y^\bullet \\ & {\scriptstyle [1]}\nwarrow \qquad \swarrow {\scriptstyle Q([v])} & \\ & Z^\bullet & \end{array}$$

is distinguished in $D(A)$. We call it the *distinguished triangle associated to the short exact sequence* (2.12).

Example 2.4.15 *Let $X^\bullet \in ObC(A)$, and let $\tau_{\leq n}X^\bullet \to X^\bullet$ be the natural inclusion. Then*

$$0 \longrightarrow \tau_{\leq n}X^\bullet \longrightarrow X^\bullet \longrightarrow X^\bullet/\tau_{\leq n}X^\bullet \longrightarrow 0$$

is short exact in $C(A)$. Thus in $D(A)$ we have the associated distinguished triangle

$$\begin{array}{ccc} \tau_{\leq n}X^\bullet & \xrightarrow{\qquad\qquad} & X^\bullet \\ & {\scriptstyle [1]}\nwarrow \qquad \swarrow & \\ & X^\bullet/\tau_{\leq n}X^\bullet & \end{array}$$

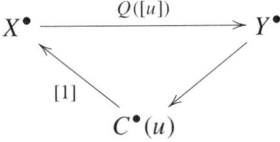

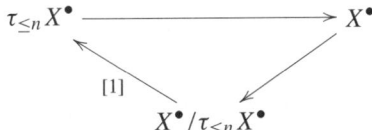

Note furthermore that the canonical map $X^\bullet/\tau_{\le n}X^\bullet \to \tau_{\ge n+1}X^\bullet$ *is a quasi-isomorphism, yielding a distinguished*

$$\begin{array}{ccc} \tau_{\le n}X^\bullet & \longrightarrow & X^\bullet \\ & {\scriptstyle [1]}\nwarrow \qquad \swarrow & \\ & \tau_{\ge n+1}X^\bullet & \end{array}$$

in $D(A)$. *Similarly, with* $\tau_{\le n-1}X^\bullet \to \tau_{\le n}X^\bullet$ *the natural inclusion, we have a short exact sequence*

$$0 \longrightarrow \tau_{\le n-1}X^\bullet \longrightarrow \tau_{\le n}X^\bullet \longrightarrow \tau_{\le n}X^\bullet/\tau_{\le n-1}X^\bullet \longrightarrow 0$$

and an associated distinguished triangle

$$\begin{array}{ccc} \tau_{\le n-1}X^\bullet & \longrightarrow & \tau_{\le n}X^\bullet \\ & {\scriptstyle [1]}\nwarrow \qquad \swarrow & \\ & \tau_{\le n}X^\bullet/\tau_{\le n-1}X^\bullet & \end{array}$$

Now $\tau_{\le n}X^\bullet/\tau_{\le n-1}X^\bullet$ *is quasi-isomorphic to*[2] $H^n(X^\bullet)[-n]$, *and we obtain the distinguished triangle*

$$\begin{array}{ccc} \tau_{\le n-1}X^\bullet & \longrightarrow & \tau_{\le n}X^\bullet \\ & {\scriptstyle [1]}\nwarrow \qquad \swarrow & \\ & H^n(X^\bullet)[-n] & \end{array}$$

in $D(A)$.

2.4.3 Derived Functors

Given an additive functor $F : A \to B$ between abelian categories A and B, we saw in Remark 2.3.5 that it induces a functor $F : K(A) \to K(B)$ between the homotopy categories. This raises the question: Does F furthermore induce a functor $F : D(A) \to D(B)$ on derived categories? The answer is no—the reason being that $F : K(A) \to K(B)$ need not take quasi-isomorphisms to quasi-isomorphisms.

Example 2.4.16 *Suppose* $F : A \to B$ *is not exact and let* $X^\bullet$ *be an acyclic complex in* $C(A)$ *with* $F(X^\bullet)$ *not acyclic. Then* $X^\bullet \xrightarrow{0} 0^\bullet$ *is a quasi-isomorphism. On the other hand* $F(0) = 0$ *as* F *is additive, and* $F(X^\bullet) \xrightarrow{F(0)=0} F(0^\bullet) = 0^\bullet$ *is not a quasi-isomorphism because* $F(X^\bullet)$ *is not acyclic.*

[2] Here we think of $H^n(X^\bullet)$ as a complex concentrated in degree 0.

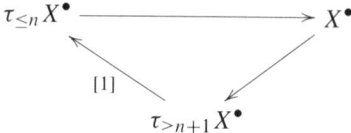

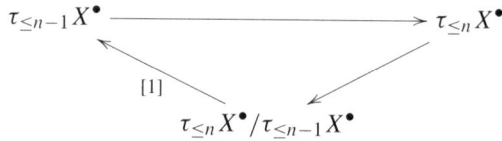

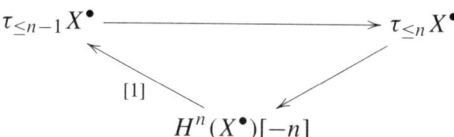

An exact functor $F : A \to B$ will indeed give rise to a functor $F : D(A) \to D(B)$, since it transforms quasi-isomorphisms to quasi-isomorphisms. This functor will then transform distinguished triangles into distinguished triangles. Unfortunately, most important functors on abelian categories, such as Hom, $\otimes$, Γ, direct image are not exact. This prompts the need for a mechanism to extend additive functors $F : A \to B$ to functors $RF : D(A) \to D(B)$. The extension RF is called the (right) derived functor of F. We shall limit our discussion to right derived functors and A an abelian category with enough injectives. Our main interest will subsequently focus on $A = Sh(X)$, the category of sheaves on a space X, and this category has enough injectives, Lemma 1.2.4. We omit a full-scale definition of derived functors via universal properties, it can be found in [Har66] or [KS90].

Lemma 2.4.17 *Let K be a category, S a multiplicative system in K, and K^0 a full subcategory of K such that $S \cap K^0$ is a multiplicative system in K^0. Suppose for every $s : I \to X$ in S with $I \in ObK^0$ there exists $f : X \to J$, $J \in ObK^0$, such that $fs \in S \cap K^0$. Then the natural functor*

$$K^0_{S \cap K^0} \longrightarrow K_S$$

is fully faithful.

Proof. To show faithful, suppose

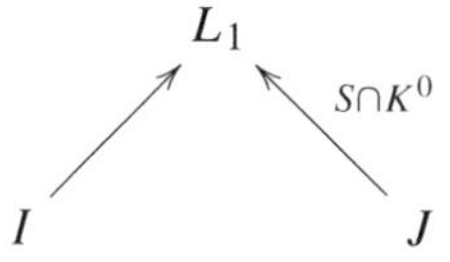

and

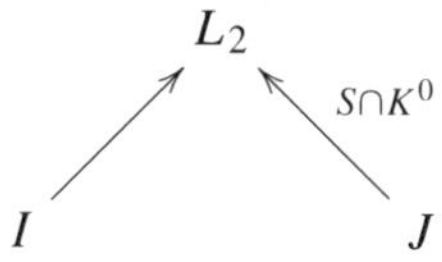

represent two morphisms in $K^0_{S \cap K^0}$ ($I, J, L_1, L_2 \in ObK^0$) which are equal in K_S, i.e. there is a diagram

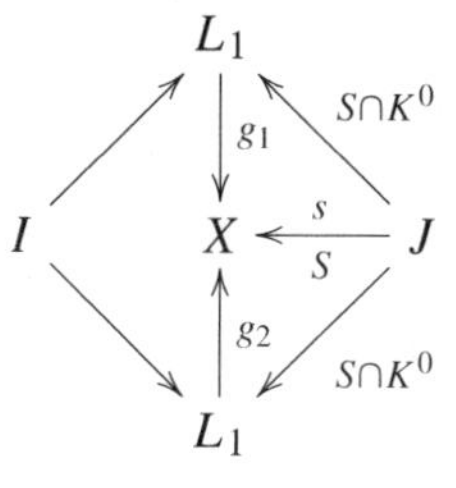

with $X \in ObK$.

For $s : J \to X$ there exists $f : X \to M$, $M \in ObK^0$, such that $fs \in S \cap K^0$. Then the diagram

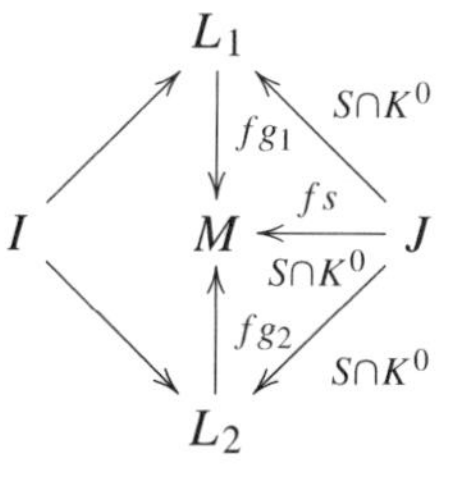

shows that the two morphisms are equal in $K^0_{S\cap K^0}$, and the functor is faithful. Now let $I, J \in ObK^0$ and

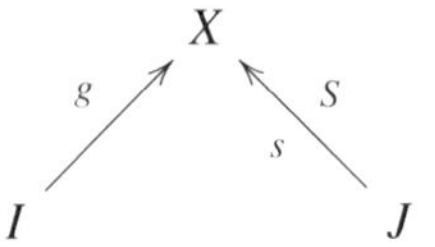

(2.13)

be a morphism in K_S, $X \in ObK$. There exists $f : X \to M$, $M \in ObK^0$, with $fs \in S \cap K^0$. Then

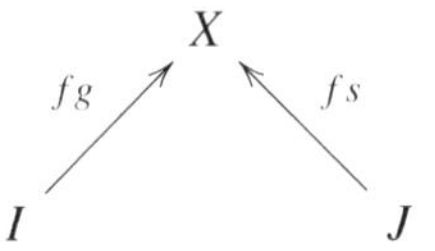

is a morphism in $K^0_{S\cap K^0}$ which is equal to (2.13) in K_S via

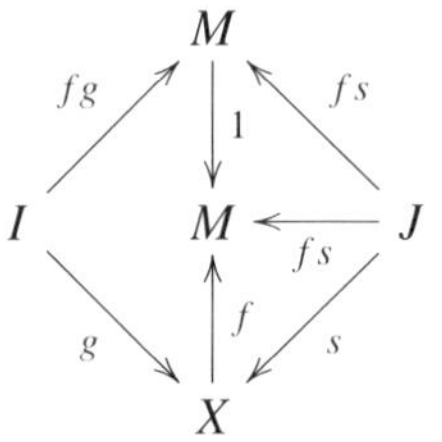

Hence the functor is full. □

Recall that $K^+(A)$ denotes the homotopy category of bounded below complexes. Similarly, let $K^-(A)$, $K^b(A)$ denote the homotopy categories of bounded above and bounded complexes, respectively. We define

$$D^+(A) = K^+(A)_{Qis}$$

and analogously $D^-(A)$, $D^b(A)$. By Lemma 2.4.17, $D^+(A)$, $D^-(A)$, $D^b(A)$ are full subcategories of $D(A)$.

Let I be the full subcategory of injective objects of A. The key observation for our construction of derived functors is the following

Lemma 2.4.18 *If $I^\bullet \in ObC^+(I)$, $X^\bullet \in ObC(A)$ and $s : I^\bullet \to X^\bullet$ is a quasi-isomorphism, then s has a homotopy inverse.*

The next result uses Lemma 2.4.18 to characterize $D^+(A)$ under the assumption that A has enough injectives.

Theorem 2.4.19 *If A has enough injectives, then the functor*

$$\iota : K^+(I) \hookrightarrow K^+(A) \xrightarrow{Q} D^+(A)$$

is an equivalence of categories.

Proof. By Remark 2.4.8, $Qis \cap K^+(I)$ is a multiplicative system in $K^+(I)$. We verify the condition in Lemma 2.4.17: Suppose $s : I^\bullet \to X^\bullet$ is a quasi-isomorphism with $I^\bullet \in ObK^+(I)$. By Lemma 2.4.18, s has a homotopy inverse $f : X^\bullet \to I^\bullet$. Then $fs = 1_{I^\bullet}$ in $K^+(I)$, so in particular $fs \in Qis \cap K^+(I)$. Therefore, we may apply Lemma 2.4.17 and conclude that

$$K^+(I)_{Qis \cap K^+(I)} \longrightarrow K^+(A)_{Qis}$$

is fully faithful. As $K^+(I)_{Qis \cap K^+(I)} = D^+(I)$ and $K^+(A)_{Qis} = D^+(A)$,

$$D^+(I) \longrightarrow D^+(A)$$

is fully faithful. Let $s : I^\bullet \to J^\bullet$ be in $Qis \cap K^+(I)$. By Lemma 2.4.18, s is a homotopy equivalence, and hence an isomorphism in $K^+(I)$. This means that every element of $Qis \cap K^+(I)$ is an isomorphism, i.e. $D^+(I) = K^+(I)$. The diagram

$$\begin{array}{ccc} K^+(I) & \longrightarrow & K^+(A) \\ \| & & \downarrow Q \\ D^+(I) & \longrightarrow & D^+(A) \end{array}$$

implies that ι is fully faithful. So far we have not used that A has enough injectives. Now using that A has enough injectives, one can show that every $X^\bullet \in ObC^+(A)$ has a quasi-isomorphism $X^\bullet \to I^\bullet$, $I^\bullet \in ObC^+(I)$ (existence of injective resolutions, see [Har66, Lemma I.4.6] or [KS90, Proposition 1.7.7]). Thus every object of $D^+(A)$ is isomorphic to one in $K^+(I)$, and ι is an equivalence of categories. □

Choose a quasi-inverse $D^+(A) \xrightarrow{\simeq} K^+(I)$ for ι (here we are invoking Freyd's theorem, using the axiom of choice; however the construction does not depend on the particular choice of quasi-inverse).

Definition 2.4.20 *Let A, B be abelian categories and $F : K^+(A) \to K(B)$ be a functor of triangulated categories. The* derived functor $RF : D^+(A) \to D(B)$ of F *is defined to be the composition*

$$RF : D^+(A) \xrightarrow{\simeq} K^+(I) \hookrightarrow K^+(A) \xrightarrow{F} K(B) \xrightarrow{Q} D(B).$$

(Frequently, $F : K^+(A) \to K(B)$ will come from an additive functor $F : A \to B$, in which case it is automatically a functor of triangulated categories.) Thus, to construct the derived functor of F, we concretely do the following: restrict F to $K^+(I)$,

$$F| : K^+(I) \longrightarrow K(B).$$

If $s \in Qis \cap K^+(I)$, then by Lemma 2.4.18, s is an isomorphism, in particular a quasi-isomorphism. So $F|$ extends:

$$\bar{F} : D^+(I) \to D(B),$$

in other words, $\bar{F}$ sends a fraction

$$I^\bullet \xrightarrow{\ f\ } L^\bullet \xleftarrow[\ s\ ]{Q_{is}} J^\bullet$$

in $D^+(I)$ to the fraction

$$F(I^\bullet) \xrightarrow{F(f)} F(L^\bullet) \xleftarrow{F(s)} F(J^\bullet)$$

in $D(B)$, where $F(s)$ is in fact an isomorphism. Then RF is

$$D^+(A) \xrightarrow{\simeq} K^+(I) = D^+(I) \xrightarrow{\bar{F}} D(B).$$

Note that RF is a functor of triangulated categories.

Remark 2.4.21 *Reversing arrows, we may consider the full subcategory of projective objects in A. Then assuming that A has enough projectives, we can use projective resolutions to construct a left derived functor*

$$LF : D^-(A) \longrightarrow D(B)$$

for $F : K^-(A) \to K(B)$ a functor of triangulated categories. Suppose that F comes from a left or right exact functor $F : A \to B$ on the level of abelian categories. If

$$DF : D^b(A) \longrightarrow D(B)$$

is any derived functor of F (e.g. RF or LF) and $X \in ObA \hookrightarrow ObD^b(A)$, then one usually requires

$$(D^0F)(X) \stackrel{def}{=} H^0(DF(X)) = F(X).$$

If F is left exact, this will be the case if $DF = RF$, as an exact sequence

$$0 \longrightarrow X \longrightarrow I^0 \longrightarrow I^1$$

will be taken to an exact sequence

$$0 \longrightarrow F(X) \longrightarrow F(I^0) \longrightarrow F(I^1).$$

If F is right exact, this will be the case if $DF = LF$, as an exact sequence

$$P^1 \longrightarrow P^0 \longrightarrow X \longrightarrow 0$$

will be taken to an exact sequence

$$F(P^1) \longrightarrow F(P^0) \longrightarrow F(X) \longrightarrow 0.$$

Thus for left exact functors F, one uses the right derived functor, and for right exact functors F, one uses the left derived functor.

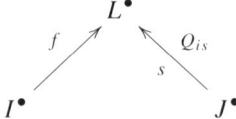

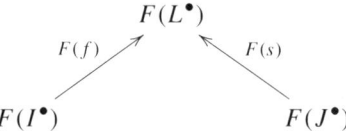

Examples 2.4.22 *For $X \in ObA$,*

$$\mathrm{Hom}(X, -) : A \longrightarrow Ab$$

is a left exact functor and we obtain a right derived functor

$$R\,\mathrm{Hom}^\bullet(X^\bullet, -) : D^+(A) \longrightarrow D(Ab)$$

($X^\bullet \in ObD^-(A)$) as well as its i-th derived functors

$$\mathrm{Ext}^i(X, Y) \stackrel{def}{=} R^i\,\mathrm{Hom}(X, Y) \stackrel{def}{=} H^i(R\,\mathrm{Hom}^\bullet(X, Y))$$

with

$$\mathrm{Ext}^0(X, Y) = \mathrm{Hom}(X, Y).$$

Unfortunately, the category $A = Sh(X)$ of sheaves on a topological space X does not have enough projectives. For example, the constant sheaf $\mathbb{Z}$ on the unit interval is not the quotient of a projective sheaf. Thus, for the right exact functor

$$\mathbf{X} \otimes - : Sh(X) \longrightarrow Sh(X),$$

where $\mathbf{X} \in Sh(X)$, a left derived functor

$$\mathbf{X}^\bullet \stackrel{L}{\otimes} - : D^-(X) \longrightarrow D(X),$$

where $\mathbf{X}^\bullet \in ObD^-(X)$, cannot be defined using projective resolutions. Instead, one defines $\mathbf{X}^\bullet \stackrel{L}{\otimes} -$ by

$$\mathbf{X}^\bullet \stackrel{L}{\otimes} \mathbf{Y}^\bullet = 3D\mathbf{X}^\bullet \otimes \mathbf{F}^\bullet,$$

where $\mathbf{F}^\bullet \to \mathbf{Y}^\bullet$ is a flat resolution of $\mathbf{Y}^\bullet$. (A sheaf $\mathbf{F}$ is called flat *if the functor $- \otimes \mathbf{F}$ is exact. Any sheaf is a quotient of a flat sheaf.) The i-th derived functors of $\mathbf{X} \otimes -$ are*

$$\mathrm{Tor}_i(\mathbf{X}, \mathbf{Y}) \stackrel{def}{=} L_i(\mathbf{X} \otimes \mathbf{Y}) \stackrel{def}{=} H^{-i}(\mathbf{X} \stackrel{L}{\otimes} \mathbf{Y})$$

with

$$\mathrm{Tor}_0(\mathbf{X}, \mathbf{Y}) = \mathbf{X} \otimes \mathbf{Y}.$$

Given a continuous map $f : X \to Y$, the direct image $f_* : Sh(X) \to Sh(Y)$ is left exact and has a right derived functor

$$Rf_* : D^+(X) \longrightarrow D(Y).$$

On the other hand, $f^* : Sh(Y) \to Sh(X)$ is exact, and so induces immediately

$$f^* : D(Y) \longrightarrow D(X).$$

3

Verdier Duality

3.1 Direct Image with Proper Support

Let X be a topological space and $\mathbf{A} \in Sh(X)$. The *support of a section* $s \in \Gamma(X; \mathbf{A})$ is the closed[1] set

$$supp(s) = \{x \in X | s(x) \neq 0\}.$$

Let $\Gamma_c(X; \mathbf{A})$ denote the subgroup of $\Gamma(X; \mathbf{A})$ consisting of all those sections that have compact support. The functor $\Gamma_c(X; -)$ is left exact, and has a right derived functor

$$R\Gamma_c(X; -) : D^+(X) \longrightarrow D^+(Ab),$$

where Ab is the category of abelian groups. The i-th derived functor of Γ_c is called *hypercohomology in degree i with compact support*, that is, we set

$$\mathcal{H}_c^i(X; \mathbf{A}^\bullet) = H^i(R\Gamma_c(X; \mathbf{A}^\bullet)).$$

Now let $f : X \to Y$ be a continuous map between topological spaces. Our goal in this section is to simultaneously generalize the functor $\Gamma_c(X; -)$ as well as the direct image functor f_*, by defining a functor $f_! : Sh(X) \to Sh(Y)$ (and its derived functor $Rf_!$). This will be carried out by defining a presheaf first. To see that this presheaf satisfies the unique gluing condition (G) for sheaves (cf. Sect. 1.1) and to calculate the stalk of the resulting sheaf, it is necessary to assume that X, Y are locally compact spaces.

Definition 3.1.1 *A topological space X is* locally compact *if it is Hausdorff and for every point $x \in X$ there exists a compact subset $K \subset X$ containing an open neighborhood of x.*

[1] Since the projection $\mathbf{A} \to X$ is a local homeomorphism, the set of points where two given global sections agree is open. Thus $supp(s)$ is closed, being the complement of the set where s and the zero section agree.

Definition 3.1.2 *Let $f : X \to Y$ be a map between topological spaces. One says that f is* proper *if f is closed and its fibers are relatively Hausdorff (two distinct points in the fiber have disjoint open neighborhoods in X) and compact.*

If X and Y are locally compact then f is proper if and only if the preimage of any compact subset of Y is compact.

From now on and throughout the rest of Sect. 3, we assume that X and Y are locally compact spaces. Let $f : X \to Y$ be a continuous map and $\mathbf{A} \in Sh(X)$. We define a presheaf on Y as follows: To an open $U \subset Y$, assign the group

$$\Gamma(U; f_!\mathbf{A}) = \{s \in \Gamma(f^{-1}(U); \mathbf{A})| \quad f| : supp(s) \to U \text{ is proper}\}.$$

Using local compactness, one shows that this presheaf satisfies the gluing condition for sheaves, see e.g. Proposition 2.2 in P.P. Grivel's paper [B+84, VI], and thus is a sheaf $f_!\mathbf{A}$.

Definition 3.1.3 *The sheaf $f_!\mathbf{A} \in Sh(Y)$ is called the* direct image with proper support of $\mathbf{A} \in Sh(X)$.

We have a monomorphism

$$f_!\mathbf{A} \hookrightarrow f_*\mathbf{A}. \tag{3.1}$$

The assignment $\mathbf{A} \mapsto f_!\mathbf{A}$ can be extended to sheaf morphisms, and we obtain a functor

$$f_! : Sh(X) \longrightarrow Sh(Y)$$

which is left exact. As an example, consider the case of $Y =$ point. Then

$$f_!\mathbf{A} = \Gamma_c(X; \mathbf{A}).$$

Proposition 3.1.4 *Let $\mathbf{A} \in Sh(X)$ and $y \in Y$. The canonical map*

$$(f_!\mathbf{A})_y \xrightarrow{\simeq} \Gamma_c(f^{-1}(y); \mathbf{A}|_{f^{-1}(y)})$$

is an isomorphism.

For a proof, consult Proposition 2.6 in [B+84, VI]. If f is proper, then $f^{-1}(y)$ is compact, so $\Gamma_c(f^{-1}(y); \mathbf{A}) = \Gamma(f^{-1}(y); \mathbf{A})$ and Proposition 3.1.4 implies that the morphism (3.1) is an isomorphism:

$$f_!\mathbf{A} \cong f_*\mathbf{A}. \tag{3.2}$$

Passing to derived categories, we have the derived functor

$$Rf_! : D^+(X) \longrightarrow D^+(Y).$$

Remark 3.1.5 *Given any functor $F : A \to B$ (A, B abelian categories), it is in applications often useful to know whether the derived functor $RF : D^+(A) \to D^+(B)$ may be computed on resolutions simpler than injective ones. Let us call a class of objects $J \subset A$* adapted *to a, say, left exact functor $F : A \to B$, if F maps any acyclic complex in $C^+(J)$ to an acyclic complex, and any object from A is a subobject of an object from J. One shows that if J is adapted to F, then $Qis \cap K^+(J)$ is a multiplicative system in $K^+(J)$ and the canonical functor*

$$K^+(J)_{Qis\cap K^+(J)} \xrightarrow{\simeq} D^+(A)$$

is an equivalence of categories (compare Lemma 2.4.17 *and Theorem* 2.4.19*). This means that to compute $RF(X^\bullet)$, we may choose a resolution $X^\bullet \to J^\bullet$, $J^\bullet \in C^+(J)$, and then*

$$RF(X^\bullet) = F(J^\bullet).$$

As an example, consider the following class of sheaves:

Definition 3.1.6 *A sheaf $\mathbf{A} \in Sh(X)$ is* c-soft, *if the restriction map $\Gamma(X; \mathbf{A}) \to \Gamma(K; \mathbf{A})$ is surjective for all compact subsets $K \subset X$.*

(Injective sheaves are c-soft.) One proves that the class of c-soft sheaves is adapted to the functor $f_!$. Thus, we may calculate $Rf_!$ on c-soft resolutions.

The *higher direct image sheaves with proper support* are

$$R^i f_!(\mathbf{A}) = \mathbf{H}^i(Rf_!\mathbf{A}).$$

Using Proposition 3.1.4, we compute their stalks as

$$\begin{aligned} R^i f_!(\mathbf{A})_y &= \mathbf{H}^i(Rf_!\mathbf{A})_y \cong H^i((Rf_!\mathbf{A})_y) \\ &= H^i(R\Gamma_c(f^{-1}(y); \mathbf{A})) = H^i_c(f^{-1}(y); \mathbf{A}). \end{aligned}$$

Example 3.1.7 *Let $i : X \hookrightarrow Y$ be an open or closed inclusion. If $y \in Y - X$, then using Proposition* 3.1.4,

$$(i_!\mathbf{A})_y \cong \Gamma_c(i^{-1}(y); \mathbf{A}) = \Gamma_c(\emptyset) = 0$$

and if $y \in X$, $(i_!\mathbf{A})_y \cong \Gamma_c(y; \mathbf{A}_y) = \mathbf{A}_y$. Thus

$$i_!\mathbf{A} = \mathbf{A}^Y$$

is extension by zero.[2] *In particular, $i_!$ is an exact functor. If $i : X \hookrightarrow Y$ is a closed inclusion, then i is proper and by* (3.2),

$$i_! \cong i_*.$$

---

[2] If X is any locally closed subspace of Y and $\mathbf{A} \in Sh(X)$ then the extension by zero, $\mathbf{A}^Y$, is defined to be the unique sheaf which restricts to $\mathbf{A}$ on X and restricts to the zero-sheaf on $Y - X$.

3.2 Inverse Image with Compact Support

In Sect. 1.1 (Proposition 1.1.11) we proved that f_* has an adjoint functor, f^*. The present section is concerned with constructing a right adjoint $f^!$ to $Rf_!$. Let $f : X \to Y$ be a continuous map between locally compact spaces, $\mathbf{A} \in Sh(X)$, $\mathbf{B} \in Sh(Y)$. Ideally, we would like to have a functor on the level of sheaves

$$f^! : Sh(Y) \longrightarrow Sh(X)$$

such that $f^! g^! = (gf)^!$ $(g : Y \to Z)$ and the adjointness formula

$$\mathbf{Hom}(f_!\mathbf{A}, \mathbf{B}) \cong f_* \mathbf{Hom}(\mathbf{A}, f^!\mathbf{B}) \tag{3.3}$$

holds in $Sh(Y)$. Unfortunately, such a functor does not exist in general. To explain why, let us explore some consequences of (3.3). Note that if $\mathbf{F} \in Sh(X)$ is any sheaf, then $\mathrm{Hom}(\mathbb{Z}_X, \mathbf{F}) = \Gamma(X; \mathbf{F})$. Consider first the case of $f = i : U \hookrightarrow X$ an open inclusion and $\mathbf{A} = \mathbb{Z}_U \in Sh(U)$ the constant sheaf. Then $(\mathbf{B} \in Sh(X))$

$$\begin{aligned}
\Gamma(U; i^!\mathbf{B}) &= \mathrm{Hom}(\mathbb{Z}_U, i^!\mathbf{B}) = \Gamma(U; i_* \mathbf{Hom}(\mathbb{Z}_U, i^!\mathbf{B})) \\
&\cong \Gamma(U; \mathbf{Hom}(i_!\mathbb{Z}_U, \mathbf{B})) = \mathrm{Hom}(i^* i_!\mathbb{Z}_U, i^*\mathbf{B}) \\
&= \mathrm{Hom}(\mathbb{Z}_U, i^*\mathbf{B}) \\
&= \Gamma(U; i^*\mathbf{B}).
\end{aligned}$$

Performing the same calculation for any open $V \subset U$, we conclude that

$$i^!\mathbf{B} \cong i^*\mathbf{B}$$

for open inclusions i. Now let $f : X \to Y$ be continuous and $i : U \hookrightarrow X$ an open inclusion as before; set $g = fi$. Using (3.3), we calculate the sections of $f^!\mathbf{B}$:

$$\begin{aligned}
\Gamma(U; f^!\mathbf{B}) &= \Gamma(U; i^* f^!\mathbf{B}) = \Gamma(U; i^! f^!\mathbf{B}) = \Gamma(U; g^!\mathbf{B}) \\
&= \mathrm{Hom}(\mathbb{Z}_U, g^!\mathbf{B}) = \Gamma(U; \mathbf{Hom}(\mathbb{Z}_U, g^!\mathbf{B})) \\
&= \Gamma(Y; g_* \mathbf{Hom}(\mathbb{Z}_U, g^!\mathbf{B})) \cong \Gamma(Y; \mathbf{Hom}(g_!\mathbb{Z}_U, \mathbf{B})) \\
&= \mathrm{Hom}(g_!\mathbb{Z}_U, \mathbf{B}) \\
&= \mathrm{Hom}(f_!(i_!\mathbb{Z}_U), \mathbf{B}),
\end{aligned}$$

where $i_!\mathbb{Z}_U$ is extension by zero. Thus, if a functor $f^!$ as described above existed, then the associated presheaf of the sheaf $f^!\mathbf{B}$ would have to be

$$\Gamma(U; f^!\mathbf{B}) = \mathrm{Hom}(f_!(\mathbb{Z}_U)^X, \mathbf{B}).$$

However, the presheaf

$$U \mapsto \mathrm{Hom}(f_!(\mathbb{Z}_U)^X, \mathbf{B})$$

is *not* a sheaf in general (it does not satisfy the gluing condition (G) for sheaves). Therefore, we cannot hope for a right adjoint $f^! : Sh(Y) \to Sh(X)$. To proceed, the idea is this: Perhaps for an appropriate class of sheaves $\mathbf{K}$, the presheaf

$$U \mapsto \mathrm{Hom}(f_!(\mathbf{K}|_U)^X, \mathbf{B})$$

will be a sheaf. The resulting sheaf $f^!_{\mathbf{K}}\mathbf{B}$ will depend on $\mathbf{K}$, so we pass to the derived category to get rid of the dependence on $\mathbf{K}$. The final product will then be a well-defined functor $f^! : D^+(Y) \to D^+(X)$ on derived categories, which is a right adjoint for $Rf_! : D^+(X) \to D^+(Y)$. This is the program that we shall carry out below.

Let $f : X \to Y$ be a continuous map.

Definition 3.2.1 *A sheaf* $\mathbf{K} \in Sh(X)$ *is* flat *if, given any monomorphism* $\iota : \mathbf{A} \hookrightarrow \mathbf{B}$, $\iota \otimes 1 : \mathbf{A} \otimes \mathbf{K} \to \mathbf{B} \otimes \mathbf{K}$ *is injective as well. A complex* $\mathbf{K}^\bullet \in C(X)$ *is* flat *if every* $\mathbf{K}^i$ *is flat.*

Fix a c-soft flat sheaf $\mathbf{K}$ on X. For two open sets $U \subset V$ in X, the natural sheaf map

$$(\mathbf{K}|_U)^X \longrightarrow (\mathbf{K}|_V)^X$$

induces

$$f_!(\mathbf{K}|_U)^X \longrightarrow f_!(\mathbf{K}|_V)^X$$

and furthermore

$$\mathrm{Hom}(f_!(\mathbf{K}|_V)^X, \mathbf{B}) \longrightarrow \mathrm{Hom}(f_!(\mathbf{K}|_U)^X, \mathbf{B})$$

for any $\mathbf{B} \in Sh(Y)$. Thus the assignment

$$U \mapsto \mathrm{Hom}(f_!(\mathbf{K}|_U)^X, \mathbf{B})$$

defines a presheaf on X. We chose $\mathbf{K}$ to be c-soft as this implies that the presheaf does satisfy unique gluing (G) and hence is a sheaf $f^!_{\mathbf{K}}\mathbf{B}$, cf. théorème 3.5 in [B+84, VI]. We chose $\mathbf{K}$ in addition to be flat because we will eventually need to use the

Lemma 3.2.2 *If* $\mathbf{K} \in Sh(X)$ *is c-soft and flat, then for any* $\mathbf{A} \in Sh(X)$, $\mathbf{A} \otimes \mathbf{K}$ *is c-soft.*

(See Proposition 6.5 [B+84, V].) For $f^!_{\mathbf{K}}\mathbf{B}$ we now have

Proposition 3.2.3 *Let* $\mathbf{A} \in Sh(X)$ *and* $\mathbf{B} \in Sh(Y)$. *There is a canonical isomorphism in* $Sh(Y)$,

$$\mathbf{Hom}(f_!(\mathbf{A} \otimes \mathbf{K}), \mathbf{B}) \cong f_* \,\mathbf{Hom}(\mathbf{A}, f^!_{\mathbf{K}}\mathbf{B}).$$

(7.14 [B+84, V].) One also checks that $f^!_{\mathbf{K}}$ transforms injective sheaves into injective sheaves. On the level of sheaves, $f^!_{\mathbf{K}}$ depends on $\mathbf{K}$, thus we wish to define $f^!$ on the derived category, where we are after all allowed to replace complexes by resolutions. The resolutions we need have to be bounded; to ensure that they exist, we require X henceforth to be of finite cohomological dimension.

Definition 3.2.4 *The* cohomological dimension $\dim X$ *of a locally compact space is the smallest integer* n *for which*

$$H^i_c(X; \mathbf{A}) = 0$$

for all $i > n$, $\mathbf{A} \in Sh(X)$.

The stratified pseudomanifolds defined in Sect. 4.1.2 have finite cohomological dimension.

Lemma 3.2.5 *If X has finite cohomological dimension, then the constant sheaf $\mathbb{Z}_X$ has a bounded c-soft flat resolution*

$$0 \to \mathbb{Z}_X \to \mathbf{K}^0 \to \mathbf{K}^1 \to \cdots \to \mathbf{K}^n \to 0,$$

$n = \dim X$.

(Proposition 6.7 [B+84, V]).

Fix a bounded c-soft flat resolution $\mathbf{K}^\bullet$ of $\mathbb{Z}_X$ as provided by the lemma. For any $\mathbf{B}^\bullet \in C^+(Y)$, we define $f^!_{\mathbf{K}^\bullet}\mathbf{B}^\bullet$ to be the simple complex associated to the double complex $\{f^!_{\mathbf{K}^i}\mathbf{B}^j\}_{i,j\in\mathbb{Z}}$. In other words, $f^!_{\mathbf{K}^\bullet}\mathbf{B}^\bullet$ has associated complex of presheaves

$$\Gamma(U; f^!_{\mathbf{K}^\bullet}\mathbf{B}^\bullet) = \mathrm{Hom}^\bullet(f_!(\mathbf{K}^\bullet|_U)^X, \mathbf{B}^\bullet),$$

$U \subset X$ open. As $\mathbf{K}^\bullet$ is bounded and $\mathbf{B}^\bullet$ is bounded below, $f^!_{\mathbf{K}^\bullet}\mathbf{B}^\bullet$ is bounded below as well. Proposition 3.2.3 implies

Proposition 3.2.6 *Let $\mathbf{A}^\bullet \in C^+(X)$ and $\mathbf{B}^\bullet \in C^+(Y)$. There is a canonical isomorphism in $C^+(Y)$,*

$$\mathbf{Hom}^\bullet(f_!(\mathbf{A}^\bullet \otimes \mathbf{K}^\bullet), \mathbf{B}^\bullet) \cong f_*\mathbf{Hom}^\bullet(\mathbf{A}^\bullet, f^!_{\mathbf{K}^\bullet}\mathbf{B}^\bullet).$$

The correspondence $\mathbf{K}^\bullet \mapsto f^!_{\mathbf{K}^\bullet}$ defines a contravariant functor from the category of c-soft flat resolutions of $\mathbb{Z}_X$ and morphisms of resolutions over $\mathbb{Z}_X$ into the category of functors $K^+(Y) \to K^+(X)$ and morphisms of functors. Moreover, if $\mathbf{K}^\bullet$ and $\mathbf{L}^\bullet$ are two c-soft flat resolutions of $\mathbb{Z}_X$ and $\mathbf{K}^\bullet \to \mathbf{L}^\bullet$ is a morphism of resolutions over $\mathbb{Z}_X$, then the morphism

$$f^!_{\mathbf{L}^\bullet} \longrightarrow f^!_{\mathbf{K}^\bullet}$$

induces a quasi-isomorphism

$$f^!_{\mathbf{L}^\bullet}(\mathbf{I}^\bullet) \longrightarrow f^!_{\mathbf{K}^\bullet}(\mathbf{I}^\bullet)$$

provided $\mathbf{I}^\bullet$ is a bounded below complex of injective sheaves. Thus the functor

$$f^! : D^+(Y) \longrightarrow D^+(X)$$

induced by $f^!_{\mathbf{K}^\bullet}$ on derived categories by setting

$$f^!\mathbf{B}^\bullet = f^!_{\mathbf{K}^\bullet}\mathbf{I}^\bullet,$$

where $\mathbf{I}^\bullet$ is an injective resolution of $\mathbf{B}^\bullet$, is independent, up to quasi-isomorphism, of the choice of $\mathbf{K}^\bullet$.

3.3 The Verdier Duality Formula

The adjointness relation between $Rf_!$ and $f^!$ on derived categories is referred to as Verdier duality [Ver66]. We shall see in Sect. 3.5 how the adjointness formula implies Poincaré duality.

Theorem 3.3.1 *Let* $\mathbf{A}^\bullet \in D^+(X)$ *and* $\mathbf{B}^\bullet \in D^+(Y)$. *There is a canonical isomorphism in* $D^+(Y)$,

$$R\mathbf{Hom}^\bullet(Rf_!\mathbf{A}^\bullet, \mathbf{B}^\bullet) \cong Rf_* R\mathbf{Hom}^\bullet(\mathbf{A}^\bullet, f^!\mathbf{B}^\bullet).$$

Proof. To compute $f^!\mathbf{B}^\bullet$, choose a bounded c-soft flat resolution $\mathbf{K}^\bullet$ of $\mathbb{Z}_X$ and bounded below injective resolution $\mathbf{I}^\bullet$ of $\mathbf{B}^\bullet$. Then

$$f^!\mathbf{B}^\bullet = f^!_{\mathbf{K}^\bullet}\mathbf{I}^\bullet.$$

As pointed out above, $f^!_{\mathbf{K}^\bullet}\mathbf{I}^\bullet$ is a complex of injective sheaves. Thus,

$$R\mathbf{Hom}^\bullet(\mathbf{A}^\bullet, f^!_{\mathbf{K}^\bullet}\mathbf{I}^\bullet) = \mathbf{Hom}^\bullet(\mathbf{A}^\bullet, f^!_{\mathbf{K}^\bullet}\mathbf{I}^\bullet).$$

One can verify that $\mathbf{Hom}^\bullet(\mathbf{A}^\bullet, f^!_{\mathbf{K}^\bullet}\mathbf{I}^\bullet)$ is a complex of flabby sheaves (a sheaf $\mathbf{X} \in Sh(X)$ is *flabby* if $\Gamma(X; \mathbf{X}) \to \Gamma(U; \mathbf{X})$ is surjective for every open $U \subset X$), and the class of flabby sheaves is adapted to the direct image functor f_* (compare Remark 3.1.5). Hence we may compute Rf_* by a flabby resolution and

$$Rf_* R\mathbf{Hom}^\bullet(\mathbf{A}^\bullet, f^!\mathbf{B}^\bullet) = f_*\mathbf{Hom}^\bullet(\mathbf{A}^\bullet, f^!_{\mathbf{K}^\bullet}\mathbf{I}^\bullet).$$

As for the left-hand side, $\mathbf{A}^\bullet \otimes \mathbf{K}^\bullet$ is a c-soft resolution of $\mathbf{A}^\bullet$ (Lemma 3.2.2) and the class of c-soft sheaves is adapted to the functor $f_!$ (Remark 3.1.5). Therefore,

$$Rf_!\mathbf{A}^\bullet = f_!(\mathbf{A}^\bullet \otimes \mathbf{K}^\bullet)$$

and

$$R\mathbf{Hom}^\bullet(Rf_!\mathbf{A}^\bullet, \mathbf{B}^\bullet) = \mathbf{Hom}^\bullet(f_!(\mathbf{A}^\bullet \otimes \mathbf{K}^\bullet), \mathbf{I}^\bullet).$$

The statement follows from Proposition 3.2.6. □

Corollary 3.3.2 *There is a canonical isomorphism*

$$\mathrm{Hom}_{D^+(Y)}(Rf_!\mathbf{A}^\bullet, \mathbf{B}^\bullet) \cong \mathrm{Hom}_{D^+(X)}(\mathbf{A}^\bullet, f^!\mathbf{B}^\bullet).$$

Proof. Let $\mathbf{X}^\bullet \in D^+(X)$ and $\mathbf{I}^\bullet \in D^+(X)$ a complex of injective sheaves. We have seen in Sect. 2.4 that the natural map

$$\mathrm{Hom}_{K^+(X)}(\mathbf{X}^\bullet, \mathbf{I}^\bullet) \longrightarrow \mathrm{Hom}_{D^+(X)}(\mathbf{X}^\bullet, \mathbf{I}^\bullet)$$

is an isomorphism. Let $\mathbf{Y}^\bullet \in D^+(X)$ and $\mathbf{I}^\bullet$ an injective resolution for $\mathbf{Y}^\bullet$. Then

$$\begin{aligned}\mathrm{Hom}_{D^+(X)}(\mathbf{X}^\bullet, \mathbf{Y}^\bullet) &\cong \mathrm{Hom}_{D^+(X)}(\mathbf{X}^\bullet, \mathbf{I}^\bullet) \cong \mathrm{Hom}_{K^+(X)}(\mathbf{X}^\bullet, \mathbf{I}^\bullet)\\ &\cong H^0\Gamma(X; \mathbf{Hom}^\bullet(\mathbf{X}^\bullet, \mathbf{I}^\bullet)) \cong H^0\Gamma(X; R\mathbf{Hom}^\bullet(\mathbf{X}^\bullet, \mathbf{Y}^\bullet))\end{aligned}$$

using Remark 1.3.7. Therefore by Theorem 3.3.1,

$$\begin{aligned}\mathrm{Hom}_{D^+(Y)}(Rf_!\mathbf{A}^\bullet, \mathbf{B}^\bullet) &\cong H^0\Gamma(Y; R\mathbf{Hom}^\bullet(Rf_!\mathbf{A}^\bullet, \mathbf{B}^\bullet)) \\ &\cong H^0\Gamma(X; R\mathbf{Hom}^\bullet(\mathbf{A}^\bullet, f^!\mathbf{B}^\bullet)) \\ &\cong \mathrm{Hom}_{D^+(X)}(\mathbf{A}^\bullet, f^!\mathbf{B}^\bullet).\end{aligned}$$

□

Examples 3.3.3 *If $i : U \hookrightarrow X$ is an open inclusion then the adjunction formula*

$$\mathbf{Hom}(i_!\mathbf{A}, \mathbf{B}) \cong i_* \mathbf{Hom}(\mathbf{A}, i^!\mathbf{B})$$

already holds on the level of sheaves and implies (see the beginning of this section) that $i^! = i^*$. Setting $\mathbf{A} = i^*\mathbf{B} = i^!\mathbf{B}$ and taking global sections yields a canonical adjunction morphism

$$i_!i^*\mathbf{B} \longrightarrow \mathbf{B}. \tag{3.4}$$

Let $j : X \hookrightarrow Y$ be a closed inclusion. Again $j_!$ has a right adjoint $j^!$ already on the category of sheaves. To describe $j^!$, let U be an open subset of Y with inclusion $i : U \hookrightarrow Y$. Applying $\mathrm{Hom}((\mathbb{Z}_{X\cap U})^Y, -)$ to the adjunction map (3.4), we obtain an isomorphism

$$\mathrm{Hom}((\mathbb{Z}_{X\cap U})^Y, i_!i^*\mathbf{B}) \xrightarrow{\simeq} \mathrm{Hom}((\mathbb{Z}_{X\cap U})^Y, \mathbf{B}).$$

As shown in the beginning of this section, the adjunction formula implies

$$\Gamma(X \cap U; j^!\mathbf{B}) = \mathrm{Hom}(j_!(\mathbb{Z}_{X\cap U})^X, \mathbf{B})$$

and thus

$$\begin{aligned}\Gamma(X \cap U; j^!\mathbf{B}) &= \mathrm{Hom}((\mathbb{Z}_{X\cap U})^Y, \mathbf{B}) \\ &\cong \mathrm{Hom}((\mathbb{Z}_{X\cap U})^Y, i_!i^*\mathbf{B}) \\ &= \mathrm{Hom}((\mathbb{Z}_{X\cap U})^Y, (\mathbf{B}|_U)^Y) \\ &= \mathrm{Hom}(\mathbb{Z}_{X\cap U}, \mathbf{B}|_U) \\ &= \Gamma_{X\cap U}(U; \mathbf{B}),\end{aligned}$$

where $\Gamma_{X\cap U}(U; \mathbf{B})$ denotes sections of $\mathbf{B}$ over U supported in $X \cap U$. Thus $j^!\mathbf{B}$ is the sheaf

$$X \cap U \mapsto \Gamma_{X\cap U}(U; \mathbf{B}).$$

3.4 The Dualizing Functor

Definition 3.4.1 *Let $f : X \to pt$ be the map to a point and regard $\mathbb{Z}$ as a sheaf $\mathbb{Z}_{pt}$ over the point. The* dualizing complex on X, *$\mathbb{D}^\bullet_X \in ObD^b(X)$, is*

$$\mathbb{D}^\bullet_X = f^!\mathbb{Z}_{pt}.$$

Fix a bounded c-soft flat resolution $\mathbf{K}^\bullet$ of $\mathbb{Z}_X$ and a bounded injective resolution $I^\bullet$ of $\mathbb{Z}$. We calculate the associated presheaf of $\mathbb{D}^\bullet_X$ ($U \subset X$ open):

$$\begin{aligned}\Gamma(U;\mathbb{D}_X^\bullet) &= \Gamma(U; f^!\mathbb{Z}_{pt}) = \Gamma(U; f^!_{\mathbf{K}^\bullet} I^\bullet) \\ &= \mathrm{Hom}^\bullet(f_!(\mathbf{K}^\bullet|_U)^X, I^\bullet) = \mathrm{Hom}^\bullet(\Gamma_c(X;(\mathbf{K}^\bullet|_U)^X), I^\bullet) \\ &= \mathrm{Hom}^\bullet(\Gamma_c(U;\mathbf{K}^\bullet), I^\bullet).\end{aligned}$$

As $f^!_{\mathbf{K}^\bullet}$ takes injectives to injectives, it follows that $\mathbb{D}_X^\bullet$ is a complex of injective sheaves.

Definition 3.4.2 *Let* $\mathbf{A}^\bullet \in D^b(X)$. *The functor*

$$\mathcal{D}_X\mathbf{A}^\bullet = R\mathbf{Hom}^\bullet(\mathbf{A}^\bullet, \mathbb{D}_X^\bullet)$$

is called the (*Borel–Moore–Verdier-*) dualizing functor.

To compute the presheaf of $\mathcal{D}_X\mathbf{A}^\bullet$, let $i : U \hookrightarrow X$ be an open inclusion and ι be the composition $\iota : U \overset{i}{\hookrightarrow} X \overset{f}{\to} pt$. The following calculation uses Proposition 3.2.6:

$$\begin{aligned}\Gamma(U;\mathcal{D}_X\mathbf{A}^\bullet) &= \Gamma(U; i^*\mathbf{Hom}^\bullet(\mathbf{A}^\bullet,\mathbb{D}_X^\bullet)) = \Gamma(U; i^*\mathbf{Hom}^\bullet(\mathbf{A}^\bullet, f^!_{\mathbf{K}^\bullet} I^\bullet)) \\ &= \mathrm{Hom}^\bullet(\mathbf{A}^\bullet|_U, i^* f^!_{\mathbf{K}^\bullet} I^\bullet) = \mathrm{Hom}^\bullet(\mathbf{A}^\bullet|_U, i^! f^!_{\mathbf{K}^\bullet} I^\bullet) \\ &= \mathrm{Hom}^\bullet(\mathbf{A}^\bullet|_U, \iota^!_{\mathbf{K}^\bullet|_U} I^\bullet) = \mathrm{Hom}^\bullet(\iota_!(\mathbf{A}^\bullet|_U \otimes \mathbf{K}^\bullet|_U), I^\bullet) \\ &= \mathrm{Hom}^\bullet(\Gamma_c(U;\mathbf{A}^\bullet \otimes \mathbf{K}^\bullet), I^\bullet).\end{aligned}$$

We need the following standard result:

Lemma 3.4.3 *Let* $C^\bullet \in C^b(Ab)$ *be a bounded complex of abelian groups. Then for each* $i \in \mathbb{Z}$, *there is a short exact sequence*

$$0 \longrightarrow \mathrm{Ext}(H^{i+1}C^\bullet, \mathbb{Z}) \longrightarrow H^{-i}DC^\bullet \longrightarrow \mathrm{Hom}(H^iC^\bullet,\mathbb{Z}) \longrightarrow 0,$$

where $DC^\bullet = R\,\mathrm{Hom}^\bullet(C^\bullet,\mathbb{Z})$.

The relation of hypercohomology with coefficients the dual of a complex $\mathbf{A}^\bullet$ to the hypercohomology with coefficients $\mathbf{A}^\bullet$ is provided by the following short exact sequence:

Theorem 3.4.4 *Let* $\mathbf{A}^\bullet \in ObD^b(X)$ *and* $U \subset X$ *open. Then for each* $i \in \mathbb{Z}$, *there is a short exact sequence*

$$0 \longrightarrow \mathrm{Ext}(\mathcal{H}_c^{i+1}(U;\mathbf{A}^\bullet),\mathbb{Z}) \longrightarrow \mathcal{H}^{-i}(U;\mathcal{D}_X\mathbf{A}^\bullet) \longrightarrow \mathrm{Hom}(\mathcal{H}_c^i(U;\mathbf{A}^\bullet),\mathbb{Z}) \longrightarrow 0.$$

Proof. As $\mathbb{D}_X^\bullet$ is injective, $\mathbf{Hom}^\bullet(\mathbf{A}^\bullet,\mathbb{D}_X^\bullet)$ is flabby. The class of flabby sheaves is adapted to Γ, whence

$$\mathcal{H}^{-i}(U;\mathcal{D}_X\mathbf{A}^\bullet) = H^{-i}\Gamma(U;\mathbf{Hom}^\bullet(\mathbf{A}^\bullet,\mathbb{D}_X^\bullet)) = H^{-i}\,\mathrm{Hom}^\bullet(\Gamma_c(U;\mathbf{A}^\bullet\otimes\mathbf{K}^\bullet), I^\bullet).$$

Put $C^\bullet = \Gamma_c(U;\mathbf{A}^\bullet\otimes\mathbf{K}^\bullet)$, a bounded complex of abelian groups. By Lemma 3.4.3, we have the exact sequence

$$0 \longrightarrow \mathrm{Ext}(H^{i+1}C^\bullet,\mathbb{Z}) \longrightarrow H^{-i}DC^\bullet \longrightarrow \mathrm{Hom}(H^iC^\bullet,\mathbb{Z}) \longrightarrow 0.$$

Now

$$H^{-i} DC^\bullet = H^{-i} \operatorname{Hom}^\bullet(C^\bullet, I^\bullet) = \mathcal{H}^{-i}(U; \mathcal{D}_X \mathbf{A}^\bullet),$$

and

$$H^i C^\bullet = H^i \Gamma_c(U; \mathbf{A}^\bullet \otimes \mathbf{K}^\bullet) = \mathcal{H}^i_c(U; \mathbf{A}^\bullet),$$

since $\mathbf{A}^\bullet \otimes \mathbf{K}^\bullet$ is a c-soft resolution of $\mathbf{A}^\bullet$. □

Thus if we take $U = X$ for a compact space X and replace the coefficient ring $\mathbb{Z}$ by a field k (in particular $\mathbb{D}^\bullet_X = f^! k_{pt}$, $f : X \to pt$), we obtain a canonical isomorphism

$$\mathcal{H}^{-i}(X; \mathcal{D}_X \mathbf{A}^\bullet) \cong \operatorname{Hom}(\mathcal{H}^i(X; \mathbf{A}^\bullet), k). \tag{3.5}$$

Duality calculations will routinely involve the following formulae:

Proposition 3.4.5 *Let $f : X \to Y$ be a continuous map and let $\mathbf{A}^\bullet \in D^+(X)$, $\mathbf{B}^\bullet \in D^+(Y)$. Then there are canonical isomorphisms*

$$\mathcal{D}_Y(Rf_! \mathbf{A}^\bullet) \cong Rf_* \mathcal{D}_X \mathbf{A}^\bullet, \qquad \mathcal{D}_X(f^* \mathbf{B}^\bullet) \cong f^! \mathcal{D}_Y \mathbf{B}^\bullet.$$

Proof. Using the commutative diagram

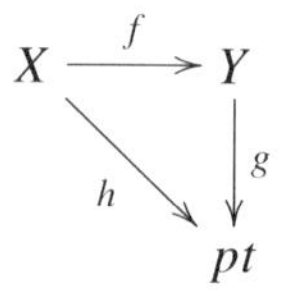

we obtain the relation

$$\mathbb{D}^\bullet_X = h^! \mathbb{Z}_{pt} \cong f^! g^! \mathbb{Z}_{pt} = f^! \mathbb{D}^\bullet_Y$$

between the dualizing complexes for X and Y. Thus, in view of Theorem 3.3.1,

$$\begin{aligned} \mathcal{D}_Y(Rf_! \mathbf{A}^\bullet) &= R\mathbf{Hom}^\bullet(Rf_! \mathbf{A}^\bullet, \mathbb{D}^\bullet_Y) \\ &\cong Rf_* R\mathbf{Hom}^\bullet(\mathbf{A}^\bullet, f^! \mathbb{D}^\bullet_Y) \\ &\cong Rf_* R\mathbf{Hom}^\bullet(\mathbf{A}^\bullet, \mathbb{D}^\bullet_X) \\ &= Rf_* \mathcal{D}_X \mathbf{A}^\bullet. \end{aligned}$$

Working with the formula

$$f^! R\mathbf{Hom}^\bullet(\mathbf{B}^\bullet, \mathbf{C}^\bullet) \cong R\mathbf{Hom}^\bullet(f^* \mathbf{B}^\bullet, f^! \mathbf{C}^\bullet)$$

(whose verification we may leave to the reader) instead of Theorem 3.3.1, we get

$$\begin{aligned} \mathcal{D}_X(f^* \mathbf{B}^\bullet) &= R\mathbf{Hom}^\bullet(f^* \mathbf{B}^\bullet, \mathbb{D}^\bullet_X) \\ &\cong R\mathbf{Hom}^\bullet(f^* \mathbf{B}^\bullet, f^! \mathbb{D}^\bullet_Y) \\ &\cong f^! R\mathbf{Hom}^\bullet(\mathbf{B}^\bullet, \mathbb{D}^\bullet_Y) \\ &= f^! \mathcal{D}_Y \mathbf{B}^\bullet. \end{aligned}$$ □

Thus the functors Rf_* and $Rf_!$ are dual to each other, as well as the functors f^* and $f^!$.

3.5 Poincaré Duality on Manifolds

On a manifold, we recognize the dualizing complex, up to quasi-isomorphism, as the orientation sheaf (shifted into the correct dimension). If the manifold is oriented, we shall recover classical Poincaré duality from Verdier duality.

Proposition 3.5.1 *Let $(M, \partial M)$ be an n-dimensional topological manifold with boundary and $\mathcal{O}_M$ be the orientation sheaf on M. Then there is a canonical isomorphism in $D^b(M)$,*

$$\mathbb{D}_M^\bullet \xrightarrow{\simeq} \mathcal{O}_M[n].$$

Proof. Let $f : M \to pt$ be the map to a point so that $\mathbb{D}_M^\bullet = f^!\mathbb{Z}_{pt}$, $\mathbf{K}^\bullet$ a bounded c-soft flat resolution of $\mathbb{Z}_M$ and $I^\bullet$ a bounded injective resolution of $\mathbb{Z}$. The cohomology sheaf $\mathbf{H}^{-i}(f^!\mathbb{Z}_{pt})$ is the sheafification of

$$U \mapsto H^{-i}\Gamma(U; f^!\mathbb{Z}_{pt})$$

($U \subset M$ open) and $\mathcal{O}_M$ is by definition the sheaf

$$U \mapsto \operatorname{Hom}(H_c^n(U; \mathbb{Z}), \mathbb{Z}).$$

In Sect. 3.4 we have seen that

$$H^{-i}\Gamma(U; f^!\mathbb{Z}_{pt}) = H^{-i}\operatorname{Hom}^\bullet(\Gamma_c(U; \mathbf{K}^\bullet), I^\bullet).$$

Put $C^\bullet = \Gamma_c(U; \mathbf{K}^\bullet)$ in Lemma 3.4.3. Then $H^iC^\bullet = H_c^i(U; \mathbb{Z})$ (as $\mathbf{K}^\bullet$ is a c-soft resolution of $\mathbb{Z}_M$) and we obtain the short exact sequence

$$\begin{aligned} 0 &\to \operatorname{Ext}(H_c^{i+1}(U; \mathbb{Z}), \mathbb{Z}) \to H^{-i}\operatorname{Hom}^\bullet(\Gamma_c(U; \mathbf{K}^\bullet), I^\bullet) \\ &\to \operatorname{Hom}(H_c^i(U; \mathbb{Z}), \mathbb{Z}) \to 0. \end{aligned}$$

When U is a small neighborhood of an interior point, use

$$H_c^i(\mathbb{R}^n; \mathbb{Z}) = \begin{cases} \mathbb{Z}, & i = n, \\ 0, & i \neq n \end{cases}$$

and when U is a small neighborhood of a boundary point, use

$$H_c^i([0,1) \times \mathbb{R}^{n-1}; \mathbb{Z}) = 0, \quad \text{all } i \in \mathbb{Z}.$$

Hence for $i = n$,

$$\operatorname{Ext}(H_c^{n+1}(U; \mathbb{Z}), \mathbb{Z}) = \operatorname{Ext}(0, \mathbb{Z}) = 0$$

and

$$H^{-n}\Gamma(U; f^!\mathbb{Z}_{pt}) \xrightarrow{\simeq} \operatorname{Hom}(H_c^n(U; \mathbb{Z}), \mathbb{Z})$$

is an isomorphism of presheaves. For $i = n - 1$, both

$$\operatorname{Hom}(H_c^{n-1}(U; \mathbb{Z}), \mathbb{Z}) = 0$$

and

$$\mathrm{Ext}(H_c^n(U;\mathbb{Z}),\mathbb{Z}) = \begin{cases} \mathrm{Ext}(\mathbb{Z},\mathbb{Z}) = 0, & U \subset \mathrm{int}\, M, \\ \mathrm{Ext}(0,\mathbb{Z}) = 0, & \text{otherwise} \end{cases}$$

vanish, and thus

$$H^{-n+1}\Gamma(U;\, f^!\mathbb{Z}_{pt}) = 0.$$

When $i \neq n, n-1$, then $H_c^i(U;\mathbb{Z}) = H_c^{i+1}(U;\mathbb{Z}) = 0$, so $H^{-i}\Gamma(U;\, f^!\mathbb{Z}_{pt}) = 0$. □

For a field k, we consider the orientation sheaf $\Gamma(U;\mathcal{O}_M) = \mathrm{Hom}(H_c^n(U;k),k)$, a sheaf of k-vector spaces.

Definition 3.5.2 *An* orientation of the manifold $(M,\partial M)$ relative to k *is an isomorphism*

$$\mathcal{O}_M|_{M-\partial M} \cong k_{M-\partial M}.$$

Assume $\partial M = \varnothing$ and M is oriented. The orientation induces an isomorphism

$$\mathcal{H}^{n-i}(M;k_M) \cong \mathcal{H}^{n-i}(M;\mathcal{O}_M).$$

By Proposition 3.5.1,

$$\mathcal{H}^{n-i}(M;\mathcal{O}_M) = \mathcal{H}^{-i}(M;\mathcal{O}_M[n]) \cong \mathcal{H}^{-i}(M;\mathbb{D}_M^\bullet)$$

and by Theorem 3.4.4,

$$\mathcal{H}^{-i}(M;\mathbb{D}_M^\bullet) \cong \mathcal{H}^{-i}(M;\mathcal{D}_M k_M) \cong \mathrm{Hom}(\mathcal{H}_c^i(M;k_M),k).$$

Composing, we get classical Poincaré duality

$$\mathcal{H}^{n-i}(M;k) \cong \mathrm{Hom}(\mathcal{H}_c^i(M;k),k).$$

4

Intersection Homology

4.1 Piecewise Linear Intersection Homology

4.1.1 Introduction

In [Poi95] and [Poi99], Poincaré defines the intersection product of two cycles of complementary dimension in a compact oriented manifold. More generally, Lefschetz [Lef26] defines the intersection product of cycles of any dimension. This product can be described as follows: Let x be an i-cycle and y a j-cycle in the compact oriented manifold M^n. If x and y are in general position, then their intersection $x \cap y$ is an $(i + j - n)$-chain. Moreover, this chain has no boundary and thus $x \cap y$ is a cycle. Next, one shows that the homology class of $x \cap y$ depends only on the homology classes of x and y. Given any two cycles x and y in M, the principle of general position in M implies that we can, by an isotopy, move one of the cycles, say y, within its homology class to a cycle y' such that x and y' are in general position. Thus, one obtains a well-defined intersection product

$$H_i(M) \times H_j(M) \stackrel{\cap}{\longrightarrow} H_{i+j-n}(M).$$

In the case of $i + j = n$, the pairing

$$H_i(M) \times H_j(M) \stackrel{\cap}{\longrightarrow} H_0(M) \stackrel{\epsilon_*}{\longrightarrow} \mathbb{Z}$$

is nondegenerate over the rationals ("Poincaré duality"; ϵ_* is the augmentation homomorphism).

For a singular space X, such an intersection pairing does not exist. As an example, consider the suspension of a 2-torus, $X^3 = \Sigma(S^1 \times S^1)$. The two isolated singularities of X^3 are the two cone-points. Let x be the 2-cycle $x = \Sigma(pt \times S^1)$ and y be the 2-cycle $y = \Sigma(S^1 \times pt)$. The intersection $x \cap y = \Sigma(pt \times pt)$ is still a $2 + 2 - 3 = 1$-chain, but it has a boundary, namely the two cone-points. This boundary does not change as we move x and y in their homology classes. The fact that Poincaré duality fails for X^3 is apparent from its list of Betti numbers:

$$b_0(X) = 1, \qquad b_1(X) = 0, \qquad b_2(X) = 2, \qquad b_3(X) = 1,$$

so that for singular spaces we do not have *"les nombres de Betti également distants des extrêmes sont égaux,"* as Poincaré asserts for manifolds.

Motivated by a question of D. Sullivan [Sul70a], Goresky and MacPherson define (in [GM80] for PL pseudomanifolds and in [GM83] for topological pseudomanifolds) a collection of groups $IH_*^{\bar{p}}(X)$, called *intersection homology groups of* X, depending on a multi-index $\bar{p}$, called a *perversity*. For these groups, a Poincaré–Lefschetz-type intersection theory can be defined, and a generalized Poincaré duality holds, but only for groups with "complementary perversities." For PL spaces, the intersection homology groups are the homology of a certain subcomplex of the complex of all PL chains. We discuss this approach in Sect. 4.1.3. However, we shall derive most of the properties of intersection homology using the more refined approach adopted in [GM83]. In that framework, $IH_*^{\bar{p}}(X)$ is realized as the hypercohomology of a complex of sheaves in the derived category on any topological pseudomanifold. Generalized Poincaré duality will be induced from a sheaf-theoretic duality statement involving the Borel–Moore–Verdier dualizing functor as defined in Sect. 3.4.

The groups $IH_*^{\bar{p}}(X)$ do not form a homology theory: If we allow arbitrary continuous maps, then intersection homology is not a homotopy invariant. Greg Friedman has however shown that intersection homology based on singular chains is invariant under stratum-preserving homotopy equivalences, see [Fri03]. Also, as $IH_*^{\bar{p}}$ can be thought of as interpolating between cohomology and homology, functoriality is an issue that has to be treated carefully (for instance $IH_*^{\bar{p}}$ is a bivariant functor on the category of topological pseudomanifolds and normally nonsingular maps).

Dual cohomology groups were discovered independently by J. Cheeger [Che79], [Che80], [Che83] by considering the L^2 De Rham complex on the open top-stratum of a space equipped with a locally conical metric. We provide a brief introduction to Cheeger's theory in Chap. 10.

4.1.2 Stratifications

We adopt the convention that the cone on the empty set is a point. The open cone on a space X is denoted by $c^\circ X$.

Definition 4.1.1 *A* 0-dimensional topological stratified pseudomanifold *is a countable set of points with the discrete topology.*

An n-dimensional topological stratified pseudomanifold *is a paracompact Hausdorff topological space* X *with a filtration by closed subspaces*

$$X = X_n \supset X_{n-1} = X_{n-2} \supset X_{n-3} \supset \cdots \supset X_1 \supset X_0 \supset X_{-1} = \varnothing$$

such that

1. *Every non-empty* $X_{n-k} - X_{n-k-1}$ *is a topological manifold of dimension* $n-k$*, called a* pure *or* open stratum of X.
2. $X - X_{n-2}$ *(called the* top stratum of X*) is dense in* X.

3. Local normal triviality: *For each point* $x \in X_{n-k} - X_{n-k-1}$, *there exists an open neighborhood* U *of* x *in* X, *a compact topological stratified pseudomanifold* L *of dimension* $k-1$ *with stratification*

$$L = L_{k-1} \supset L_{k-3} \supset \cdots \supset L_0 \supset L_{-1} = \varnothing$$

and a homeomorphism

$$\phi : U \xrightarrow{\simeq} \mathbb{R}^{n-k} \times c^\circ L$$

which is stratum-preserving, i.e. ϕ *restricts to homeomorphisms* $\phi| : U \cap X_{n-l} \xrightarrow{\simeq} \mathbb{R}^{n-k} \times c^\circ L_{k-l-1}$ *(see Fig. 4.1).*

The definition implies that topological stratified pseudomanifolds are locally compact. In the local normal triviality condition, we think of the factor $\mathbb{R}^{n-k}$ as the "horizontal direction" along the manifold $X_{n-k} - X_{n-k-1}$, and we think of the factor $c^\circ L$ as the "normal slice" to the stratum at x. The space L is called the *link* of the stratum $X_{n-k} - X_{n-k-1}$ at the point x, and U is a *distinguished neighborhood* of x. We refer to X_{n-2} as the *singular set of* X and will often write $\Sigma = X_{n-2}$.

A *PL space* X is a topological space together with a class $\mathcal{T}$ of locally finite simplicial triangulations T of X. The class $\mathcal{T}$ is required to be closed under the operation of linear subdivision, and if $T, T' \in \mathcal{T}$, then T and T' are required to have a common linear subdivision. A *closed PL subspace of* X is a subcomplex of a suitable triangulation $T \in \mathcal{T}$ of X.

Definition 4.1.2 *An* n-dimensional PL stratified pseudomanifold *is a PL space* X *of dimension* n *with a filtration by closed PL subspaces*

$$X = X_n \supset X_{n-1} = X_{n-2} \supset X_{n-3} \supset \cdots \supset X_1 \supset X_0 \supset X_{-1} = \varnothing$$

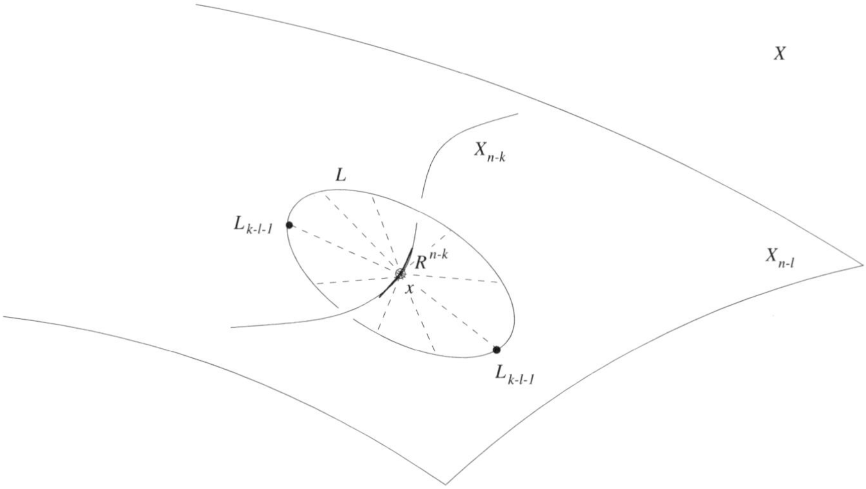

Fig. 4.1. Local normal triviality and the link of a stratum.

such that

1. *Every nonempty* $X_{n-k} - X_{n-k-1}$ *is a PL manifold of dimension* $n - k$,
2. $X - X_{n-2}$ *is dense in* X,
3. Local normal triviality: *Same as in Definition* 4.1.1(3), *with* L *a PL stratified pseudomanifold and* ϕ *a PL isomorphism.*

Examples 4.1.3

- *Let* X *be a PL pseudomanifold of dimension* n, *i.e. a PL space* X *with an admissible triangulation* $T \in \mathcal{T}$ *such that* X *is the union of the* n*-simplices in* T *and each* $(n-1)$*-simplex is a face of exactly two* n*-simplices. Then* X *can be filtered so as to acquire the structure of a PL stratified pseudomanifold: For* $i \neq n-1$, *take* X_i *to be the* i*-skeleton of* T *and set* $X_{n-1} = X_{n-2}$ *(of course this is not in general an "intrinsic" stratification).*
- *All irreducible complex algebraic or analytic varieties can be filtered so as to be topological stratified pseudomanifolds.*
- *The set of points* x *in a normal* n*-dimensional real algebraic variety such that every neighborhood of* x *has topological dimension* n *can be filtered so as to be a topological stratified pseudomanifold.*
- *The framework of stratifications is frequently useful even when the underlying space is not singular. This point of view is sometimes adopted in high-dimensional knot theory. If a manifold is stratified by submanifolds, then the links are all spheres.*

4.1.3 Piecewise Linear Intersection Homology

Let X be a PL space and $T \in \mathcal{T}$ be an admissible triangulation of X. A *simplicial* i*-chain* ξ is a function

$$\xi : \{\sigma \in T | \sigma \text{ an oriented } i\text{-simplex } \} \longrightarrow \mathbb{Z}$$

with $\xi(-\sigma) = -\xi(\sigma)$, where $-\sigma$ denotes σ with the opposite orientation. In particular, we allow infinite chains. The support $|\xi|$ of ξ is the union of the closed i-simplices σ of T such that $\xi(\sigma) \neq 0$; $|\xi|$ is a closed PL subspace of X since T is locally finite. Let $C_i^T(X)$ denote the abelian group of all simplicial i-chains.

Let T' be a subdivision of T and $\xi \in C_i^T(X)$. Define an i-chain $\xi' \in C_i^{T'}(X)$ by (σ' an i-simplex of T')

$$\xi'(\sigma') = \begin{cases} 0, & \text{if } \sigma' \text{ is not contained in an } i\text{-simplex of } T, \\ \xi(\sigma), & \text{if } \sigma' \text{ is contained in the compatibly oriented } i\text{-simplex } \sigma \in T. \end{cases}$$

The assignment $\xi \mapsto \xi'$ defines a canonical map

$$C_i^T(X) \longrightarrow C_i^{T'}(X);$$

note that $|\xi'| = |\xi|$.

Now let T, T' be any two admissible triangulations of X and $\xi \in C_i^T(X)$, $\xi' \in C_i^{T'}(X)$. We call ξ and ξ' equivalent if T and T' have a common subdivision T'' such that the images of ξ and ξ' under the canonical maps

$$\xi \in C_i^T(X) \longrightarrow C_i^{T''}(X) \longleftarrow C_i^{T'}(X) \ni \xi'$$

coincide in $C_i^{T''}(X)$. Define the *group of PL i-chains* $C_i(X)$ to be the set of equivalence classes of simplicial i-chains. In other words,

$$C_i(X) = \varinjlim_{T \in \mathcal{T}} C_i^T(X).$$

Elements $\xi \in C_i(X)$ have a well-defined support $|\xi|$. The reason for allowing infinite chains here is that we shall eventually need restriction maps $C_i(X) \to C_i(U)$, $U \subset X$ open, which would not exist if we insisted on finite simplicial chains.

Definition 4.1.4 *The homology of the complex $C_\bullet(X)$ of PL chains on X,*

$$H_i(X) = H_i(C_\bullet(X)),$$

is called homology with closed supports of X *or* Borel–Moore homology of X.

Remark 4.1.5 *Let $\hat{X} = X \cup \{\infty\}$ be the one-point compactification of the locally compact space X. At least when X is a forward tame ANR, we may think of Borel–Moore homology as $H_*(X) = H_*^{ord}(\hat{X}, \infty)$, where H_*^{ord} is the ordinary relative homology of the pair $(\hat{X}, \infty)$, see [HR96], Chap. 7. If X is compact, then the Borel–Moore homology of X agrees with the ordinary homology of X.*

Example 4.1.6 *We have*

$$H_i(\mathbb{R}^n; \mathbb{Z}) = \begin{cases} \mathbb{Z}, & i = n, \\ 0, & i \neq n. \end{cases}$$

Let X^n be a PL stratified pseudomanifold. In computing $H_*(X)$ from the chain complex $C_\bullet(X)$, we allow arbitrary PL chains. Call a PL i-chain ξ *transverse to the stratification of* X if

$$\dim(|\xi| \cap X_{n-k}) = i + (n - k) - n = i - k,$$

for all $k \geq 2$. If we define a chain complex by allowing only transverse PL chains and compute its i-th homology, then, by a theorem of McCrory (see e.g. [McC75], Theorem 5.2), we obtain the cohomology $H^{n-i}(X)$ of X. Hence, if we could move every chain to be transverse to the stratification, then Poincaré duality would hold. However, as we have seen in Sect. 4.1.1, classical Poincaré duality fails for general pseudomanifolds, thus so does transversality. The idea, then, is to introduce a parameter, a "perversity," that specifies the allowable "deviation" from full transversality, and to associate a group to each value of the parameter, thereby obtaining a whole spectrum of groups ranging from cohomology to homology.

Definition 4.1.7 *A* perversity $\bar{p}$ *is a function associating to each integer* $k \geq 2$ *an integer* $\bar{p}(k)$ *such that* $\bar{p}(2) = 0$ *and* $\bar{p}(k) \leq \bar{p}(k+1) \leq \bar{p}(k) + 1$.

Examples 4.1.8 *Several perversities play a distinguished role*: *The* zero perversity $\bar{0}$ *is the function* $\bar{0}(k) = 0$, *the* top perversity $\bar{t}$ *is* $\bar{t}(k) = k - 2$. *The* lower middle perversity $\bar{m}$ *is* $(0, 0, 1, 1, 2, 2, \ldots)$ *and the* upper middle perversity $\bar{n}$ *is* $(0, 1, 1, 2, 2, 3, \ldots)$. *Observe that* $\bar{m} + \bar{n} = \bar{t}$.

We call two perversities $\bar{p}, \bar{q}$ *complementary* if $\bar{p} + \bar{q} = \bar{t}$.

Definition 4.1.9 *Let X be an n-dimensional PL pseudomanifold with stratification*

$$X = X_n \supset X_{n-2} \supset \cdots \supset X_0 \supset \varnothing.$$

The group $IC_i^{\bar{p}}(X)$ of i-dimensional intersection chains of perversity $\bar{p}$ *is defined to be*

$$IC_i^{\bar{p}}(X) = \{\xi \in C_i(X) |\ \dim(|\xi| \cap X_{n-k}) \leq i - k + \bar{p}(k),$$
$$\dim(|\partial\xi| \cap X_{n-k}) \leq i - 1 - k + \bar{p}(k),\ \textit{for all } k \geq 2\}.$$

The boundary operators

$$\partial_i : IC_i^{\bar{p}}(X) \longrightarrow IC_{i-1}^{\bar{p}}(X)$$

are well-defined by the condition imposed on $|\partial\xi|$. Thus $\{(IC_i^{\bar{p}}(X), \partial_i)\}_{i \geq 0}$ forms a subcomplex $IC_\bullet^{\bar{p}}(X)$ of the PL chain complex $C_\bullet(X)$.

Definition 4.1.10 *The homology groups of the PL intersection chain complex,*

$$IH_i^{\bar{p}}(X) = H_i(IC_\bullet^{\bar{p}}(X)),$$

are called the perversity $\bar{p}$ intersection homology groups *of the PL stratified pseudomanifold X*.

If $\bar{p}$ and $\bar{q}$ are two perversities, we write $\bar{p} \leq \bar{q}$ if $\bar{p}(k) \leq \bar{q}(k)$ for all $k \geq 2$. Note that $\bar{0} \leq \bar{p} \leq \bar{t}$ for any $\bar{p}$. If $\bar{p} \leq \bar{q}$, we have an obvious inclusion

$$IC_\bullet^{\bar{p}}(X) \hookrightarrow IC_\bullet^{\bar{q}}(X)$$

which induces a canonical morphism

$$IH_*^{\bar{p}}(X) \longrightarrow IH_*^{\bar{q}}(X).$$

We say that a PL pseudomanifold X^n is *orientable* if for some admissible triangulation T of X one can orient each n-simplex so that the chain that assigns $1 \in \mathbb{Z}$ to each n-simplex with the chosen orientation is an n-cycle. A choice of such an n-cycle is called an *orientation for* X, and its homology class

$$[X] \in H_n(X)$$

is the *fundamental class of* X.

Proposition 4.1.11 *The cap product with the fundamental class*

$$H^{n-k}(X) \xrightarrow{\cap[X]} H_k(X)$$

factors through $IH_k^{\bar{0}}(X)$:

$$H^{n-k}(X) \xrightarrow{\cap[X]} IH_k^{\bar{0}}(X).$$

Proof. Let T be an admissible triangulation of X, T' its first barycentric subdivision, and $C_T^\bullet(X)$ the simplicial cochain complex of T. Let σ be an $(n-k)$-simplex in T. The cap product

$$C_T^{n-k}(X) \xrightarrow{\cap[X]} C_k^{T'}(X)$$

maps the cochain 1_σ,

$$1_\sigma(\sigma') = \begin{cases} 1, & \sigma' = \sigma, \\ 0, & \sigma' \neq \sigma \end{cases}$$

($\sigma' \in T$ any $(n-k)$-simplex) to the dual block $D_X\sigma$ in X, a k-dimensional subcomplex of T'.

There exists a unique open stratum $X_s - X_{s-1}$ such that $\mathrm{int}(\sigma) \subset X_s - X_{s-1}$, $s \geq n-k$. We have $|D_X\sigma| \subset X - X_{s-1}$, and if $n-l \geq s$ then

$$D_X\sigma \cap (T'|_{X_{n-l}}) = D_{X_{n-l}}\sigma.$$

Let X_{n-l} be any stratum. If $n-l \geq s$, then

$$\dim(|D_X\sigma| \cap X_{n-l}) = \dim|D_{X_{n-l}}\sigma| = (n-l)-(n-k) = k-l,$$

and if $n-l < s$, then

$$|D_X\sigma| \cap X_{n-l} \subset (X - X_{s-1}) \cap X_{n-l} = \varnothing.$$

Thus (arguing similarly for $\partial D_X\sigma$),

$$D_X\sigma \in IC_k^{\bar{0}}(X). \qquad \square$$

The natural inclusion

$$IC_\bullet^{\bar{t}}(X) \hookrightarrow C_\bullet(X)$$

induces

$$IH_*^{\bar{t}}(X) \longrightarrow H_*(X).$$

Summarizing, we have canonical maps ($\bar{p}$ any perversity)

$$H^{n-k}(X) \to IH_k^{\bar{0}}(X) \to IH_k^{\bar{p}}(X) \to IH_k^{\bar{t}}(X) \to H_k(X),$$

whose composition is $\cap[X]$.

Let us consider some hands-on examples.

Example 4.1.12 *Let $(M^n, \partial M)$ be a compact manifold and let X^n be the singular space obtained by coning off the boundary of M:*

$$X = M \cup_{\partial M} c(\partial M).$$

The space X has one singular stratum $X_0 = \{c\}$, the cone-point. Its codimension is n, thus in computing $IH_i^{\bar{p}}(X)$, only the value $\bar{p}(n)$ is relevant. We distinguish three cases:

- $i - n + \bar{p}(n) < -1$: *An i-cycle ξ is not allowed to pass through X_0. If ξ bounds, $\xi = \partial \xi'$, then ξ' is also not allowed to pass through X_0.*

$$\begin{aligned}
IH_i^{\bar{p}}(X) &= \ker(\partial_i : IC_i^{\bar{p}}(X) \to IC_{i-1}^{\bar{p}}(X))/\partial_{i+1} IC_{i+1}^{\bar{p}}(X) \\
&= \frac{\{\xi \in C_i(X) \mid \partial\xi = 0, \dim(|\xi| \cap X_0) \le i - n + \bar{p}(n)\}}{\begin{array}{r}\partial_{i+1}\{\xi' \in C_{i+1}(X) \mid \dim(|\xi'| \cap X_0) \le i + 1 - n + \bar{p}(n), \\ \dim(|\partial\xi'| \cap X_0) \le i - n + \bar{p}(n)\}\end{array}} \\
&= \frac{\{\xi \mid \partial\xi = 0, |\xi| \cap X_0 = \varnothing\}}{\partial_{i+1}\{\xi' \mid |\xi'| \cap X_0 = \varnothing\}} \\
&= \frac{\ker(\partial_i : C_i(M) \to C_{i-1}(M))}{\partial_{i+1} C_{i+1}(M)} \\
&= H_i(M).
\end{aligned}$$

- $i - n + \bar{p}(n) = -1$: *An i-cycle ξ is not allowed to pass through X_0. If ξ bounds, $\xi = \partial\xi'$, then ξ' may pass through X_0.*

$$\begin{aligned}
IH_i^{\bar{p}}(X) &= \frac{\{\xi \mid \partial\xi = 0, \dim(|\xi| \cap X_0) \le -1\}}{\partial_{i+1}\{\xi' \mid \dim(|\xi'| \cap X_0) \le 0, \dim(|\partial\xi'| \cap X_0) \le -1\}} \\
&= \frac{\{\xi \mid \partial\xi = 0, |\xi| \cap X_0 = \varnothing\}}{\partial_{i+1}\{\xi' \mid |\partial\xi'| \cap X_0 = \varnothing\}} \\
&= \frac{\ker(\partial_i : C_i(M) \to C_{i-1}(M))}{(\partial C_{i+1}(X)) \cap C_i(M)} \\
&= \operatorname{im}(H_i(M) \longrightarrow H_i(X)).
\end{aligned}$$

- $i - n + \bar{p}(n) \ge 0$: *An i-cycle ξ may pass through X_0. If ξ bounds, $\xi = \partial\xi'$, then ξ' may pass through X_0 as well.*

$$\begin{aligned}
IH_i^{\bar{p}}(X) &= \frac{\{\xi \mid \partial\xi = 0, |\xi| \cap X_0 \text{ arbitrary}\}}{\partial_{i+1}\{\xi' \mid |\xi'| \cap X_0 \text{ arbitrary}\}} \\
&= \frac{\ker(\partial_i : C_i(X) \to C_{i-1}(X))}{\partial_{i+1} C_{i+1}(X)} \\
&= H_i(X).
\end{aligned}$$

Summarizing, we have

$$IH_i^{\bar{p}}(X) = \begin{cases} H_i(M), & i < n - 1 - \bar{p}(n), \\ \operatorname{im}(H_i(M) \longrightarrow H_i(X)), & i = n - 1 - \bar{p}(n), \\ H_i(X), & i > n - 1 - \bar{p}(n). \end{cases}$$

Example 4.1.13 *Consider the space $X^4 = (\Sigma T^2) \times S^1$, stratified as $X = X_4 \supset X_1 = \{c_+\} \times S^1 \sqcup \{c_-\} \times S^1 \supset \varnothing$, where c_+ and c_- are the two suspension points of the unreduced suspension ΣT^2 of the 2-torus; we shall give a table of its lower and upper middle perversity intersection homology. Write the 2-torus as $T^2 = S_1^1 \times S_2^1$. Note that $\bar{m}(3) = 0, \bar{n}(3) = 1$. The following table shows the various generating cycles in X (left column), generators of the lower middle perversity intersection homology groups (middle column), and generators of the upper middle perversity intersection homology groups (right column). The groups $IH_*^{\bar{m}}$ and $IH_*^{\bar{n}}$ are not self-dual, but they are dual to each other. The entries beneath the columns record the transversality condition that chains have to satisfy, depending on the perversity.*

	Cycles		$IH_*^{\bar{m}}(X)$		$IH_*^{\bar{n}}(X)$	
$*=4$		$\Sigma T^2 \times S^1$		$\Sigma T^2 \times S^1$		$\Sigma T^2 \times S^1$
$*=3$	$\Sigma(S_1 \times S_2)$	$S_1 \times S_2 \times S^1$ $(\Sigma S_1) \times S^1$ $(\Sigma S_2) \times S^1$	$\Sigma(S_1 \times S_2)$		$\Sigma(S_1 \times S_2)$	$(\Sigma S_1) \times S^1$ $(\Sigma S_2) \times S^1$
$*=2$	$S_1 \times S_2$ ΣS_1 ΣS_2	$S_1 \times S^1$ $S_2 \times S^1$		$S_1 \times S^1$ $S_2 \times S^1$	ΣS_1 ΣS_2	
$*=1$	S_1 S_2	S^1	S_1 S_2	S^1		S^1
$*=0$	*pt.*		*pt.*		*pt.*	
			$\dim(\xi_i \cap X_1) \leq i - 3$ $\dim(\partial\xi_i \cap X_1) \leq i - 4$		$\dim(\xi_i \cap X_1) \leq i - 2$ $\dim(\partial\xi_i \cap X_1) \leq i - 3$	

Example 4.1.14 *Let X^n be a PL stratified pseudomanifold and stratify the PL space $\mathbb{R} \times X$ by*

$$(\mathbb{R} \times X)_{i+1} = \mathbb{R} \times X_i.$$

Given an intersection i-chain $\xi \in IC_i^{\bar{p}}(X)$, the $(i+1)$-chain $\mathbb{R} \times \xi$ satisfies

$$\begin{aligned} \dim(|\mathbb{R} \times \xi| \cap (\mathbb{R} \times X)_{(n+1)-k}) &= \dim(\mathbb{R} \times (|\xi| \cap X_{n-k})) \\ &\leq 1 + \dim(|\xi| \cap X_{n-k}) \\ &\leq (i+1) - k + \bar{p}(k) \end{aligned}$$

(similarly for $\partial(\mathbb{R} \times \xi)$) and thus is an element $\mathbb{R} \times \xi \in IC_{i+1}^{\bar{p}}(\mathbb{R} \times X)$, called the suspension *of ξ. Suspension induces a map of complexes*

$$IC_{\bullet}^{\bar{p}}(X) \longrightarrow IC_{\bullet}^{\bar{p}}(\mathbb{R} \times X)[1],$$

and it turns out that the corresponding map on homology,

$$IH_i^{\bar{p}}(X) \xrightarrow{\simeq} IH_{i+1}^{\bar{p}}(\mathbb{R} \times X) \tag{4.1}$$

is an isomorphism; for details cf. N. Habegger's notes [B⁺84, II].

Example 4.1.15 (*The intersection homology of cones and punctured cones*) *Let X^{k-1} be a compact PL stratified pseudomanifold. We consider the open cone $c^\circ X$ on X with stratification*

$$(c^\circ X)_i = \begin{cases} c^\circ(X_{i-1}), & i > 0, \\ \{c\}, & i = 0, \end{cases}$$

where c denotes the cone-point. Suppose ξ is an intersection chain $\xi \in IC_{i-1}^{\bar{p}}(X)$, so that by definition

$$\begin{aligned} \dim(|\xi| \cap X_{k-1-j}) &\leq i - 1 - j + \bar{p}(j), \\ \dim(|\partial\xi| \cap X_{k-1-j}) &\leq i - 2 - j + \bar{p}(j). \end{aligned}$$

Let $c^\circ\xi$ denote the cone on the chain ξ, $c^\circ\xi \in C_i(c^\circ X)$. When $j < k$, then

$$\begin{aligned} \dim(|c^\circ\xi| \cap (c^\circ X)_{k-j}) &= \dim(c^\circ(|\xi| \cap X_{k-j-1})) \\ &\leq 1 + \dim(|\xi| \cap X_{k-j-1}) \\ &\leq 1 + (i-1) - j + \bar{p}(j) \\ &= i - j + \bar{p}(j) \end{aligned}$$

(similarly for $\partial\xi$). When $j = k$,

$$\dim(|c^\circ\xi| \cap (c^\circ X)_0) = \dim(|c^\circ\xi| \cap \{c\}) = \dim(\{c\}) = 0,$$

thus we need

$$0 \leq i - k + \bar{p}(k)$$

in order that $c^\circ\xi \in IC_i(c^\circ X)$. Furthermore, if $\partial\xi \neq 0$, we need even

$$0 \leq i - 1 - k + \bar{p}(k).$$

The situation is summarized as follows:

$$\begin{aligned} i > k - \bar{p}(k) &\Rightarrow c^\circ\xi \in IC_i(c^\circ X) \\ i = k - \bar{p}(k) &\Rightarrow c^\circ\xi \in IC_i(c^\circ X) \quad \textit{precisely when } \xi \textit{ is a cycle} \\ i < k - \bar{p}(k) &\Rightarrow c^\circ\xi \notin IC_i(c^\circ X). \end{aligned}$$

This suggests that we should regard coning $\xi \mapsto c^\circ\xi$ as a map of complexes

$$
\begin{array}{ccc}
\tau_{\geq k-\bar{p}(k)-1} IC_{\bullet}^{\bar{p}}(X) & \xrightarrow{c^\circ} & IC_{\bullet}^{\bar{p}}(c^\circ X)[1] \\
\| & & \| \\
\vdots & & \vdots \\
\downarrow & & \downarrow \\
IC_{k-\bar{p}(k)+1}(X) & \xrightarrow{c^\circ} & IC_{k-\bar{p}(k)+2}(c^\circ X) \\
\downarrow & & \downarrow \\
IC_{k-\bar{p}(k)}(X) & \xrightarrow{c^\circ} & IC_{k-\bar{p}(k)+1}(c^\circ X) \\
\downarrow & & \downarrow \\
\ker \partial & \xrightarrow{c^\circ} & IC_{k-\bar{p}(k)}(c^\circ X) \\
\downarrow & & \downarrow \\
0 & \xrightarrow{c^\circ} & IC_{k-\bar{p}(k)-1}(c^\circ X) \\
\downarrow & & \downarrow \\
\vdots & & \vdots
\end{array}
$$

We claim that this map induces an isomorphism on homology. A sketch of the proof, following [B⁺84, §3, II], runs as follows: c° *is injective, so let us consider the short exact sequence*

$$
\begin{aligned}
0 \to \tau_{\geq k-\bar{p}(k)-1} IC_{\bullet}(X) &\overset{c^\circ}{\to} IC_{\bullet}(c^\circ X)[1] \\
&\to IC_{\bullet}(c^\circ X)[1]/c^\circ \tau_{\geq k-\bar{p}(k)-1} IC_{\bullet}(X) \to 0.
\end{aligned}
$$

We have to check that the quotient complex has no homology, i.e. given $\xi \in IC_i(c^\circ X)$ *with* $\partial \xi = c^\circ \eta$, $\eta \in IC_{i-2}(X)$, *we have to find* $\gamma \in IC_{i-1}(X)$ *such that* $\xi - c^\circ \gamma$ *bounds an element of* $IC_{i+1}(c^\circ X)$. *Let* $N = \pi^{-1}[0, \epsilon]$, $\pi : c^\circ X \to \mathbb{R}_+$ *the projection, be a closed neighborhood of the cone-point, chosen so small that it contains no vertex of some triangulation of* $|\xi|$, *except* c. *Then* $\xi \cap N = c^\circ \gamma \cap N$ *with* $\gamma \in IC_{i-1}(X)$, $\partial \gamma = \eta$, *and* $\xi - c^\circ \gamma$ *is a cycle supported in* $\pi^{-1}[\epsilon, \infty)$, *which is PL-isomorphic to* $\mathbb{R}_+ \times X$. *Now it is easy to see that any intersection cycle in* $IC_i(\mathbb{R} \times X)$ *supported in* $\mathbb{R}_+ \times X$ *bounds an element of* $IC_{i+1}(\mathbb{R} \times X)$, *supported in* $\mathbb{R}_+ \times X$.

The intersection homology of a cone is thus

$$
IH_i^{\bar{p}}(c^\circ X) = \begin{cases} IH_{i-1}^{\bar{p}}(X), & i \geq k - \bar{p}(k), \\ 0, & i < k - \bar{p}(k). \end{cases} \tag{4.2}
$$

As for the punctured cone $c^\circ X - \{c\} \cong \mathbb{R} \times X$, *the map*

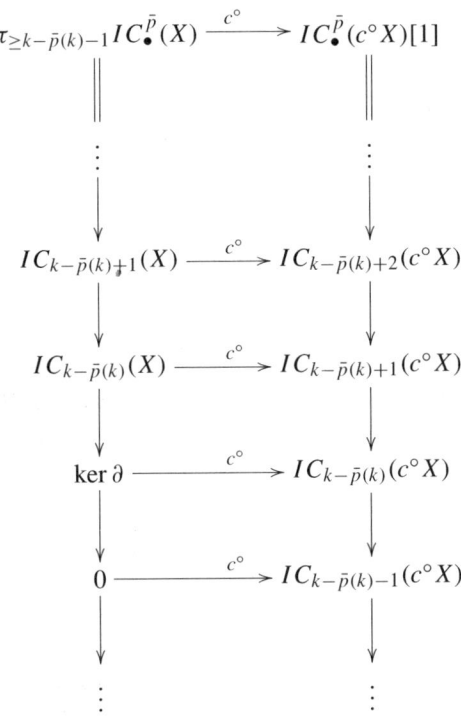

$$\begin{array}{ccc} IC^{\bar{p}}_{\bullet}(X) & \to & IC^{\bar{p}}_{\bullet}(c^{\circ}X - \{c\})[1] \\ \xi & \mapsto & (c^{\circ}\xi) - \{c\} \end{array}$$

induces an isomorphism on homology:

$$IH^{\bar{p}}_i(c^{\circ}X - \{c\}) \cong IH^{\bar{p}}_{i-1}(X). \tag{4.3}$$

Example 4.1.16 (*The intersection homology of distinguished neighborhoods*) *We consider PL stratified pseudomanifolds that have the form of a distinguished neighborhood*

$$\mathbb{R}^{n-k} \times c^{\circ}L,$$

where L is a $(k-1)$-dimensional compact PL stratified pseudomanifold, to be interpreted as a link of a stratum of codimension k. Composing $\xi \mapsto c^{\circ}\xi$ with $n-k$ suspensions, we obtain the chain map

$$\tau_{\geq k-\bar{p}(k)-1} IC^{\bar{p}}_{\bullet}(L) \longrightarrow IC^{\bar{p}}_{\bullet}(\mathbb{R}^{n-k} \times c^{\circ}L)[n-k+1]. \tag{4.4}$$

Composing $\xi \mapsto (c^{\circ}\xi) - \{c\}$ with $n-k$ suspensions, we get

$$IC^{\bar{p}}_{\bullet}(L) \longrightarrow IC^{\bar{p}}_{\bullet}(\mathbb{R}^{n-k} \times (c^{\circ}L - \{c\}))[n-k+1] \tag{4.5}$$

and the diagram

$$\begin{array}{ccc} \tau_{\geq k-\bar{p}(k)-1} IC^{\bar{p}}_{\bullet}(L) & \longrightarrow & IC^{\bar{p}}_{\bullet}(\mathbb{R}^{n-k} \times c^{\circ}L)[n-k+1] \\ {\scriptstyle inclusion}\downarrow & & \downarrow{\scriptstyle restriction} \\ IC^{\bar{p}}_{\bullet}(L) & \longrightarrow & IC^{\bar{p}}_{\bullet}(\mathbb{R}^{n-k} \times (c^{\circ}L - \{c\}))[n-k+1] \end{array} \tag{4.6}$$

commutes. Combining (4.1) *of Example* 4.1.14 *and* (4.2) *of Example* 4.1.15, *we compute the intersection homology of a distinguished neighborhood as*

$$IH^{\bar{p}}_i(\mathbb{R}^{n-k} \times c^{\circ}L) \cong \begin{cases} IH^{\bar{p}}_{i-(n-k+1)}(L), & i \geq n - \bar{p}(k) \\ 0, & i < n - \bar{p}(k) \end{cases} \tag{4.7}$$

(*induced by* (4.4)). *Using* (4.1) *of Example* 4.1.14 *and* (4.3) *of Example* 4.1.15, *we obtain*

$$IH^{\bar{p}}_i(\mathbb{R}^{n-k} \times (c^{\circ}L - \{c\})) \cong IH^{\bar{p}}_{i-(n-k+1)}(L) \tag{4.8}$$

(*induced by* (4.5)).

4.1.4 The Sheafification of the Intersection Chain Complex

Let X^n be a PL stratified pseudomanifold and $U \subset X$ an open subset. Then U has an induced PL structure: S is an admissible triangulation of U if there exist an admissible triangulation T of X and a linear subdivision S' of S such that each simplex of S' is linearly contained in a simplex of T.

Given a PL i-chain $\xi \in C_i(X)$, choose a triangulation T of X and a triangulation S of U such that ξ is simplicial with respect to T and every simplex $\sigma' \in S$ is contained in some simplex $\sigma \in T$. Define $\xi|_U$ to be the PL i-chain on U represented by the simplicial i-chain

$$(\xi|_U)(\sigma') = \begin{cases} 0, & \text{if } \sigma' \text{ is not contained in an } i\text{-simplex of } T, \\ \xi(\sigma), & \text{if } \sigma' \text{ is contained in the compatibly oriented } i\text{-simplex } \sigma \in T \end{cases}$$

($\sigma' \in S$ an i-simplex). Thus we have a restriction map

$$C_i(X) \longrightarrow C_i(U).$$

Note that $|\xi|_U| = |\xi| \cap U$. In particular for open subsets $V \subset U$, we have restrictions

$$C_i(U) \longrightarrow C_i(V),$$

which are functorial with respect to inclusions of open subsets. Now U has an induced filtration

$$U \supset U \cap X_{n-2} \supset U \cap X_{n-3} \supset \cdots \supset U \cap X_0 \supset \varnothing,$$

and we can consider the group of intersection i-chains on U, $IC_i^{\bar{p}}(U) \subset C_i(U)$. For open subsets $V \subset U$, the composition

$$IC_i^{\bar{p}}(U) \hookrightarrow C_i(U) \longrightarrow C_i(V)$$

factors through $IC_i^{\bar{p}}(V)$, and we have functorial restrictions

$$IC_i^{\bar{p}}(U) \longrightarrow IC_i^{\bar{p}}(V).$$

Therefore the assignment

$$U \mapsto IC_i^{\bar{p}}(U)$$

defines a presheaf on X. If $\{U_\alpha\}$ is an open cover of $U \subset X$ and $\xi_\alpha \in IC_i^{\bar{p}}(U_\alpha)$ chains such that $\xi_\alpha|_{U_\alpha \cap U_\beta} = \xi_\beta|_{U_\alpha \cap U_\beta}$, all α, β, then there exists a unique intersection chain $\xi \in IC_i^{\bar{p}}(U)$ with $\xi|_{U_\alpha} = \xi_\alpha$, all α. Hence, the presheaf satisfies the unique gluing condition (G) (Sect. 1.1) and is a sheaf.

Definition 4.1.17 *The sheaf*

$$U \mapsto IC_i^{\bar{p}}(U)$$

is called the perversity $\bar{p}$ intersection chain sheaf on X *and denoted by* $\mathbf{IC}_{\bar{p}}^{-i}(X) \in Sh(X)$. *The* sheaf-complex $\mathbf{IC}_{\bar{p}}^{\bullet}(X) \in C^b(X)$ of perversity $\bar{p}$ intersection chains on *X is given in degree j by*

$$(\mathbf{IC}_{\bar{p}}^{\bullet}(X))^j = \mathbf{IC}_{\bar{p}}^{j}(X)$$

and has differentials

$$\mathbf{IC}_{\bar{p}}^{-i}(X) \longrightarrow \mathbf{IC}_{\bar{p}}^{-i+1}(X)$$

induced by

$$\partial_i : IC_i^{\bar{p}}(U) \longrightarrow IC_{i-1}^{\bar{p}}(U).$$

Our indexing convention is thus homology superscripts and agrees with the notation used in the original paper [GM83]. We have

$$\Gamma(U; \mathbf{IC}_{\bar{p}}^{-i}(X)) = IC_i^{\bar{p}}(U).$$

Our immediate next goal is to show that $\mathbf{IC}_{\bar{p}}^{\bullet}(X)$ is soft. If $K \subset X$ is a closed PL subspace and $\xi \in IC_i^{\bar{p}}(K)$, then we can triangulate X so that K is triangulated by a subcomplex and moreover ξ is a simplicial chain with respect to the subcomplex. This simplicial chain may of course be regarded as an element of $C_i(X)$, but unfortunately it will not in general be in $IC_i^{\bar{p}}(X)$, since, e.g., the intersection conditions imply that an intersection chain $\xi \in IC_i^{\bar{p}}(X)$ is never contained in the singular set Σ. Thus we have to use a slightly more elaborate argument.

Godement [God58, Chapitre II, §3.3] proves

Theorem 4.1.18 *If X is a paracompact topological space, $\mathbf{A} \in Sh(X)$, and $K \subset X$ a closed subspace, then*

$$\Gamma(K; \mathbf{A}) = \varinjlim_{U \supset K} \Gamma(U; \mathbf{A})$$

(U ranges over all open neighborhoods of K in X).

If T is any simplicial complex and $v \in T$ a vertex, let $St(v, T)$ denote the closed star of v in T.

Proposition 4.1.19 $\mathbf{IC}_{\bar{p}}^{\bullet}(X)$ *is soft.*

Proof. Let $K \subset X$ be a closed subspace and $\xi \in \Gamma(K; \mathbf{IC}_{\bar{p}}^{-i}(X))$ a section over K. By Theorem 4.1.18, ξ is represented by a section

$$\hat{\xi} \in \Gamma(U; \mathbf{IC}_{\bar{p}}^{-i}(X)) = IC_i^{\bar{p}}(U),$$

over some open neighborhood $U \supset K$ of K. For an admissible triangulation T of U, let $K(T)$ be the set of vertices of simplices in T that intersect K. Choose a triangulation T of U such that $\hat{\xi}$ is simplicial with respect to T, every $U \cap X_i$ is covered by a subcomplex of T, and

$$N = \bigcup_{v \in K(T)} St(v, T')$$

(T' the first barycentric subdivision of T) is closed in X. For each vertex $v \in T$, $\hat{\xi} \cap St(v, T')$ is in $IC_i^{\bar{p}}(X)$ because $St(v, T')$ is transverse to the strata of X, that is, $St(v, T') \in IC_n^{\bar{0}}(X)$, so that $\hat{\xi} \cap St(v, T') \in IC_i^{\bar{p}+\bar{0}}(X) = IC_i^{\bar{p}}(X)$. Thus

$$\hat{\xi} \cap N \in \Gamma(X; \mathbf{IC}_{\bar{p}}^{-i}(X)) = IC_i^{\bar{p}}(X)$$

is a global section which restricts to $\xi \in \Gamma(K; \mathbf{IC}_{\bar{p}}^{-i}(X))$. □

Corollary 4.1.20 $IH_i^{\bar{p}}(X) \cong \mathcal{H}^{-i}(X; \mathbf{IC}^{\bullet}_{\bar{p}}(X))$.

In the derived category $D^b(X)$, the complex $\mathbf{IC}^{\bullet}_{\bar{p}}(X)$ is characterized by a certain set of axioms, which correspond to key properties of the incarnation of $\mathbf{IC}^{\bullet}_{\bar{p}}(X)$ constructed by using PL chains on X. A good starting point is to understand where on X and in which range the derived sheaves of $\mathbf{IC}^{\bullet}_{\bar{p}}(X)$ can have nonzero stalks.

Proposition 4.1.21 *Let $x \in X$ and U be a distinguished neighborhood of x in X. Then the restriction map induces an isomorphism*

$$IH_i^{\bar{p}}(U) \xrightarrow{\simeq} \mathbf{H}^{-i}(\mathbf{IC}^{\bullet}_{\bar{p}}(X))_x.$$

Proof. Let $X_{n-k} - X_{n-k-1}$ be the pure stratum that contains x and let L^{k-1} be the link of that stratum at x. Write the open cone $c^{\circ}L$ on L as $L \times [0,1)/\sim$, $(z_1, 0) \sim (z_2, 0)$, all $z_1, z_2 \in L$. As U is distinguished, we can choose a stratum preserving PL isomorphism

$$\phi : (-1, +1)^{n-k} \times (L \times [0,1)/\sim) \xrightarrow{\simeq} U.$$

Construct a fundamental system of neighborhoods for x as follows: For $1 \geq \epsilon > 0$, let

$$N_{\epsilon} = (-\epsilon, \epsilon)^{n-k} \times (L \times [0, \epsilon)/\sim)$$

and put $U_{\epsilon} = \phi(N_{\epsilon}) \subset U$, so that $U = U_1$. The chain map (4.4) from Example 4.1.16 gives, for each $\epsilon > 0$, a quasi-isomorphism

$$\tau_{\geq k-\bar{p}(k)-1} IC_{\bullet}^{\bar{p}}(L) \longrightarrow IC_{\bullet}^{\bar{p}}(N_{\epsilon})[n-k+1].$$

If $\epsilon > \delta > 0$, then $N_{\delta} \subset N_{\epsilon}$ and we have a restriction map

$$IC_{\bullet}^{\bar{p}}(N_{\epsilon}) \longrightarrow IC_{\bullet}^{\bar{p}}(N_{\delta}).$$

Restricting commutes with suspending and we obtain a commutative diagram

$$\begin{array}{ccc} & \tau_{\geq k-\bar{p}(k)-1} IC_{\bullet}^{\bar{p}}(L) & \\ \swarrow & & \searrow \\ IC_{\bullet}^{\bar{p}}(N_{\epsilon})[n-k+1] & \xrightarrow{\text{restr}} & IC_{\bullet}^{\bar{p}}(N_{\delta})[n-k+1] \end{array} \tag{4.9}$$

For each $\epsilon > 0$, $\phi| : N_{\epsilon} \xrightarrow{\simeq} U_{\epsilon}$ induces

$$IC_{\bullet}^{\bar{p}}(N_{\epsilon}) \xrightarrow{\simeq} IC_{\bullet}^{\bar{p}}(U_{\epsilon})$$

such that the square

$$\begin{array}{ccc} IC_{\bullet}^{\bar{p}}(N_{\epsilon})[n-k+1] & \xrightarrow{\text{restr}} & IC_{\bullet}^{\bar{p}}(N_{\delta})[n-k+1] \\ \simeq \downarrow \phi|_{*} & & \phi|_{*} \downarrow \simeq \\ IC_{\bullet}^{\bar{p}}(U_{\epsilon})[n-k+1] & \xrightarrow{\text{restr}} & IC_{\bullet}^{\bar{p}}(U_{\delta})[n-k+1] \end{array} \tag{4.10}$$

commutes ($\epsilon > \delta > 0$). Taking homology in diagram (4.9), we see that

$$IH_{*}^{\bar{p}}(N_{\epsilon}) \xrightarrow{\simeq} IH_{*}^{\bar{p}}(N_{\delta})$$

is an isomorphism. Thus, taking homology in diagram (4.10) confirms that

$$IH_{*}^{\bar{p}}(U_{\epsilon}) \xrightarrow{\simeq} IH_{*}^{\bar{p}}(U_{\delta})$$

is an isomorphism. As restrictions are functorial with respect to inclusions of open subsets, we have the commutative diagram

$$\begin{array}{ccc} & IH_{i}^{\bar{p}}(U) & \\ \simeq\swarrow & & \searrow\simeq \\ IH_{i}^{\bar{p}}(U_{\epsilon}) & \xrightarrow{\simeq} & IH_{i}^{\bar{p}}(U_{\delta}) \end{array}$$

and hence an isomorphism

$$IH_{i}^{\bar{p}}(U) \underset{\text{restr}}{\xrightarrow{\simeq}} \lim_{\epsilon \to 0} IH_{i}^{\bar{p}}(U_{\epsilon}). \tag{4.11}$$

Now

$$\mathbf{H}^{-i}(\mathbf{IC}_{\bar{p}}^{\bullet}(X)) = \mathbf{Sheaf}(V \mapsto H^{-i}\Gamma(V; \mathbf{IC}_{\bar{p}}^{\bullet}(X)))$$

($V \subset X$ open), and the stalk at x is given by

$$\begin{aligned} \mathbf{H}^{-i}(\mathbf{IC}_{\bar{p}}^{\bullet}(X))_{x} &= \varinjlim_{V \ni x} H^{-i}\Gamma(V; \mathbf{IC}_{\bar{p}}^{\bullet}(X)) \\ &= \lim_{\epsilon \to 0} H^{-i}\Gamma(U_{\epsilon}; \mathbf{IC}_{\bar{p}}^{\bullet}(X)) \\ &= \lim_{\epsilon \to 0} H^{-i}(IC_{-\bullet}^{\bar{p}}(U_{\epsilon})) \\ &= \lim_{\epsilon \to 0} IH_{i}^{\bar{p}}(U_{\epsilon}). \end{aligned}$$

Thus (4.11) is an isomorphism

$$IH_{i}^{\bar{p}}(U) \xrightarrow{\simeq} \mathbf{H}^{-i}(\mathbf{IC}_{\bar{p}}^{\bullet}(X))_{x}.$$ □

There is an analogous statement for deleted distinguished neighborhoods. Set $U_k = X - X_{n-k}$, $k \geq 2$, and let $i_k : U_k \hookrightarrow U_{k+1}$ be the inclusions.

Proposition 4.1.22 *Let $x \in X_{n-k} - X_{n-k-1}$ and U be a distinguished neighborhood of x in X. Then restriction from the deleted neighborhood $U \cap U_k = U - X_{n-k}$ induces an isomorphism*

$$IH_{i}^{\bar{p}}(U \cap U_k) \xrightarrow{\simeq} \mathbf{H}^{-i}(i_{k*}\mathbf{IC}_{\bar{p}}^{\bullet}(X)|_{U_k})_{x}.$$

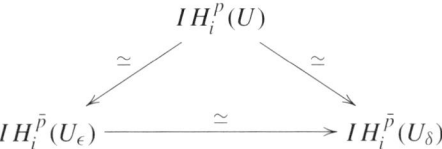

Proof. We use the notation from the proof of Proposition 4.1.21. Using chain map (4.5) in Example 4.1.16, we have for each $1 > \epsilon > 0$ a quasi-isomorphism

$$IC_{\bullet}^{\bar{p}}(L) \longrightarrow IC_{\bullet}^{\bar{p}}(N_{\epsilon} - ((-\epsilon, \epsilon)^{n-k} \times \{c\}))[n-k+1]$$

which we employ as in the proof of Proposition 4.1.21 to see that restriction induces an isomorphism

$$IH_{*}^{\bar{p}}(U_{\epsilon} \cap U_k) \xrightarrow{\simeq} IH_{*}^{\bar{p}}(U_{\delta} \cap U_k),$$

$\epsilon > \delta > 0$, and thus

$$IH_i^{\bar{p}}(U \cap U_k) \underset{\text{restr}}{\xrightarrow{\simeq}} \lim_{\epsilon \to 0} IH_i^{\bar{p}}(U_{\epsilon} \cap U_k).$$

The derived sheaf is given by

$$\begin{aligned} \mathbf{H}^{-i}(i_{k*}\mathbf{IC}_{\bar{p}}^{\bullet}(X)|_{U_k}) &= \mathbf{Sheaf}(V \mapsto H^{-i}\Gamma(V; i_{k*}\mathbf{IC}_{\bar{p}}^{\bullet}(X)|_{U_k})) \\ &= \mathbf{Sheaf}(V \mapsto H^{-i}\Gamma(V \cap U_k; \mathbf{IC}_{\bar{p}}^{\bullet}(X)|_{U_k})) \end{aligned}$$

and its stalk over x is

$$\begin{aligned} \mathbf{H}^{-i}(i_{k*}\mathbf{IC}_{\bar{p}}^{\bullet}(X)|_{U_k})_x &= \varinjlim_{V \ni x} H^{-i}\Gamma(V \cap U_k; \mathbf{IC}_{\bar{p}}^{\bullet}(X)|_{U_k}) \\ &= \lim_{\epsilon \to 0} IH_i^{\bar{p}}(U_{\epsilon} \cap U_k). \end{aligned}$$

□

Combining Proposition 4.1.21 with our calculation of the intersection homology of distinguished neighborhoods in Example 4.1.16, we obtain

Proposition 4.1.23 *The derived sheaves of the complex* $\mathbf{IC}_{\bar{p}}^{\bullet}(X)$ *of intersection chains satisfy the vanishing condition*

$$\mathbf{H}^j(\mathbf{IC}_{\bar{p}}^{\bullet}(X)|_{U_{k+1}}) = 0, \quad \textit{for } j > \bar{p}(k) - n, k \geq 2.$$

Proof. Suppose $x \in X_{n-l} - X_{n-l-1}$, $l \leq k$, and let U be a distinguished neighborhood of x in X. Then

$$\begin{aligned} \mathbf{H}^j(\mathbf{IC}_{\bar{p}}^{\bullet}(X))_x &\cong IH_{-j}^{\bar{p}}(U) && \text{(by Proposition 4.1.21)} \\ &\cong IH_{-j}^{\bar{p}}(\mathbb{R}^{n-l} \times c^{\circ}L) && \text{(since } U \text{ is distinguished)} \\ &\cong \begin{cases} IH_{-j-(n-l+1)}^{\bar{p}}(L), & -j \geq n - \bar{p}(l) \\ 0, & -j < n - \bar{p}(l) \end{cases} && \text{((4.7) in Example 4.1.16).} \end{aligned}$$

So if $j > \bar{p}(k) - n$, then $-j < n - \bar{p}(l)$ (as $l \leq k$ implies $\bar{p}(l) \leq \bar{p}(k)$) and

$$\mathbf{H}^j(\mathbf{IC}_{\bar{p}}^{\bullet}(X))_x = 0.$$

□

Recall that if $\mathbf{A} \in Sh(U_{k+1})$ is any sheaf on U_{k+1}, then there is the canonical adjunction morphism

$$\mathbf{A} \longrightarrow i_{k*}i_k^*\mathbf{A}$$

induced by the restriction of sections ($U \subset U_{k+1}$ open)

$$\Gamma(U;\mathbf{A}) \to \Gamma(U \cap U_k; \mathbf{A}|_{U_k}) = \Gamma(U; i_{k*}i_k^*\mathbf{A}),$$

see the proof of Proposition 1.1.11. Let j_k denote the inclusion $j_k : U_{k+1} - U_k \hookrightarrow U_{k+1}$. The following "attaching property" measures in what range $\mathbf{IC}^\bullet_{\bar{p}}(X)|_{U_{k+1}}$ can be reconstructed (up to quasi-isomorphism) from $\mathbf{IC}^\bullet_{\bar{p}}(X)|_{U_k}$ by pushing forward to U_{k+1}.

Proposition 4.1.24 *The map on derived sheaves*

$$\mathbf{H}^j(j_k^*\mathbf{IC}^\bullet_{\bar{p}}(X)|_{U_{k+1}}) \to \mathbf{H}^j(j_k^* i_{k*}i_k^*\mathbf{IC}^\bullet_{\bar{p}}(X)|_{U_{k+1}})$$

induced by the canonical adjunction

$$\mathbf{IC}^\bullet_{\bar{p}}(X)|_{U_{k+1}} \to i_{k*}i_k^*(\mathbf{IC}^\bullet_{\bar{p}}(X)|_{U_{k+1}})$$

is an isomorphism for $j \le \bar{p}(k) - n$.

Proof. If $-j \ge n - \bar{p}(k)$ then by (4.7), Example 4.1.16,

$$IH^{\bar{p}}_{-j}(U) \cong IH^{\bar{p}}_{-j-(n-k+1)}(L),$$

where U is a distinguished neighborhood of a point $x \in U_{k+1} - U_k = X_{n-k} - X_{n-k-1}$, so that there is a stratum preserving isomorphism $U \cong \mathbb{R}^{n-k} \times c^\circ L$. By (4.8), Example 4.1.16,

$$IH^{\bar{p}}_{-j}(U \cap U_k) \cong IH^{\bar{p}}_{-j-(n-k+1)}(L),$$

and we have the commutative diagram

$$\begin{array}{ccc} & IH^{\bar{p}}_{-j-(n-k+1)}(L) & \\ {\scriptstyle\simeq}\swarrow & & \searrow{\scriptstyle\simeq} \\ IH^{\bar{p}}_{-j}(U) & \xrightarrow{\text{restr}} & IH^{\bar{p}}_{-j}(U \cap U_k) \end{array}$$

whence the restriction

$$IH^{\bar{p}}_{-j}(U) \xrightarrow{\simeq} IH^{\bar{p}}_{-j}(U \cap U_k)$$

is an isomorphism, provided $-j \ge n - \bar{p}(k)$. In the commutative diagram of restrictions

$$\begin{array}{ccc} IH^{\bar{p}}_{-j}(U) & \xrightarrow{\simeq} & IH^{\bar{p}}_{-j}(U \cap U_k) \\ \simeq\downarrow & & \downarrow\simeq \\ \mathbf{H}^j(\mathbf{IC}^\bullet_{\bar{p}}(X))_x & \longrightarrow & \mathbf{H}^j(i_{k*}i_k^*\mathbf{IC}^\bullet_{\bar{p}}(X))_x \end{array}$$

the left vertical map is an isomorphism by Proposition 4.1.21, and the right vertical map is an isomorphism by Proposition 4.1.22. Thus

$$\mathbf{H}^j(\mathbf{IC}^\bullet_{\bar{p}}(X))_x \xrightarrow{\simeq} \mathbf{H}^j(i_{k*}i_k^*\mathbf{IC}^\bullet_{\bar{p}}(X))_x$$

is an isomorphism. □

Definition 4.1.25 *Let X be a topological space and Ξ a filtration $X = X_n \supset X_{n-1} \supset \cdots \supset X_0 \supset X_{-1} = \emptyset$ by closed subsets. A complex of sheaves $\mathbf{A}^\bullet \in C(X)$ on X is called* cohomologically locally constant with respect to Ξ, *if for all i and all $0 \le j \le n$, $\mathbf{H}^i(\mathbf{A}^\bullet)|_{X_j - X_{j-1}}$ is locally constant. We say that $\mathbf{A}^\bullet$ is* constructible with respect to the filtration Ξ, *if $\mathbf{A}^\bullet$ is cohomologically locally constant with respect to Ξ and its cohomology sheaves have finitely generated stalks.*

We denote the full subcategory of $D^b(X)$ consisting of all constructible bounded complexes of sheaves by $D^b_c(X)$.

If X is a topological pseudomanifold and $\mathbf{A}^\bullet \in D^b_c(X)$, then evaluation induces an isomorphism

$$\mathbf{A}^\bullet \xrightarrow{\simeq} \mathcal{D}_X\mathcal{D}_X\mathbf{A}^\bullet.$$

Proposition 4.1.26 *The complex $\mathbf{IC}^\bullet_{\bar{p}}(X)$ of intersection PL-chains on the PL stratified pseudomanifold X is constructible.*

Motivated by our study of the properties of $\mathbf{IC}^\bullet_{\bar{p}}(X)$ (Propositions 4.1.23 and 4.1.24), we adopt the following set of axioms:

Definition 4.1.27 *We say that a complex of sheaves $\mathbf{A}^\bullet \in D^b_c(X)$ in the constructible derived bounded category of X* satisfies the set of axioms [AX], *if*

(AX0): *(Normalization)* $\mathbf{A}^\bullet|_{X-\Sigma} \cong \mathbb{R}_{X-\Sigma}[n]$, *the constant sheaf on $X - \Sigma$.*
(AX1): *(Lower Bound)* $\mathbf{H}^i(\mathbf{A}^\bullet) = 0$ *for all $i < -n$,*
(AX2): *(Vanishing Condition)* $\mathbf{H}^i(\mathbf{A}^\bullet|_{U_{k+1}}) = 0$ *for all $i > \bar{p}(k) - n$, $k \ge 2$.*
(AX3): *(Attaching Condition) The maps*

$$\mathbf{H}^i(j_k^*\mathbf{A}^\bullet|_{U_{k+1}}) \to \mathbf{H}^i(j_k^*Ri_{k*}i_k^*\mathbf{A}^\bullet|_{U_{k+1}})$$

are isomorphisms for $i \le \bar{p}(k) - n$ and $k \ge 2$.

Remarks 4.1.28

1. *We may more generally replace* (AX0) *by the requirement that $\mathbf{A}^\bullet|_{X-\Sigma} \cong \mathcal{S}[n]$, where $\mathcal{S}$ is a local coefficient system, i.e. a locally constant sheaf, on $X - \Sigma$.*
2. *Ri_{k*} is the right derived functor $Ri_{k*} : D^b(U_k) \longrightarrow D^b(U_{k+1})$ of i_{k*}, see* Sect. *2.4.3. In the preceding discussion it was permissible to use i_{k*} because $\mathbf{IC}^\bullet_{\bar{p}}(X)$ is soft.*

3. *We will see later that the assumption that* $\mathbf{A}^\bullet$ *be constructible is redundant. Constructibility follows from* (AX0)–(AX3).

Let k be a field. By considering simplicial i-chains

$$\xi : \{\sigma \in T | \sigma \text{ an oriented } i\text{-simplex}\} \longrightarrow k$$

(T an admissible triangulation of X), we obtain Borel–Moore homology $H_i(X;k)$ with coefficients in k, as well as intersection homology $IH_i^{\bar{p}}(X;k)$ with coefficients in k. The corresponding complex of sheaves $\mathbf{IC}^\bullet_{\bar{p}}(X;k)$ is a complex of sheaves of k-vector spaces.

From now on, we shall take $k = \mathbb{R}$ and write $\mathbf{IC}^\bullet_{\bar{p}}(X) = \mathbf{IC}^\bullet_{\bar{p}}(X;\mathbb{R})$ unless otherwise specified.

Theorem 4.1.29 *Let* $\mathbf{IC}^\bullet_{\bar{p}}(X)$ *be the sheaf-complex of intersection PL-chains on the oriented PL stratified pseudomanifold* X^n. *Then* $\mathbf{IC}^\bullet_{\bar{p}}(X)$ *satisfies* [AX].

Proof. $\mathbf{IC}^\bullet_{\bar{p}}(X)$ is bounded by definition and constructible by Proposition 4.1.26. We show that it satisfies the normalization (AX0) over the top-stratum $X - \Sigma$: Let $\mathbf{C}^{-i}(X)$ denote the sheaf

$$U \mapsto C_i(U)$$

of all ($\mathbb{R}$-valued) PL i-chains on X so that there is a canonical morphism

$$\mathbf{IC}^\bullet_{\bar{p}}(X) \longrightarrow \mathbf{C}^\bullet(X)$$

induced by the inclusion of chains. If $U \subset X - \Sigma$, then

$$IC_i^{\bar{p}}(U) \longrightarrow C_i(U)$$

is an isomorphism (since chains on U do not intersect the singular set Σ at all), and thus

$$\mathbf{IC}^\bullet_{\bar{p}}(X)|_{X-\Sigma} \xrightarrow{\simeq} \mathbf{C}^\bullet(X)|_{X-\Sigma}.$$

Therefore, it suffices to prove

$$\mathbf{H}^{-i}(\mathbf{C}^\bullet(X)|_{X-\Sigma}) \cong \begin{cases} \mathbb{R}_{X-\Sigma}, & i = n, \\ 0, & i \neq n. \end{cases}$$

The derived sheaf is given by ($U \subset X - \Sigma$ open)

$$\begin{aligned} \mathbf{H}^{-i}(\mathbf{C}^\bullet(X)|_{X-\Sigma}) &= \mathbf{Sheaf}(U \mapsto H^{-i}\Gamma(U;\mathbf{C}^\bullet(X))) \\ &= \mathbf{Sheaf}(U \mapsto H_i(U)). \end{aligned}$$

If $i \neq n$, then for U a small Euclidean neighborhood, $H_i(U) = H_i(\mathbb{R}^n) = 0$. Now let $i = n$. By the universal coefficient sequence for Borel–Moore homology

$$0 \to \mathrm{Ext}(H_c^{n+1}(U;\mathbb{R}),\mathbb{R}) \to H_n(U;\mathbb{R}) \to \mathrm{Hom}(H_c^n(U;\mathbb{R}),\mathbb{R}) \to 0,$$

cf. [Ive86, Proposition VIII.1.6], we have

$$H_n(U) \cong \mathrm{Hom}(H_c^n(U), \mathbb{R}).$$

Hence $\mathbf{H}^{-n}(\mathbf{C}^\bullet(X)|_{X-\Sigma})$ is the orientation sheaf $\mathcal{O}_{X-\Sigma}$ on $X - \Sigma$, given by

$$U \mapsto \mathrm{Hom}(H_c^n(U), \mathbb{R}),$$

see Sect. 3.5. As X is oriented, we have an identification

$$\mathcal{O}_{X-\Sigma} \cong \mathbb{R}_{X-\Sigma},$$

and so

$$\mathbf{H}^{-n}(\mathbf{C}^\bullet(X)|_{X-\Sigma}) \cong \mathcal{O}_{X-\Sigma} \cong \mathbb{R}_{X-\Sigma}.$$

The lower bound condition (AX1) is clearly satisfied by $\mathbf{IC}^\bullet_{\bar{p}}(X)$. The vanishing condition (AX2) is Proposition 4.1.23, and the attaching condition (AX3) is Proposition 4.1.24. □

We discuss an alternative description of the attaching axiom (AX3). Let $K \subset X$ be a closed subset with inclusion $j : K \hookrightarrow X$ and complementary inclusion $i : X - K \hookrightarrow X$. Let $\mathbf{B} \in Sh(X)$ be an injective sheaf on X. We would first like to obtain a functorial description of the kernel of the adjunction morphism

$$\mathbf{B} \longrightarrow i_* i^* \mathbf{B}.$$

If $U \subset X$ is any open subset, then this morphism is the restriction of sections

$$\Gamma(U; \mathbf{B}) \stackrel{\text{restr}}{\longrightarrow} \Gamma(U - K; \mathbf{B}).$$

This map is surjective, since injectivity of $\mathbf{B}$ implies injectivity of $\mathbf{B}|_U$, which in turn implies that $\mathbf{B}|_U$ is flabby. The kernel of the restriction is

$$\{s \in \Gamma(U; \mathbf{B}) |\ s|_{U-K} = 0\} = \Gamma_{U\cap K}(U; \mathbf{B}).$$

Consequently, there is a short exact sequence

$$0 \to \Gamma_{U\cap K}(U; \mathbf{B}) \longrightarrow \Gamma(U; \mathbf{B}) \longrightarrow \Gamma(U - K; \mathbf{B}) \to 0.$$

As taking direct limits is an exact functor, this induces a short exact sequence

$$\begin{array}{ccccccccc}
0 & \longrightarrow & \varinjlim_U \Gamma_{U\cap K}(U; \mathbf{B}) & \longrightarrow & \varinjlim \Gamma(U; \mathbf{B}) & \longrightarrow & \varinjlim \Gamma(U - K; \mathbf{B}) & \longrightarrow & 0 \\
 & & \| & & \| & & \| & & \\
0 & \longrightarrow & (j_* j^! \mathbf{B})_x & \longrightarrow & \mathbf{B}_x & \longrightarrow & (i_* i^* \mathbf{B})_x & \longrightarrow & 0
\end{array}$$

So in $D^b(X)$, we get a distinguished triangle

$$j_* j^! \mathbf{B}^\bullet \longrightarrow \mathbf{B}^\bullet \longrightarrow Ri_* i^* \mathbf{B}^\bullet \xrightarrow{[1]} j_* j^! \mathbf{B}^\bullet$$

Suppose $\mathbf{A}^\bullet$ satisfies [AX] and consider the distinguished triangle

$$j_k^! \mathbf{A}^\bullet \longrightarrow j_k^* \mathbf{A}^\bullet \longrightarrow j_k^* R i_{k*} i_k^* \mathbf{A}^\bullet \xrightarrow{[1]} j_k^! \mathbf{A}^\bullet$$

Then

$$\text{(AX3)} \quad \mathbf{H}^i(j_k^* \mathbf{A}^\bullet|_{U_{k+1}}) \xrightarrow{\cong} \mathbf{H}^i(j_k^* R i_{k*} i_k^* \mathbf{A}^\bullet|_{U_{k+1}})$$

for $i \leq \bar{p}(k) - n$ iff

$$\text{(AX3}'\text{)} \quad \mathbf{H}^i(j_k^! \mathbf{A}^\bullet) = 0, \quad i \leq \bar{p}(k) - n + 1.$$

If M^m is a manifold of dimension m and $f_x : \{x\} \hookrightarrow M$ denotes the inclusion of a point, then the identity

$$f_x^! \mathbf{B}^\bullet = f_x^* \mathbf{B}^\bullet[-m] \tag{4.12}$$

holds for cohomologically locally constant $\mathbf{B}^\bullet$. Let $j_x : \{x\} \hookrightarrow X$ be the inclusion of a point into X. Factoring j_x as

$$\{x\} \overset{f_x}{\hookrightarrow} X_{n-k} - X_{n-k-1} \overset{j_k}{\hookrightarrow} X$$

and using (4.12), we see that (AX3′) is furthermore equivalent to

$$\text{(AX3}''\text{)} \quad \mathbf{H}^i(j_x^! \mathbf{A}^\bullet) = 0, \quad i \leq \bar{p}(k) - k + 1, x \in X_{n-k} - X_{n-k-1}.$$

The complex $j_x^! \mathbf{A}^\bullet$ is often referred to as the *costalk* of $\mathbf{A}^\bullet$ at x, and (AX3″) is a *costalk vanishing condition.*

4.2 Deligne's Sheaf

We have defined the intersection chain complex $\mathbf{IC}^\bullet_{\bar{p}}(X)$ for a PL stratified pseudomanifold X, and have verified that it satisfies the system of axioms [AX]. Deligne's sheaf as constructed in [GM83] is an object in the (constructible, bounded) derived category, which is defined on any topological stratified pseudomanifold (not assuming a piecewise linear structure), and which, by its very definition, satisfies [AX]. The key advantage of introducing [AX] is that it uniquely characterizes a sheaf complex up to isomorphism in the derived category, see Theorem 4.2.1 below. It follows that Deligne's sheaf is isomorphic to $\mathbf{IC}^\bullet_{\bar{p}}(X)$, as constructed in Sects. 4.1.3 and 4.1.4 for

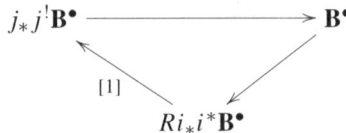

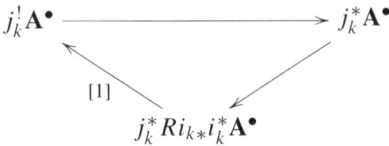

PL spaces. After this has been established, we will in fact use the notation $\mathbf{IC}^\bullet_{\bar{p}}(X)$ for any incarnation of the isomorphism class characterized by [AX], and still call it the *perversity $\bar{p}$ intersection chain sheaf*. The purely sheaf theoretic axiomatic approach via Deligne's functorial construction has thus several advantages: It is defined on any topological pseudomanifold, and does not assume any additional geometric structure.[1] Even more seriously, the efficient characterization by axioms in the derived category allows for dramatically simpler proofs. For example, proving generalized Poincaré duality is fairly challenging using combinatorial chains (see [GM80]), but reduces to a relatively straightforward check of axioms for the Verdier-dual complex in the framework of derived categories (see Theorem 4.4.1 below).

Let X be any n-dimensional topological stratified pseudomanifold. *Deligne's sheaf* $\mathbf{S}^\bullet$ for perversity $\bar{p}$ is defined by the formula

$$\mathbf{S}^\bullet = \tau_{\leq \bar{p}(n)-n} Ri_{n*} \cdots \tau_{\leq \bar{p}(3)-n} Ri_{3*} \tau_{\leq \bar{p}(2)-n} Ri_{2*} \mathbb{R}_{X-\Sigma}[n] \tag{4.13}$$

as an object in $D^b_c(X)$. Here, the i_k are the open inclusions of Sect. 4.1.4 (that is, $U_k = X - X_{n-k}$, $k \geq 2$, $i_k : U_k \hookrightarrow U_{k+1}$), $X - \Sigma = U_2$ is the top stratum, Ri_{k*} is the derived functor of i_{k*} (see Examples 2.4.22), and $\tau_{\leq l}$ is the truncation from Sect. 1.3 which is well-defined on the derived category by Remark 2.4.13. Clearly, $\mathbf{S}^\bullet$ satisfies [AX] (the truncations $\tau_{\leq \bar{p}(k)-n}$ are designed precisely so that the stalk vanishing condition (AX2) holds).

Theorem 4.2.1 *If* $\mathbf{A}^\bullet \in D^b(X)$ *satisfies* [AX], *then* $\mathbf{A}^\bullet \cong \mathbf{S}^\bullet$.

Proof. We proceed by induction on the codimension of strata. On $U_2 = X - \Sigma$,

$$\mathbf{A}^\bullet|_{U_2} \cong \mathbb{R}_{U_2}[n] \cong \mathbf{S}^\bullet|_{U_2}$$

by the normalization (AX0). Suppose inductively we have constructed $\mathbf{A}^\bullet|_{U_k} \cong \mathbf{S}^\bullet|_{U_k}$. We shall first extend it to a morphism over U_{k+1}, then check that it is an isomorphism. The adjunction morphism

$$\mathbf{A}^\bullet|_{U_{k+1}} \longrightarrow Ri_{k*} i_k^* (\mathbf{A}^\bullet|_{U_{k+1}})$$

induces a morphism

$$\tau_{\leq \bar{p}(k)-n} \mathbf{A}^\bullet|_{U_{k+1}} \longrightarrow \tau_{\leq \bar{p}(k)-n} Ri_{k*} i_k^* (\mathbf{A}^\bullet|_{U_{k+1}}).$$

Using the stalk condition (AX2) and the induction hypothesis, we construct the composition

$$\begin{aligned} \mathbf{A}^\bullet|_{U_{k+1}} \cong \tau_{\leq p(k)-n} \mathbf{A}^\bullet|_{U_{k+1}} \longrightarrow\ & \tau_{\leq \bar{p}(k)-n} Ri_{k*} i_k^* (\mathbf{A}^\bullet|_{U_{k+1}}) \\ \cong\ & \tau_{\leq \bar{p}(k)-n} Ri_{k*} (\mathbf{A}^\bullet|_{U_k}) \\ \cong\ & \tau_{\leq \bar{p}(k)-n} Ri_{k*} \mathbf{S}^\bullet|_{U_k} \\ =\ & \mathbf{S}^\bullet|_{U_{k+1}}, \end{aligned}$$

[1] It should be pointed out however that using appropriate "singular intersection chains" it is possible to define intersection homology of topological pseudomanifolds without employing sheaf machinery [Kin85].

defining the morphism $\mathbf{A}^\bullet|_{U_{k+1}} \to \mathbf{S}^\bullet|_{U_{k+1}}$. It extends the isomorphism over U_k, whence it remains to be shown that it is an isomorphism over $U_{k+1} - U_k$:

$$j_k^* \mathbf{A}^\bullet \cong j_k^* \tau_{\leq \bar{p}(k)-n} \mathbf{A}^\bullet \cong j_k^* \tau_{\leq \bar{p}(k)-n} R i_{k*} i_k^* \mathbf{A}^\bullet|_{U_{k+1}},$$

by the attaching condition (AX3). Thus we have $\mathbf{A}^\bullet|_{U_{k+1}} \xrightarrow{\cong} \mathbf{S}^\bullet|_{U_{k+1}}$. □

Corollary 4.2.2 *For a PL stratified pseudomanifold X, $\mathbf{IC}^\bullet_{\bar{p}}(X) \cong \mathbf{S}^\bullet$ in $D^b_c(X)$.*

Remark 4.2.3 *In Theorem* 4.2.1 *it was not necessary to assume that $\mathbf{A}^\bullet$ be constructible. Since $\mathbf{S}^\bullet$ is constructible by its definition via Deligne's formula, Theorem* 4.2.1 *shows that any $\mathbf{A}^\bullet$ satisfying* (AX0)–(AX3) *is automatically constructible. This was claimed in Remark* 4.1.28(3).

Henceforth, we shall write $\mathbf{IC}^\bullet_{\bar{p}}(X)$ to denote any incarnation of the isomorphism class uniquely characterized by [AX]. The intersection homology groups ("with infinite chains") are defined as the hypercohomology of such a complex:

$$IH_i^{\bar{p}}(X) = \mathcal{H}^{-i}(X; \mathbf{IC}^\bullet_{\bar{p}}(X)).$$

Let $\bar{p} \leq \bar{q}$. If we think of $\mathbf{IC}^\bullet_{\bar{p}}(X)$ and $\mathbf{IC}^\bullet_{\bar{q}}(X)$ as being defined by the formula (4.13), then the canonical inclusion of complexes $\tau_{\leq a} \to \tau_{\leq b}$, $a \leq b$, induces a canonical morphism

$$\mathbf{IC}^\bullet_{\bar{p}}(X) \longrightarrow \mathbf{IC}^\bullet_{\bar{q}}(X). \tag{4.14}$$

4.3 Topological Invariance of Intersection Homology

Theorem 4.3.1 *Let X^n be a topologically stratifiable pseudomanifold, let Ξ_1 and Ξ_2 be two stratifications of X. If $\mathbf{IC}^\bullet_{\Xi_1}(X)$ and $\mathbf{IC}^\bullet_{\Xi_2}(X)$ denote the perversity $\bar{p}$ intersection chain sheaves as constructed by the Deligne formula applied to Ξ_1 and Ξ_2, respectively, then*

$$\mathbf{IC}^\bullet_{\Xi_1}(X) \cong \mathbf{IC}^\bullet_{\Xi_2}(X)$$

in $D^b(X)$.

Proof. We proceed as in [B$^+$84, V]. The axiomatics [AX] of Definition 4.1.27 depends on a fixed stratification Ξ of X. Throughout this proof, we will indicate this dependence by writing $[\mathrm{AX}]_\Xi$. The argument will be structured as follows:

- Step 1. Introduce a more intrinsic axiomatics $[\mathrm{AX}_{\mathrm{intr}}]_\Xi$, which depends on the stratification Ξ only very weakly.
- Step 2. $[\mathrm{AX}]_\Xi \Leftrightarrow [\mathrm{AX}_{\mathrm{intr}}]_\Xi$.
- Step 3. Construct an *intrinsic* sheaf complex $\mathbf{IC}^\bullet_{\mathrm{intr}}(X)$, which satisfies $[\mathrm{AX}_{\mathrm{intr}}]_\Xi$ for every topological stratification Ξ of X.
- Step 4. Establish $\mathbf{IC}^\bullet_{\Xi_1}(X) \cong \mathbf{IC}^\bullet_{\Xi_2}(X)$ via $\mathbf{IC}^\bullet_{\mathrm{intr}}(X)$.

Step 1. For $j = 0, 1, 2, \ldots$, set

$$\bar{p}^{-1}(j) = \begin{cases} \min\{k : \bar{p}(k) \geq j\}, & j \leq \bar{p}(n), \\ \infty, & j > \bar{p}(n). \end{cases}$$

Thus

$$\bar{p}(k) \geq j \iff k \geq \bar{p}^{-1}(j) \quad \text{for } 2 \leq k \leq n.$$

The axiomatic system $[\mathrm{AX}_{\mathrm{intr}}]_{\Xi}$ involves the j-th *cohomological supports*

$$\{x \in X : H^j(\iota_x^* \mathbf{A}^\bullet) \neq 0\}$$

and the j-th *cohomological cosupports*

$$\{x \in X : H^j(\iota_x^! \mathbf{A}^\bullet) \neq 0\},$$

where $\iota_x : \{x\} \hookrightarrow X$ and $\mathbf{A}^\bullet \in D(X)$. We say that a complex of sheaves $\mathbf{A}^\bullet$ satisfies the set of axioms $[\mathrm{AX}_{\mathrm{intr}}]_{\Xi}$ with respect to a filtration Ξ of X by closed subsets

$$X = X_n \supset \Sigma = X_{n-2} \supset X_{n-3} \supset \cdots \supset X_0,$$

if

$(\mathrm{AX0})_{\mathrm{intr}}$: (Normalization) $\mathbf{A}^\bullet$ is bounded, cohomologically locally constant with respect to Ξ, and $\mathbf{A}^\bullet|_{X-\Sigma} \cong \mathbb{R}_{X-\Sigma}[n]$,
$(\mathrm{AX1})_{\mathrm{intr}}$: (Lower Bound) $\mathbf{H}^j(\mathbf{A}^\bullet) = 0$ for all $j < -n$,
$(\mathrm{AX2})_{\mathrm{intr}}$: (Support Dimensions)

$$\dim\{x \in X : H^j(\iota_x^* \mathbf{A}^\bullet) \neq 0\} \leq n - \bar{p}^{-1}(j+n) \quad \text{for all } j > -n,$$

$(\mathrm{AX3})_{\mathrm{intr}}$: (Cosupport Dimensions)

$$\dim\{x \in X : H^j(\iota_x^! \mathbf{A}^\bullet) \neq 0\} \leq n - \bar{q}^{-1}(-j) \quad \text{for all } j < 0,$$

where $\bar{q}$ is the complementary perversity to $\bar{p}$.

Here, dim may be taken to be the topological dimension of Hurewicz and Wallman [HW41]. Note that only $(\mathrm{AX0})_{\mathrm{intr}}$ involves the filtration Ξ. It is not assumed that $\mathbf{A}^\bullet$ is constructible with respect to Ξ.

Step 2. $(\mathrm{AX2}) \Rightarrow (\mathrm{AX2})_{\mathrm{intr}}$: Fix $j > -n$. Let x be a point in the j-th cohomological support set S of $\mathbf{A}^\bullet$. If k is such that $\bar{p}(k) - n < j$, then, by (AX2), $H^j(\mathbf{A}_y^\bullet) = 0$ for all $y \in U_{k+1} = X - X_{n-(k+1)}$. Hence $x \notin U_{k+1}$, i.e. $x \in X_{n-(k+1)}$. Since $j > -n$, $\bar{p}(2) - n = -n < j$, so that $\{k : \bar{p}(k) - n < j\} \neq \varnothing$. Let

$$K = \max\{k : \bar{p}(k) - n < j\}.$$

The maximality implies that $\bar{p}(K+1) \geq j + n$, that is, $K + 1 \geq \bar{p}^{-1}(j+n)$. As $S \subset X_{n-(K+1)}$, the monotonicity of dimension implies

$$\dim S \leq \dim X_{n-(K+1)} \leq n - (K+1) \leq n - \bar{p}^{-1}(j+n).$$

$(AX0)_{intr}\&(AX2)_{intr} \Rightarrow (AX2)$: We must show that if $x \in X_{n-k} - X_{n-k-1}$ is a point such that $H^j(\mathbf{A}^\bullet_x) \neq 0$, then $j \leq \bar{p}(k) - n$. Let S again be the j-th cohomological support set of $\mathbf{A}^\bullet$. By $(AX0)_{intr}$, $\mathbf{A}^\bullet$ is cohomologically locally constant with respect to Ξ, whence $\mathbf{H}^j(\mathbf{A}^\bullet)|_{X_{n-k}-X_{n-k-1}}$ is locally constant. Thus $x \in S \cap (X_{n-k} - X_{n-k-1})$ has an open neighborhood $U \subset X_{n-k} - X_{n-k-1}$ with $H^j(\mathbf{A}^\bullet_y) \neq 0$ for all $y \in U$. Consequently $U \subset S$ and

$$n - k = \dim U \leq \dim S \leq n - \bar{p}^{-1}(j+n)$$

by $(AX2)_{intr}$. We conclude that $k \geq \bar{p}^{-1}(j+n)$, i.e. $\bar{p}(k) \geq j+n$.

The equivalence of the cosupport dimension axiom and the costalk vanishing condition (AX3″) is shown similarly. As we have seen earlier, the attaching condition (AX3) is equivalent to (AX3″).

Step 3. Given a filtration Ξ of X consisting of closed subsets

$$X = X_n(\Xi) \supset X_{n-2}(\Xi) \supset X_{n-3}(\Xi) \supset \cdots \supset X_0(\Xi),$$

we write $U_k(\Xi) = X - X_{n-k}(\Xi)$, $i_k(\Xi): U_k(\Xi) \hookrightarrow U_{k+1}(\Xi)$, $j_k(\Xi): X_{n-k}(\Xi) - X_{n-k-1}(\Xi) \hookrightarrow U_{k+1}(\Xi)$. Let

$$\mathbf{IC}^\bullet_\Xi(X) = \tau_{\leq \bar{p}(n)-n} Ri_n(\Xi)_* \cdots \tau_{\leq \bar{p}(3)-n} Ri_3(\Xi)_* \tau_{\leq \bar{p}(2)-n} Ri_2(\Xi)_* \mathbb{R}_{U_2(\Xi)}[n]$$

and

$$\mathbf{IC}^\bullet_{\Xi,k}(X) = \mathbf{IC}^\bullet_\Xi(X)|_{U_k(\Xi)}.$$

By induction on k, one can construct an intrinsic filtration Ξ_{intr} of X such that

(i) every nonempty $X_{n-k}(\Xi_{intr}) - X_{n-k-1}(\Xi_{intr})$ is a manifold of dimension $n-k$,
(ii) $j_k(\Xi_{intr})^* \mathbf{IC}^\bullet_{\Xi_{intr},k+1}(X)$ is cohomologically locally constant,
(iii) $j_k(\Xi_{intr})^! \mathbf{IC}^\bullet_{\Xi_{intr},k+1}(X)$ is cohomologically locally constant, and
(iv) for every topological stratification Ξ of X, $X_{n-k}(\Xi_{intr}) - X_{n-k-1}(\Xi_{intr})$ is a union of connected components of pure strata of Ξ, and $U_k(\Xi) \subset U_k(\Xi_{intr})$.

(For details of this construction, cf. [B⁺84, V]. For $k = 2$, start by taking $U_2(\Xi_{intr})$ to be the largest open submanifold of X, that is, the union of all open submanifolds of X.) Put $\mathbf{IC}^\bullet_{intr}(X) = \mathbf{IC}^\bullet_{\Xi_{intr}}(X)$. By construction, $\mathbf{IC}^\bullet_{intr}(X)$ satisfies $[AX]_{\Xi_{intr}}$. By step 2, it satisfies $[AX_{intr}]_{\Xi_{intr}}$. Let Ξ be any topological stratification of X. Since $U_2(\Xi)$ is an open submanifold of X, we have $U_2(\Xi) \subset U_2(\Xi_{intr})$ and therefore $(AX0_{intr})$ holds for Ξ. The other three axioms are stratification independent. Thus $\mathbf{IC}^\bullet_{intr}(X)$ satisfies $[AX_{intr}]_\Xi$.

Step 4. Let Ξ_1 and Ξ_2 be two topological stratifications of X. By construction, $\mathbf{IC}^\bullet_{\Xi_1}(X)$ satisfies $[AX]_{\Xi_1}$. By step 3, $\mathbf{IC}^\bullet_{intr}(X)$ satisfies $[AX_{intr}]_{\Xi_1}$. By step 2,

$\mathbf{IC}^\bullet_{\mathrm{intr}}(X)$ satisfies $[\mathrm{AX}]_{\Xi_1}$. Theorem 4.2.1 applied to Ξ_1 asserts that $\mathbf{IC}^\bullet_{\Xi_1}(X) \cong \mathbf{IC}^\bullet_{\mathrm{intr}}(X)$ in $D(X)$. The same argument applied to Ξ_2 yields an isomorphism $\mathbf{IC}^\bullet_{\Xi_2}(X) \cong \mathbf{IC}^\bullet_{\mathrm{intr}}(X)$. Thus $\mathbf{IC}^\bullet_{\Xi_1}(X) \cong \mathbf{IC}^\bullet_{\Xi_2}(X)$. □

Let X be a pseudomanifold with topological stratification Ξ, let Y be a pseudomanifold with topological stratification Υ, and let $f : X \to Y$ be any homeomorphism. Then the pullback filtration $f^*\Upsilon$, given by

$$f^{-1}(Y_n(\Upsilon)) \supset f^{-1}(Y_{n-2}(\Upsilon)) \supset f^{-1}(Y_{n-3}(\Upsilon)) \supset \cdots \supset f^{-1}(Y_0(\Upsilon)),$$

is a topological stratification of X and

$$f^*\mathbf{IC}^\bullet_\Upsilon(Y) \cong \mathbf{IC}^\bullet_{f^*\Upsilon}(X).$$

By Theorem 4.3.1, $\mathbf{IC}^\bullet_{f^*\Upsilon}(X) \cong \mathbf{IC}^\bullet_\Xi(X)$. These isomorphisms induce an isomorphism of the intersection homology groups of (X, Ξ) and (Y, Υ):

$$\mathcal{H}^j(Y; \mathbf{IC}^\bullet_\Upsilon(Y)) \cong \mathcal{H}^j(X; f^*\mathbf{IC}^\bullet_\Upsilon(Y)) \cong \mathcal{H}^j(X; \mathbf{IC}^\bullet_\Xi(X)).$$

4.4 Generalized Poincaré Duality on Singular Spaces

We will establish a nondegenerate intersection pairing between complementary intersection homology groups of a closed oriented topological stratified pseudomanifold. The term "complementary" is understood to mean complementary dimensions and complementary perversities. In fact, the pairing will be induced on hypercohomology by a Verdier-duality statement on the sheaf level.

We will use the following identities for the stalks and costalks of a constructible complex of sheaves: Let $\mathbf{A}^\bullet \in D^b_c(X)$, $x \in X$, and U_x a small distinguished open neighborhood of x. Then an argument involving the hypercohomology spectral sequence (Proposition 1.3.5) shows

$$\mathbf{H}^i(\mathbf{A}^\bullet)_x \cong \mathcal{H}^i(U_x; \mathbf{A}^\bullet)$$
$$\mathbf{H}^i(j_x^!\mathbf{A}^\bullet) \cong \mathcal{H}^i_c(U_x; \mathbf{A}^\bullet).$$

Recall that two perversities $\bar{p}, \bar{q}$ are *complementary* if $\bar{p} + \bar{q} = \bar{t}$, the top perversity. In the following theorem, $\mathcal{D}$ denotes the Borel–Moore–Verdier dualizing functor from Sect. 3.4 (Definition 3.4.2, but use real coefficients).

Theorem 4.4.1 *Let X be an oriented topological pseudomanifold and $\bar{p}, \bar{q}$ complementary perversities. Then there exists an isomorphism*

$$\mathcal{D}\mathbf{IC}^\bullet_{\bar{q}}(X)[n] \cong \mathbf{IC}^\bullet_{\bar{p}}(X)$$

extending the orientation over $X - \Sigma$.

Proof. By Theorem 4.2.1, we only have to show that $\mathcal{D}\mathbf{IC}^\bullet_{\bar{q}}[n]$ satisfies [AX] for $\bar{p}$.

(AX0): Since i_2 is an open inclusion, we have $i_2^* = i_2^!$ (see Examples 3.3.3). By Proposition 3.4.5, $\mathcal{D}i_2^* \cong i_2^!\mathcal{D} = i_2^*\mathcal{D}$. Hence

$$\begin{aligned} i_2^*\mathcal{D}\mathbf{IC}^\bullet_{\bar{q}}[n] &\cong \mathcal{D}(i_2^*\mathbf{IC}^\bullet_{\bar{q}})[n] \cong \mathcal{D}(\mathbb{R}_{U_2}[n])[n] \cong \mathbb{D}^\bullet_{U_2} \\ &\cong \mathcal{O}_{U_2}[n] \cong \mathbb{R}_{U_2}[n]. \end{aligned}$$

Here, we are using Proposition 3.5.1, and the last isomorphism is given by the orientation.

(AX2): Let $x \in X_{n-k} - X_{n-k-1}$. We know that

$$H^i(j_x^!\mathbf{IC}^\bullet_{\bar{q}}) = 0, \quad i \le \bar{q}(k) - k + 1,$$

by (AX3$''$) for $\bar{q}$. In the following calculations, we will repeatedly make use of Theorem 3.4.4. We rewrite the stalk of the dual of $\mathbf{IC}^\bullet_{\bar{q}}$:

$$\begin{aligned} \mathbf{H}^j(\mathcal{D}\mathbf{IC}^\bullet_{\bar{q}}[n])_x &\cong \mathcal{H}^{j+n}(U_x; \mathcal{D}\mathbf{IC}^\bullet_{\bar{q}}) \\ &\cong \mathrm{Hom}(\mathcal{H}_c^{-j-n}(U_x; \mathbf{IC}^\bullet_{\bar{q}}), \mathbb{R}) \\ &\cong \mathrm{Hom}(H^{-j-n}(j_x^!\mathbf{IC}^\bullet_{\bar{q}}), \mathbb{R}) \\ &= 0 \end{aligned}$$

for $-j-n \le \bar{q}(k)-k+1$, which is equivalent to $j > \bar{p}(k)-n$ as $\bar{p}(k)+\bar{q}(k) = k-2$.

(AX3): Similarly, the $\bar{q}$-stalk condition for $\mathbf{IC}^\bullet_{\bar{q}}$ implies the $\bar{p}$-costalk condition for its dual:

$$\begin{aligned} H^j(j_x^!\mathcal{D}\mathbf{IC}^\bullet_{\bar{q}}[n]) &\cong \mathcal{H}_c^{j+n}(U_x; \mathcal{D}\mathbf{IC}^\bullet_{\bar{q}}) \\ &\cong \mathrm{Hom}(\mathcal{H}^{-j-n}(U_x; \mathbf{IC}^\bullet_{\bar{q}}), \mathbb{R}) \\ &\cong \mathrm{Hom}(\mathbf{H}^{-j-n}(\mathbf{IC}^\bullet_{\bar{q}})_x, \mathbb{R}) \\ &= 0 \end{aligned}$$

for $-j-n > \bar{q}(k)-n$, which is equivalent to $j \le \bar{p}(k) - k + 1$.

(AX1): Use the calculation from (AX2) and the fact that $H^i(j_x^!\mathbf{IC}^\bullet_{\bar{q}}) = 0$ when $i > 0$. □

Generalized Poincaré duality for intersection homology groups follows readily from Theorem 4.4.1 by taking hypercohomology. Assume X compact.

$$\begin{aligned} IH_i^{\bar{p}}(X) &= \mathcal{H}^{-i}(X; \mathbf{IC}^\bullet_{\bar{p}}) \\ &\cong \mathcal{H}^{-i}(X; \mathcal{D}\mathbf{IC}^\bullet_{\bar{q}}[n]) \\ &\cong \mathcal{H}^{-i+n}(X; \mathcal{D}\mathbf{IC}^\bullet_{\bar{q}}) \\ &\cong \mathrm{Hom}(\mathcal{H}^{i-n}(X; \mathbf{IC}^\bullet_{\bar{q}}), \mathbb{R}) \\ &= \mathrm{Hom}(IH^{\bar{q}}_{n-i}(X), \mathbb{R}). \end{aligned}$$

That is, we obtain a nondegenerate bilinear pairing

$$IH_i^{\bar{p}}(X) \otimes IH^{\bar{q}}_{n-i}(X) \longrightarrow \mathbb{R}. \tag{4.15}$$

5

Characteristic Classes and Smooth Manifolds

5.1 Introduction

In order to emphasize the importance of the signature invariant and characteristic classes such as the L-classes, it is helpful to place them in the context of various classification schemes for smooth manifolds. Such schemes include various bordism theories (such as smooth oriented bordism Ω_*^{SO}, to be reviewed in Sect. 5.2) and the rather refined method of the surgery program, to be reviewed in Sect. 5.5. These reviews illustrate the central position which the signature $\sigma(M) \in \mathbb{Z}$ of a manifold M and its L-classes $L_i(M) \in H^{4i}(M; \mathbb{Q})$ occupy in both schemes. Consequently, it is desirable to extend such classes to singular spaces. Among their applications are then the development of classification frameworks in analogy with frameworks available for manifolds. As we shall see, these classes share many properties with their nonsingular counterparts, but there are important differences as well.[1] We emphasize that this chapter is not an introduction to differential topology, but merely a *repetitorium*. For more details, we recommend [MS74,Ran02] and [Lüc02]. Throughout this chapter, $H_*(-)$ denotes ordinary homology, not Borel–Moore homology.

If M is a *closed* manifold, then $\sigma(M)$ is a homotopy invariant of M. This is not true for open manifolds.

Example 5.1.1 *Let B^{2k} be a base-manifold and E, E' the total spaces of $2k$-plane vector bundles ξ, ξ' over B. The Euler number $e(\xi)$ is the self-intersection $B \cap B$ of B in E as the zero section. Thus $\sigma(E) = \operatorname{sign} e(\xi)$ and $\sigma(E') = \operatorname{sign} e(\xi')$ depends on the vector bundle, whereas E and E' are homotopy equivalent, since they are both homotopy equivalent to B.*

Even if M is a closed manifold, the $L_i(M)$ are not homotopy invariants of M, and are therefore much more refined. In particular, they can help to distinguish manifolds in a given homotopy type.

[1] Of course, the class $L(X)$ of a singular space X coincides with the classical definition of the L-class when X is nonsingular.

Example 5.1.2 *There exist infinitely many manifolds $M_i, i = 1, 2, \ldots$ in the homotopy type of $S^2 \times S^4$, distinguished by the first Pontrjagin class of their tangent bundle $p_1(TM_i) \in H^4(S^2 \times S^4) \cong \mathbb{Z}$, namely $p_1(TM_i) = Ki$, K a fixed integer $\neq 0$.*

In fact, we have the following result:

Theorem 5.1.3 (*Browder–Novikov*) *The homotopy type and L-classes determine the diffeomorphism class of a smooth simply connected manifold of dimension ≥ 5 up to a finite number of possibilities.*

5.2 Smooth Oriented Bordism

For a smooth oriented manifold M, let $-M$ denote the same underlying smooth manifold with the opposite orientation. We say that two smooth oriented closed manifolds M and M' are *bordant* if there exists a smooth oriented compact manifold W such that $\partial W \cong M \sqcup -M'$, where ∂W is the boundary of W with the induced orientation, $\cong$ denotes orientation preserving diffeomorphism, and $\sqcup$ is disjoint union. This relation is indeed an equivalence relation, and we denote the equivalence class of M by $[M]$. Let Ω_n^{SO} be the set of bordism classes $[M^n]$ of n-dimensional smooth oriented closed manifolds M^n. We make Ω_n^{SO} into an abelian group by setting

$$[M_1] + [M_2] = [M_1 \sqcup M_2].$$

The zero element is the bordism class of the empty set (and any manifold which bounds is a representative for that class). The cartesian product $(M_1, M_2) \mapsto M_1 \times M_2$ induces an associative bilinear product

$$\Omega_m^{SO} \times \Omega_n^{SO} \to \Omega_{m+n}^{SO}$$

which makes

$$\Omega_*^{SO} = \bigoplus_{n \geq 0} \Omega_n^{SO}$$

into a graded ring. The one-element $1 \in \Omega_0^{SO}$ is the bordism class of a point. We have the identity $[M_1^m \times M_2^n] = (-1)^{mn}[M_2 \times M_1]$. In Sect. 5.4 we shall illustrate how characteristic classes can be employed to gain insight into the structure of Ω_*^{SO}. In fact, we will outline the rational computation of this ring, leaving aside the more subtle torsion questions.

5.3 The Characteristic Classes of Chern and Pontrjagin

In [Sti36], Stiefel introduced characteristic homology classes for the tangent bundle of a closed differentiable manifold as those cycles where a given number of vector fields becomes linearly dependent. Simultaneously, Whitney [Whi35] considered sphere bundles over simplicial complexes and constructed ($\mathbb{Z}$- or $\mathbb{Z}/_2$-valued) functions on the homology of the complex (later to be phrased as cohomology classes) that are invariants of the bundle. Pontrjagin obtained new characteristic classes for orientable smooth manifolds in [Pon47] by investigating the homology of oriented Grassmann manifolds and, among other remarks, pointed out the bordism invariance of the associated characteristic numbers. Characteristic classes for complex vector bundles were developed by Chern in [Che46], where a description in terms of differential forms constructed from a hermitian metric is given as well.

Let us briefly review the formal axiomatics of Chern and Pontrjagin classes. For any complex vector bundle ξ over a base space B there are associated *Chern classes* $c_i(\xi) \in H^{2i}(B;\mathbb{Z})$ (and we define the *total Chern class* $c(\xi)$ by $c(\xi) = \sum_0^\infty c_i(\xi) \in H^*(B)$) such that

1. $c_0(\xi) = 1$.
2. $c_i(f^*\xi) = f^*c_i(\xi)$ for a map $f : B' \to B$.
3. $c(\xi \oplus \eta) = c(\xi)c(\eta)$.
4. $c(\gamma) = 1 - g$.

In (4), the generator $g \in H^2(\mathbb{C}P^n;\mathbb{Z})$ is the Poincaré dual of $\mathbb{C}P^{n-1} \subset \mathbb{C}P^n$ and γ is the tautological line bundle over $\mathbb{C}P^n$ whose fiber over a point is the line in $\mathbb{C}^{n+1}$ which represents that point. The c_i are uniquely determined by (1)–(4). We have $c_i(\xi) = 0$ if $i > \operatorname{rank}\xi$. With $n = \operatorname{rank}\xi$, the top Chern class $c_n(\xi) = e(\xi_\mathbb{R})$ is the *Euler class*, where $\xi_\mathbb{R}$ is the underlying real vector bundle of ξ. If M is a smooth compact oriented manifold with tangent bundle TM then

$$\langle e(TM), [M]\rangle = \chi(M),$$

where χ denotes the Euler characteristic (and $[M]$ the fundamental class of M).

Now let ξ be a real vector bundle over B. We define the i-th *Pontrjagin class* of ξ to be

$$p_i(\xi) = (-1)^i c_{2i}(\xi \otimes \mathbb{C}) \in H^{4i}(B;\mathbb{Z}).$$

(Recall that the Chern classes of odd index of a complexified bundle have order 2.) The *total Pontrjagin class* of ξ is $p(\xi) = \sum_0^\infty p_i(\xi) \in H^*(B;\mathbb{Z})$. These classes satisfy axioms similar to the ones above:

1. $p_0(\xi) = 1$.
2. $p_i(f^*\xi) = f^*p_i(\xi)$ for a map $f : B' \to B$.
3. $p(\xi \oplus \eta) = p(\xi)p(\eta)$ modulo 2-torsion (thus we have actual equality e.g. in $H^*(B;\mathbb{Q})$).
4. $p(\gamma_\mathbb{R}) = 1 + g^2$.

The last statement for example is deduced from

$$\begin{aligned} p_0(\gamma_\mathbb{R}) + p_1(\gamma_\mathbb{R}) &= 1 - c_2(\gamma_\mathbb{R} \otimes \mathbb{C}) = 1 - c_2(\gamma \oplus \bar\gamma) \\ &= 1 - (c_2\gamma + c_1\gamma c_1\bar\gamma + c_2\bar\gamma) = 1 + (c_1\gamma)^2. \end{aligned}$$

5.4 The Rational Calculation of Ω_*^{SO}

The investigation of the rational structure of the Ω_i^{SO} falls into two parts: In the first part, we shall construct a set of generators and use characteristic classes to verify their linear independence. This gives in particular a lower bound on the rank of the bordism groups. In the second part, we use Thom spaces and Serre's theorem to establish the same number as an upper bound for the rank.

5.4.1 The Lower Bound

Our goal is to see that

$$\{[\mathbb{C}P^{2i_1} \times \cdots \times \mathbb{C}P^{2i_r}] | (i_1, \ldots, i_r) \text{ a partition of } k\} \tag{5.1}$$

is a set of linearly independent elements in Ω_{4k} (we abbreviate $\Omega_i = \Omega_i^{SO}$). Thus a first basic problem is even: how to prove that a given element is nonzero? To attack this problem, we introduce *characteristic numbers*: Given a partition $I = (i_1, \ldots, i_r)$ of k, we define

$$p_I[M] = \langle p_{i_1}(TM) \cdots p_{i_r}(TM), [M^{4k}] \rangle.$$

Example 5.4.1 *We have* $c(\mathbb{C}P^n) = (1+g)^{n+1}$ *and* $p(\mathbb{C}P^n) = (1+g^2)^{n+1}$, *where here* g *is the generator* $g = -c_1(\gamma) \in H^2(\mathbb{C}P^n)$. *Hence* $p_i(\mathbb{C}P^n) = \binom{n+1}{i} g^{2i}$ *and*

$$p_I[\mathbb{C}P^{2n}] = \binom{2n+1}{i_1} \cdots \binom{2n+1}{i_r}.$$

Suppose now that M^{4k} is a boundary, $M = \partial W$. Then $TW|_M = TM \oplus \epsilon^1$ (ϵ^m denotes the trivial m-plane bundle) and $p_i(M) = p_i(TM \oplus \epsilon^1) = p_i(TW|_M) = p_i(W)|_M$. Thus, with $\iota : M \hookrightarrow W$ the inclusion of the boundary, we have

$$\langle p_I(TM), [M] \rangle = \langle \iota^* p_I(TW), [M] \rangle = \langle p_I(TW), \iota_*[M] \rangle.$$

Now $\iota_*[M] = 0$, as M bounds W. We have obtained Pontrjagin's theorem: If M^{4k} is a boundary, then every $p_I[M^{4k}] = 0$. Using this fact and the above example, we conclude that

$$[\mathbb{C}P^{2n}] \neq 0 \in \Omega_{4n}.$$

We remark that $\mathbb{C}P^{2n+1}$ bounds, as it can be written as the total space of a 2-sphere bundle over quaternionic space $\mathbb{H}P^n$ (which bounds the corresponding disk bundle). Pontrjagin's theorem also implies that $M^{4k} \mapsto p_I[M^{4k}]$ induces a well-defined group homomorphism $\Omega_{4k} \to \mathbb{Z}$.

Now let us investigate the linear independence of the elements of (5.1). We will write $\mathbb{C}P^I = \mathbb{C}P^{2i_1} \times \cdots \times \mathbb{C}P^{2i_r}$, where I is a partition of k. Suppose there were integers λ_I such that

$$\sum_I \lambda_I [\mathbb{C}P^I] = 0 \in \Omega_{4k}$$

(summation over all partitions I of k). Then for each partition J,

$$\sum_I \lambda_I p_J[\mathbb{C}P^I] = 0,$$

i.e. the matrix

$$(p_J[\mathbb{C}P^I])_{I,J}$$

is singular.

Example 5.4.2 *We calculate this matrix for $k = 4$, and verify that it is nonsingular. Hence the rank of Ω_{16} is at least 5. In the present example, cohomology will be taken with rational coefficients. With the notation*

$$\begin{aligned}
H^*(\mathbb{C}P^2) &= \mathbb{Q}[x]/(x^3 = 0),\ x \in H^2(\mathbb{C}P^2) : x = -c_1(\gamma),\ \langle x^2, [\mathbb{C}P^2]\rangle = +1,\\
H^*(\mathbb{C}P^4) &= \mathbb{Q}[y]/(y^5 = 0),\ y \in H^2(\mathbb{C}P^4) : y = -c_1(\gamma),\ \langle y^4, [\mathbb{C}P^4]\rangle = +1,\\
H^*(\mathbb{C}P^6) &= \mathbb{Q}[z]/(z^7 = 0),\ z \in H^2(\mathbb{C}P^6) : z = -c_1(\gamma),\ \langle z^6, [\mathbb{C}P^6]\rangle = +1,\\
H^*(\mathbb{C}P^8) &= \mathbb{Q}[u]/(u^9 = 0),\ u \in H^2(\mathbb{C}P^8) : u = -c_1(\gamma),\ \langle u^8, [\mathbb{C}P^8]\rangle = +1,
\end{aligned}$$

we have the following Pontrjagin classes:

$$\begin{aligned}
p(\mathbb{C}P^2) &= 1 + 3x^2,\\
p(\mathbb{C}P^4) &= 1 + 5y^2 + 10y^4,\\
p(\mathbb{C}P^6) &= 1 + 7z^2 + 21z^4 + 35z^6,\\
p(\mathbb{C}P^8) &= 1 + 9u^2 + 36u^4 + 84u^6 + 126u^8.
\end{aligned}$$

The partitions of 4 are:

$$(4), (1,3), (2,2), (1,1,2), (1,1,1,1)$$

and we claim that the manifolds

$$\begin{aligned}
M_1 &= \mathbb{C}P^8,\\
M_2 &= \mathbb{C}P^2 \times \mathbb{C}P^6,\\
M_3 &= \mathbb{C}P^4 \times \mathbb{C}P^4,\\
M_4 &= \mathbb{C}P^2 \times \mathbb{C}P^2 \times \mathbb{C}P^4,\\
M_5 &= \mathbb{C}P^2 \times \mathbb{C}P^2 \times \mathbb{C}P^2 \times \mathbb{C}P^2
\end{aligned}$$

are linearly independent as elements of Ω_{16}. Indeed, the corresponding 5×5 matrix of Pontrjagin numbers is

$$\begin{pmatrix}
p_4[M_1] & p_1p_3[M_1] & p_2^2[M_1] & p_1^2p_2[M_1] & p_1^4[M_1]\\
p_4[M_2] & p_1p_3[M_2] & p_2^2[M_2] & p_1^2p_2[M_2] & p_1^4[M_2]\\
p_4[M_3] & p_1p_3[M_3] & p_2^2[M_3] & p_1^2p_2[M_3] & p_1^4[M_3]\\
p_4[M_4] & p_1p_3[M_4] & p_2^2[M_4] & p_1^2p_2[M_4] & p_1^4[M_4]\\
p_4[M_5] & p_1p_3[M_5] & p_2^2[M_5] & p_1^2p_2[M_5] & p_1^4[M_5]
\end{pmatrix}$$

$$= \begin{pmatrix}
126 & 756 & 1296 & 2916 & 6561\\
105 & 546 & 882 & 1911 & 4116\\
100 & 500 & 825 & 1750 & 3750\\
90 & 405 & 630 & 1305 & 2700\\
81 & 324 & 486 & 972 & 1944
\end{pmatrix}$$

with determinant 17, 222, 625. *To find, say,* $p_1^2 p_2[M_4]$, *we calculate the first and second Pontrjagin class of* M_4 *(where* $\pi_1 : M_4 \to \mathbb{C}P^2$, $\pi_2 : M_4 \to \mathbb{C}P^2$, $\pi_3 : M_4 \to \mathbb{C}P^4$ *are the first, second, and third factor projections, respectively):*

$$\begin{aligned} p_1(\mathbb{C}P^2 \times \mathbb{C}P^2 \times \mathbb{C}P^4) &= p_1(\pi_1^* T\mathbb{C}P^2 \oplus \pi_2^* T\mathbb{C}P^2 \oplus \pi_3^* T\mathbb{C}P^4) \\ &= \pi_1^* p_1(\mathbb{C}P^2) + \pi_2^* p_1(\mathbb{C}P^2) + \pi_3^* p_1(\mathbb{C}P^4) \\ &= \pi_1^*(3x^2) + \pi_2^*(3x^2) + \pi_3^*(5y^2) \\ &= 3x^2 \times 1 \times 1 + 1 \times 3x^2 \times 1 + 1 \times 1 \times 5y^2, \end{aligned}$$

$$\begin{aligned} p_2(\mathbb{C}P^2 \times \mathbb{C}P^2 \times \mathbb{C}P^4) &= \pi_1^* p_1(\mathbb{C}P^2)\pi_2^* p_1(\mathbb{C}P^2) + \pi_1^* p_1(\mathbb{C}P^2)\pi_3^* p_1(\mathbb{C}P^4) \\ &\quad + \pi_2^* p_1(\mathbb{C}P^2)\pi_3^* p_1(\mathbb{C}P^4) + \pi_3^* p_2(\mathbb{C}P^4) \\ &= \pi_1^*(3x^2)\pi_2^*(3x^2) + \pi_1^*(3x^2)\pi_3^*(5y^2) \\ &\quad + \pi_2^*(3x^2)\pi_3^*(5y^2) + \pi_3^*(10y^4) \\ &= 3x^2 \times 3x^2 \times 1 + 3x^2 \times 1 \times 5y^2 \\ &\quad + 1 \times 3x^2 \times 5y^2 + 1 \times 1 \times 10y^4 \end{aligned}$$

and evaluate the product $p_1^2 p_2$ *on the fundamental class:*

$$\begin{aligned} \langle p_1^2(M_4)p_2(M_4), [M_4]\rangle &= \langle(3x^2 \times 1 \times 1 + 1 \times 3x^2 \times 1 + 1 \times 1 \times 5y^2)^2 \\ &\quad (3x^2 \times 3x^2 \times 1 + 3x^2 \times 1 \times 5y^2 \\ &\quad + 1 \times 3x^2 \times 5y^2 + 1 \times 1 \times 10y^4), [M_4]\rangle \\ &= \langle(3x^2 \times 3x^2 \times 1 + 3x^2 \times 1 \times 5y^2 \\ &\quad + 3x^2 \times 3x^2 \times 1 + 1 \times 3x^2 \times 5y^2 \\ &\quad + 3x^2 \times 1 \times 5y^2 + 1 \times 3x^2 \times 5y^2 \\ &\quad + 1 \times 1 \times 25y^4) \\ &\quad (3x^2 \times 3x^2 \times 1 + 3x^2 \times 1 \times 5y^2 \\ &\quad + 1 \times 3x^2 \times 5y^2 + 1 \times 1 \times 10y^4), [M_4]\rangle \\ &= \langle 3x^2 \times 3x^2 \times 25y^4 + 3x^2 \times 6x^2 \times 25y^4 \\ &\quad + 6x^2 \times 3x^2 \times 25y^4 + 6x^2 \times 3x^2 \times 10y^4, \\ &\quad [\mathbb{C}P^2] \times [\mathbb{C}P^2] \times [\mathbb{C}P^4]\rangle \\ &= 9 \cdot 25 + 18 \cdot 25 + 18 \cdot 25 + 180 \\ &= 1305. \end{aligned}$$

To see that in fact for any k, the matrix of Pontrjagin numbers is nonsingular, the trick is to change the basis to one in which the matrix has triangular form. Then we compute the diagonal elements and check that they are all nonzero, establishing Thom's theorem below. Changing basis means the introduction of a new set of characteristic classes s_I. Consider the polynomial ring $\mathbb{Z}[t_1, \ldots, t_k]$ and the subring $S \subset \mathbb{Z}[t_1, \ldots, t_k]$ of symmetric polynomials. A well known algebraic fact asserts that $S = \mathbb{Z}[\sigma_1, \dots, \sigma_k]$, where σ_i is the i-th elementary symmetric polynomial, $1 + \sigma_1 + \cdots + \sigma_k = (1 + t_1) \cdots (1 + t_k)$. Thus, for every partition $I = (i_1, \dots, i_r)$ of k, there exists a polynomial s_I defined by

$$s_I(\sigma_1, \dots, \sigma_k) = \sum t_1^{i_1} \cdots t_r^{i_r},$$

where the summation makes the right-hand side symmetric, that is, we sum over all permutations of $t_1, \dots, t_r$ that give different monomials.

Example 5.4.3 *For $k = 2$ and the partition* (2), *we obtain*

$$s_2(\sigma_1, \sigma_2) = \sum t_1^2 = t_1^2 + t_2^2 = (t_1 + t_2)^2 - 2t_1t_2 = \sigma_1^2 - 2\sigma_2.$$

Let $S^k \subset S$ denote the additive subgroup of homogeneous symmetric polynomials of degree k. Obviously, $\{\sigma_{i_1} \cdots \sigma_{i_r} \mid I \text{ a partition of } k\}$ is an additive basis of S^k, but so is $\{s_I(\sigma_1, \dots, \sigma_k) \mid I\}$. We define new characteristic numbers

$$s_I[M^{4k}] = \langle s_I(p_1(M), \dots, p_k(M)), [M] \rangle.$$

Then proving that

$$(p_J[\mathbb{C}P^I])_{I,J}$$

is nonsingular is equivalent to proving that

$$(s_J[\mathbb{C}P^I])_{I,J}$$

is nonsingular. The new numbers have the advantage that they satisfy a product formula

$$s_I[M^{4k} \times N^{4l}] = \sum_{I_1 I_2 = I} s_{I_1}[M^{4k}] s_{I_2}[N^{4l}],$$

where $I_1 I_2$ denotes the juxtaposition of I_1, a partition of k, and I_2, a partition of l, giving a partition of $k + l$. Consequently, if $I = (i_1, \dots, i_r)$ and $J = (j_1, \dots, j_q)$ are partitions of k then

$$s_I[\mathbb{C}P^J] = \sum_{I_1 \cdots I_q = I} s_{I_1}[\mathbb{C}P^{2j_1}] \cdots s_{I_q}[\mathbb{C}P^{2j_q}] = 0,$$

unless I is a refinement of J (in particular, $s_I[\mathbb{C}P^J] = 0$ if $r < q$). Thus if we order all partitions of k by increasing length, then $(s_I[\mathbb{C}P^J])_{J,I}$ is triangular.

Example 5.4.4 *We continue Example* 5.4.2. *Note that*

$$\begin{aligned}
s_4 &= \sigma_1^4 - 4\sigma_1^2\sigma_2 + 2\sigma_2^2 + 4\sigma_1\sigma_3 - 4\sigma_4 \\
s_{1,3} &= \sigma_1^2\sigma_2 - 2\sigma_2^2 - \sigma_1\sigma_3 + 4\sigma_4 \\
s_{2,2} &= \sigma_2^2 - 2\sigma_1\sigma_3 + 2\sigma_4 \\
s_{1,1,2} &= \sigma_1\sigma_3 - 4\sigma_4 \\
s_{1,1,1,1} &= \sigma_4
\end{aligned}$$

whence

$$\begin{pmatrix} s_4[M_1] & s_{1,3}[M_1] & s_{2,2}[M_1] & s_{1,1,2}[M_1] & s_{1,1,1,1}[M_1] \\ s_4[M_2] & s_{1,3}[M_2] & s_{2,2}[M_2] & s_{1,1,2}[M_2] & s_{1,1,1,1}[M_2] \\ s_4[M_3] & s_{1,3}[M_3] & s_{2,2}[M_3] & s_{1,1,2}[M_3] & s_{1,1,1,1}[M_3] \\ s_4[M_4] & s_{1,3}[M_4] & s_{2,2}[M_4] & s_{1,1,2}[M_4] & s_{1,1,1,1}[M_4] \\ s_4[M_5] & s_{1,3}[M_5] & s_{2,2}[M_5] & s_{1,1,2}[M_5] & s_{1,1,1,1}[M_5] \end{pmatrix}$$

$$= \begin{pmatrix} p_4[M_1] & p_1p_3[M_1] & p_2^2[M_1] & p_1^2p_2[M_1] & p_1^4[M_1] \\ p_4[M_2] & p_1p_3[M_2] & p_2^2[M_2] & p_1^2p_2[M_2] & p_1^4[M_2] \\ p_4[M_3] & p_1p_3[M_3] & p_2^2[M_3] & p_1^2p_2[M_3] & p_1^4[M_3] \\ p_4[M_4] & p_1p_3[M_4] & p_2^2[M_4] & p_1^2p_2[M_4] & p_1^4[M_4] \\ p_4[M_5] & p_1p_3[M_5] & p_2^2[M_5] & p_1^2p_2[M_5] & p_1^4[M_5] \end{pmatrix}$$

$$\times \begin{pmatrix} -4 & 4 & 2 & -4 & 1 \\ 4 & -1 & -2 & 1 & 0 \\ 2 & -2 & 1 & 0 & 0 \\ -4 & 1 & 0 & 0 & 0 \\ 1 & 0 & 0 & 0 & 0 \end{pmatrix}$$

$$= \begin{pmatrix} 9 & 72 & 36 & 252 & 126 \\ 0 & 21 & 0 & 126 & 105 \\ 0 & 0 & 25 & 100 & 100 \\ 0 & 0 & 0 & 45 & 90 \\ 0 & 0 & 0 & 0 & 81 \end{pmatrix}.$$

It remains to calculate the diagonal elements in general. To find $s_k[\mathbb{C}P^{2k}]$, we consider $\mathbb{Z}[t_1, \ldots, t_{2k+1}]$ and $1 + \sigma_1 + \cdots + \sigma_{2k+1} = (1 + t_1) \cdots (1 + t_{2k+1})$. Comparing with $1 + p_1(\mathbb{C}P^{2k}) + \cdots + p_{2k+1}(\mathbb{C}P^{2k}) = (1 + g^2)^{2k+1}$ yields

$$s_k(\mathbb{C}P^{2k}) = t_1^k + \cdots + t_{2k+1}^k = g^{2k} + \cdots + g^{2k} = (2k + 1)g^{2k}$$

and $s_k[\mathbb{C}P^{2k}] = 2k + 1$. The product formula then shows that all diagonal entries are non-zero. We have outlined the demonstration of the following result:

Theorem 5.4.5 (*Thom*) *For each k, the matrix $(p_J[\mathbb{C}P^I])_{I,J}$ of Pontrjagin numbers is nonsingular. In particular,*

$$\text{rank}\, \Omega_{4k}^{SO} \geq p(k),$$

where $p(k)$ denotes the number of partitions of k.

Example 5.4.6 (*Continuing Example* 5.4.4) *Using the product formula for the s_I, we check the diagonal entries of the triangular matrix obtained in* 5.4.4:

$$\begin{aligned}
&s_4[\mathbb{C}P^8] &&= 9, \\
&s_{1,3}[\mathbb{C}P^2 \times \mathbb{C}P^6] &&= s_1[\mathbb{C}P^2]s_3[\mathbb{C}P^6] = 3 \cdot 7 = 21, \\
&s_{2,2}[\mathbb{C}P^4 \times \mathbb{C}P^4] &&= s_2^2[\mathbb{C}P^4] = 5^2 = 25, \\
&s_{1,1,2}[\mathbb{C}P^2 \times \mathbb{C}P^2 \times \mathbb{C}P^4] &&= s_1^2[\mathbb{C}P^2]s_2[\mathbb{C}P^4] = 3^2 \cdot 5 = 45, \\
&s_{1,1,1,1}[\mathbb{C}P^2 \times \mathbb{C}P^2 \times \mathbb{C}P^2 \times \mathbb{C}P^2] &&= s_1^4[\mathbb{C}P^2] = 3^4 = 81.
\end{aligned}$$

5.4.2 The Upper Bound

The program for showing that $p(k)$ is an upper bound for $\Omega_{4k}^{SO} \otimes \mathbb{Q}$ is to show that the latter groups are the image of the homotopy groups of certain spaces (the *Thom spaces*), and to calculate those homotopy groups rationally, using Serre's theorem. We organize the discussion into five steps:

1. Thom Spaces: Let ξ be a k-plane bundle over the CW-complex B. We denote the disk bundle by $D(\xi)$ and the sphere bundle by $S(\xi)$. The *Thom space* associated to ξ is the quotient space $T(\xi) = D(\xi)/S(\xi)$. Let $t_0 \in T(\xi)$ be the image of $S(\xi)$.
   Fact 1: $T(\xi)$ is $(k-1)$-connected (n-cells of B correspond to $(n+k)$-cells of T).
   Fact 2: If ξ is oriented, then the Thom isomorphism $H_{k+i}(D, S) \cong H_i(B)$ implies $H_{k+i}(T, t_0) \cong H_i(B)$.
2. Serre's Theorem: Quoting Gromov, "The rational homotopy invariants of many simply connected spaces are essentially homological ones (and so there is nothing new and unexpected down there hidden from our eyes in the depths of homotopies)." Precisely, we have the following statement:

   Theorem 5.4.7 (*Serre*) *Let X be a finite, $(k-1)$-connected complex, $k \geq 2$. Then the Hurewicz map is an isomorphism*

   $$\pi_i(X) \otimes \mathbb{Q} \xrightarrow{\cong} H_i(X; \mathbb{Q})$$

   for $i < 2k - 1$.

3. Application to Universal Bundles: Let γ be the canonical bundle over the Grassmannian $Gr_k(\mathbb{R}^{k+p})$ of oriented k-planes in $\mathbb{R}^{k+p}$. Then by fact 1, $MSO_p^k := T(\gamma)$ is $(k-1)$-connected. Therefore, using Serre's theorem and fact 2,

   $$\pi_{i+k}(MSO_p^k) \otimes \mathbb{Q} \cong H_{i+k}(MSO_p^k; \mathbb{Q}) \cong H_i(Gr_k)$$

   for $k > i + 1, i \geq 0$. Now the homology of the Grassmannian is well known:

   $$\operatorname{rank} H_i(Gr_k; \mathbb{Q}) = \begin{cases} p(j), & i = 4j \text{ (generated by } p_1(\gamma), p_2(\gamma), \ldots), \\ 0, & i \not\equiv 0(4). \end{cases}$$

   Hence for $i \geq 0, k > i + 1$,

   $$\operatorname{rank} \pi_{i+k}(MSO_p^k) \otimes \mathbb{Q} = \begin{cases} p(j), & i = 4j, \\ 0, & i \not\equiv 0(4). \end{cases}$$

4. The map $\pi_{i+k}(T(\xi)) \to \Omega_i$: Let ξ be an oriented k-plane bundle over the smooth base B. Given a continuous map $f : S^{i+k} \to T(\xi)$, there exists a homotopy $f \simeq \tilde{f}$ to a map $\tilde{f}$ which is smooth in $\tilde{f}^{-1}(T - t_0)$ and transverse to the zero section B (i.e. on $\tilde{f}^{-1}(B)$, the differential of $\tilde{f}$ is onto the normal space

of B in T). The implicit function theorem implies that $\tilde{f}^{-1}(B)$ is a smooth i-manifold and so defines a bordism class in Ω_i. Thus we propose the assignment

$$f \mapsto \tilde{f}^{-1}(B).$$

In order for this to be well-defined, we need to see that the bordism class $[\tilde{f}^{-1}(B)]$ depends only on the homotopy class $[f]$. Suppose $f_1, f_2 : S^{i+k} \to T(\xi)$ are smooth on the inverse images of $T - t_0$, transverse to B, and $f_1 \simeq f_2$. Then there exists a homotopy $H : S^{i+k} \times [0, 1] \to T$ such that

1. H smooth in $H^{-1}(T - t_0)$,
2. H transverse to B,
3. $H(-, t) = f_1, t \in [0, \epsilon]$; $H(-, t) = f_2, t \in [1 - \epsilon, 1]$.

Consequently, $H^{-1}(B)$ is an oriented bordism from $f_1^{-1}(B)$ to $f_2^{-1}(B)$.

5. If $k, p \geq i$, then $\pi_{i+k}(MSO_p^k) \to \Omega_i$ is onto: Given $[M^i] \in \Omega_i$, we embed into Euclidean space, $M^i \hookrightarrow \mathbb{R}^{i+k}$. Take a tube neighborhood N of M such that $N \cong E(\nu^k)$, where ν^k is the normal bundle of the embedding. The composition

$$N \cong E(\nu^k) \xrightarrow{\text{Gauss map}} E(\gamma_i^k) \hookrightarrow E(\gamma_p^k) \xrightarrow{\text{collapse}} T(\gamma_p^k) = MSO_p^k$$

defines a map $f : N \to MSO_p^k$ which is transverse to the zero section $Gr_k(\mathbb{R}^{k+p})$ and $f^{-1}(Gr_k) = M$. Extend f to $\mathbb{R}^{i+k} \cup \{\infty\} = S^{i+k}$ by sending $S^{i+k} - N \mapsto \{t_0\}$. Then $\bar{f} : S^{i+k} \to MSO_p^k$ gives rise to the bordism class of M.

We have established the following result:

Theorem 5.4.8 (*Thom*)

$$\Omega_i^{SO} \begin{cases} \textit{is finite}, & i \not\equiv 0(4), \\ \textit{has rank } p(j), & i = 4j \end{cases}$$

and $\{[\mathbb{C}P^{2i_1} \times \cdots \times \mathbb{C}P^{2i_r}] | (i_1, \ldots, i_r)$ *a partition of* $j\}$ *is a set of generators for* $\Omega_{4j} \otimes \mathbb{Q}$. *Thus*

$$\Omega_*^{SO} \otimes \mathbb{Q} \cong \mathbb{Q}[\mathbb{C}P^2, \mathbb{C}P^4, \mathbb{C}P^6, \ldots]$$

as an algebra.

Remark 5.4.9 *It follows from step 5 above that each* Ω_i^{SO} *is finitely generated: Taking* $k \geq 2$, MSO_p^k *is simply connected (indeed* $(k-1)$*-connected). A simply connected space has all its homotopy groups finitely generated if and only if all its homology groups are finitely generated. The Grassmannian* $Gr_k(\mathbb{R}^{k+p})$ *is a finite CW-complex, so* MSO_p^k *is a finite CW-complex, whence all* $H_i(MSO_p^k)$ *are finitely generated. Therefore* $\pi_{i+k}(MSO_p^k)$, *and* a fortiori Ω_i^{SO} *are finitely generated. As for the torsion, C. T. C. Wall has proved that M is an oriented boundary if and only if all Pontrjagin numbers and all Stiefel–Whitney numbers of M vanish. Every* Ω_i *is hence of the form* $\Omega_i = \mathbb{Z} \oplus \cdots \oplus \mathbb{Z} \oplus \mathbb{Z}/2 \oplus \cdots \oplus \mathbb{Z}/2$.

5.5 Surgery Theory

Surgery theory is a framework for classifying manifolds within a given homotopy type. We briefly summarize its early history (in the 1960's). It was first developed by Kervaire and Milnor to classify homotopy spheres in dimension ≥ 6, up to h-cobordism (and hence, by Smale's theorem, up to diffeomorphism). Browder worked out simply connected surgery theory, starting with the question: When is a simply connected Poincaré complex homotopy equivalent to a differentiable manifold? Novikov initiated the use of surgery in the study of the nonuniqueness of manifold structures in the homotopy type of a manifold. Wall constructed a comprehensive surgery obstruction theory in addressing the non-simply connected case.

The two fundamental problems we are trying to solve are:

1. Given a space X, when is X homotopy equivalent to a manifold?
2. How to classify manifolds M in a given homotopy type?

In dealing with question 1, an obvious condition on X immediately presents itself: Since manifolds satisfy Poincaré duality, X has to satisfy Poincaré duality as well. We make the following definition[2]:

Definition 5.5.1 *A finite CW-complex X is called a* Poincaré complex *if $H_n(X) \cong \mathbb{Z}$ for some n, and cap product with a generator $[X] \in H_n(X)$ is an isomorphism $H^q(X) \xrightarrow{\cap [X]} H_{n-q}(X)$ for all q. We call $[X]$ an* orientation *and n the (Poincaré-)* dimension *of X.*

Suppose X is a Poincaré complex. Assume we already had a homotopy equivalence $f : M \to X$. Let $g : X \to M$ be its homotopy inverse, and let ν_M be the normal bundle[3] of M. Then $\xi^k = g^* \nu_M$ is a vector bundle over X such that $f^*\xi \cong f^*g^*\nu_M \cong 1^*\nu_M \cong \nu_M$, since $gf \simeq 1$. A map $f : X \to Y$ between Poincaré complexes X and Y has *degree* 1 if $f_*[X] = [Y]$.

Definition 5.5.2 *A* normal map (*or* normal invariant) *is a degree 1 map $f : M \to X$, where M is a (smooth) manifold, together with a vector bundle ξ over X and a bundle isomorphism $f^*\xi \cong \nu_M$. That is, we have a diagram*

$$\begin{array}{ccc} \nu & \longrightarrow & \xi \\ \downarrow & & \downarrow \\ M & \xrightarrow{f} & X \end{array}$$

---

[2] We will mainly emphasize the simply-connected case, although we state the surgery structure sequence for the general case.

[3] We obtain the normal bundle by choosing an embedding $M^n \hookrightarrow \mathbb{R}^{n+k}$. Any two such embeddings are isotopic (k large). A more intrinsic approach is often preferable, and one may also work with the tangent bundle. The convenience of the normal bundle comes from its natural occurrence in the Thom–Pontrjagin construction, to be used later on in this section.

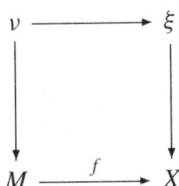

Thus we discover a second obstruction in resolving question 1: The given space X has to possess a normal map. This is a genuine obstruction since there are finite Poincaré complexes X for which no normal maps exist, see [MM79, page 32f]. Suppose then that we have X together with a normal map f. The final step in producing a manifold homotopy equivalent to X is to modify f until it is a homotopy equivalence (keeping X fixed). Note that if f is of degree 1 then the diagram

$$\begin{array}{ccc} H^*(M) & \xleftarrow{f^*} & H^*(X) \\ {\scriptstyle \cap[M]}\big\downarrow{\scriptstyle \cong} & & {\scriptstyle \cap[X]}\big\downarrow{\scriptstyle \cong} \\ H_*(M) & \xrightarrow{f_*} & H_*(X) \end{array}$$

commutes, providing an "Umkehrmap" $H_*(X) \to H_*(M)$ for $f_* : H_*(M) \to H_*(X)$. This implies that f_* is surjective on homology. Consequently, in modifying f into a homotopy equivalence, we have to kill the kernel of f_*. We take our clue from the world of CW-complexes, where elements can be killed by attaching cells. This does not quite work here as it would destroy the manifold structure of M. In the world of manifolds, "elementary surgeries" replace the operation of attaching cells—under surgery, the resulting space is again a closed manifold. The goal is thus: Perform a finite sequence of surgeries after which we end up with a manifold M' and a homotopy equivalence $f' : M' \to X$. Unfortunately, this process is obstructed. The obstruction is one of the most central invariants of high-dimensional topology: the signature.

Definition 5.5.3 *Let X be a Poincaré complex of dimension $n = 4k$. Define a symmetric, bilinear pairing $(-,-) : H^{2k}(X) \times H^{2k}(X) \to \mathbb{Z}$ by*

$$(x, y) = \langle x \cup y, [X] \rangle.$$

Over $\mathbb{Q}$, represent this pairing by a symmetric matrix (after having chosen a basis), then diagonalize the matrix. Define the signature *$\sigma(X)$ of X to be the number of positive diagonal entries minus the number of negative diagonal entries.*[4]

The role of the signature is explained as follows: The manifold resulting from surgery is bordant to the original M. By Theorem 6.1.2 of Sect. 6.1, the signature is a bordism invariant. This means in particular that the signature of M never changes during the modification of f. Now if M' is homotopy equivalent to X, then $\sigma(M') = \sigma(X)$, since the signature is a homotopy invariant. But by bordism invariance, $\sigma(M') = \sigma(M)$. Therefore, we can modify a given normal map $f : M \to X$ into a homotopy equivalence only if $\sigma(M) - \sigma(X)$ vanishes.

We summarize: Given a Poincaré complex X, the attempt to find a manifold M and a homotopy equivalence $f : M \to X$ will be divided into two steps: First find any manifold M' and a normal map $g : M' \to X$ (a "surgery problem"). Then perform surgeries on M' to change g to a homotopy equivalence f.

[4] If the dimension of X is not divisible by 4, set $\sigma(X) = 0$.

Next, we have to discuss how to cook up a surgery problem for a given X, i.e. how to decide whether X has a normal map and how to construct one if it exists. The idea is to define a weak notion of normal bundle, the "Spivak normal fibration." Every Poincaré complex has such a normal fibration. Then try to lift this to an actual vector bundle ξ over X (we shall see that this cannot always be done). Then X sits nicely (as the zero section) in the total space of ξ and transversality works. The Thom–Pontrjagin construction will yield the normal map.

Theorem 5.5.4 (*Spivak*) *Given any simply connected Poincaré complex X^n, there exists a spherical fiber space ν with total space $E(\nu)$ and projection $\pi : E(\nu) \to X$ with fiber a homotopy-S^{k-1}, and an element $\alpha \in \pi_{n+k}(T(\nu))$ such that*

$$\alpha_*[S^{n+k}] \cap U = [X] \in H_n(X).$$

Here $T(\nu)$ is the Thom space $T(\nu) = cyl(\pi) \cup_E cE$ and $U \in H^k(T\nu)$ the Thom class.

Roughly, the theorem is obtained by embedding $X \hookrightarrow \mathbb{R}^{n+k} \subset S^{n+k}$ and taking a regular neighborhood N. The pair $(N, \partial N)$ is a manifold with boundary. Take $E(\nu) = \partial N$; note $T(\nu) \cong N/\partial N$. We have $\alpha_*[S^{n+k}] \in H_{n+k}(T\nu) \cong H_{n+k}(N, \partial N)$, $U \in H^k(T\nu) \cong H^k(N, \partial N)$ and $\cap U$ is a map $H_{n+k}(N, \partial N) \to H_n(N) \cong H_n(X)$, as N is homotopy equivalent to X. The element α is the homotopy class of the natural collapse $S^{n+k} \to N/\partial N$. The Spivak normal fiber space for X is unique in the following sense:

Definition 5.5.5 *Let $p : E \to B$ be a fibration. Two maps $f_0, f_1 : X \to E$ are* fiber homotopic *($f_0 \simeq_p f_1$) if there exists a homotopy $F : X \times I \to E$ ($F(-, 0) = f_0$, $F(-, 1) = f_1$) such that*

$$pF(x, t) = pf_0(x)$$

for all x, t, that is, the tracks of the homotopy stay in the same fiber.

Definition 5.5.6 *Two fibrations $p_i : E_i \to B$ are* fiber homotopy equivalent *if there exist $E_1 \underset{g}{\overset{f}{\rightleftarrows}} E_2$ such that $gf \simeq_{p_1} 1_{E_1}$, $fg \simeq_{p_2} 1_{E_2}$.*

Theorem 5.5.7 *Let ξ_1, ξ_2 be $(k-1)$-spherical fiber spaces over a Poincaré complex X of dimension n ($k \gg n$). Let $\alpha_i \in \pi_{n+k}(T\xi_i)$, $i = 1, 2$, be such that $\alpha_{i*}[S^{n+k}] \cap U_i = [X]$. Then there exists a fiber homotopy equivalence $b : \xi_1 \to \xi_2$ such that $T(b)_*(\alpha_1) = \alpha_2$.*

Now that we know that every Poincaré complex X has a well-defined normal spherical fibration, we would make progress if we knew that this fibration really is the sphere bundle of some vector bundle over X. To decide this question, we reformulate it as a lifting problem using classifying spaces:

Theorem 5.5.8 (*Stasheff*) *There exists a classifying space BG for spherical fiber spaces. (Think of G as the "group" of self-homotopy equivalences of spheres.)*

We remark that BG is a small space—$BG \otimes \mathbb{Q} \simeq pt$—in particular, it carries no rational characteristic classes. Given a Poincaré complex X, its Spivak normal fibration ν_X has a classifying map $X \to BG$ and ν_X is the sphere bundle of a vector bundle if and only if there exists a lift

$$\begin{array}{ccc} & & BO \\ & \nearrow & \downarrow \\ X & \longrightarrow & BG \end{array}$$

$J : BO \to BG$ is the map induced from regarding an orthogonal transformation as a self-homotopy equivalence of the sphere. We may also define it as follows: Let $J_k : BO(k) \to BG(k)$ be the classifying map for the universal k-plane bundle over $BO(k)$ viewed as a spherical fibration. The stabilization map $O(k) \to O(k+1)$ given by direct sum with the identity $1 : \mathbb{R} \to \mathbb{R}$ induces $BO(k) \to BO(k+1)$, and the map $G(k) \to G(k+1)$ given by suspending a self-homotopy equivalence of S^k induces $BG(k) \to BG(k+1)$. Then let $J = \operatorname{colim}_{k\to\infty} J_k : BO = \operatorname{colim}_{k\to\infty} BO(k) \to \operatorname{colim}_{k\to\infty} BG(k) = BG$. On homotopy groups, J induces the much-studied *J-homomorphism* of G. W. Whitehead, now completely calculated by J. F. Adams and others. Using the fibration $BO \to BG \to BG/O$, the existence of the lift is equivalent to the composition $X \to BG \to BG/O$ being homotopic to the constant map. Let us assume that for a given X this is indeed the case. Then the lift $X \to BO$ defines a vector bundle ξ over X, which we think of as the normal bundle of X. One employs the Thom–Pontrjagin construction to complete the construction of a normal map for X: α (see Theorem 5.5.4) gives a degree 1 map $F : S^{n+k} \to T(\xi)$. We can change F in its homotopy class so that it is transverse to X. An important point here is that, although X is not a manifold, transversality still works, since we only need that X is the zero section in a vector bundle. With F transverse to X, set $M = F^{-1}(X)$ and $f = F|_M : M \to X$. Then f is of degree 1 and by transversality M is a closed manifold. The transversality construction also gives us a bundle map $\nu_M \to \xi$ covering f. This concludes the construction of a normal map for X.

Step two consists of trying to surger f to a homotopy equivalence. The obstruction theory for this problem can be very elegantly formulated, especially having question 2 above in mind as well, by the *surgery structure sequence*. Let us briefly discuss the terms that appear in the sequence:

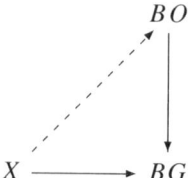

- As we mentioned above, performing a geometric surgery does not change the bordism class of a manifold M. Indeed, in killing a class in the kernel of the normal map, one tries to represent it by an embedded sphere S^k with trivial normal bundle, that is, we have an embedded $S^k \times D^{n-k}$ (where n is the dimension of M). Then $M - S^k \times \operatorname{int} D^{n-k}$ is a manifold with boundary $S^k \times S^{n-k-1}$. If we attach $D^{k+1} \times S^{n-k-1}$ along the boundary, we obtain a closed manifold M', the *effect of surgery on* $S^k \hookrightarrow M$. Now M and M' are bordant via $W = D^{k+1} \times D^{n-k} \cup M \times [0, 1]$ (attaching a handle to the cylinder). The class of S^k is now zero in M'. Moreover, the normal map extends over W. In particular, we also have a normal map $M' \to X$. It is thus natural to consider the set $\mathcal{N}(X)$ of normal bordism classes of normal maps to X:

$$\mathcal{N}(X) = \{f : M \to X |\ \deg f = 1,\ f^*\xi \cong \nu_M\}/\sim,$$

where the equivalence relation $\sim$ is given by *normal bordism*: $f : M \to X$ and $f' : M' \to X$ are normally bordant, if there exists $F : W \to X$, $\partial W = M \sqcup -M'$, $F|_{\partial W} = f \sqcup f'$ and $F^*\xi \cong \nu_W$ extending the bundle isomorphisms over the boundary.

- The *Structure Set*: Let

$$\mathcal{S}^{Diff}(X) = \{M \xrightarrow{f} X |\ f \text{ a homotopy equivalence}\}/\sim,$$

where M is a smooth, closed manifold and the equivalence relation $\sim$ is $f \sim f'$ iff there is a homotopy-commutative diagram

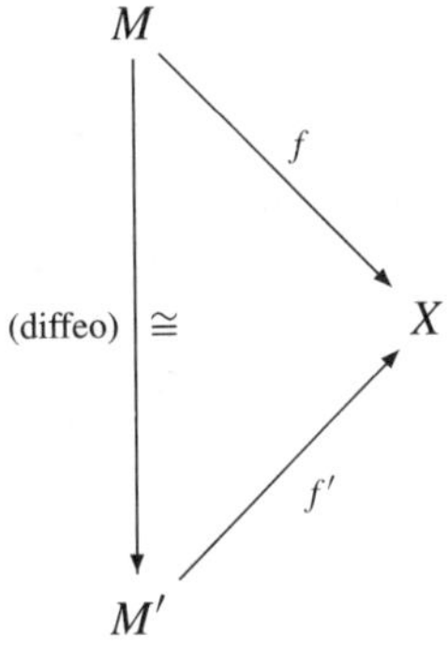

Answering question 2 precisely means computing $\mathcal{S}^{Diff}(X)$.[5]

- The surgery obstruction groups: The obstruction to changing a normal map to a homotopy equivalence takes values in certain abelian groups $L_n(\pi)$ (the *L-groups* of C. T. C. Wall), $n \in \mathbb{Z}$, which depend only on the fundamental group $\pi = \pi_1(X)$. In fact, L_n is a covariant functor from finitely presented groups π to abelian groups. For fixed π, the sequence of L-groups is 4-periodic,

$$L_{n+4}(\pi) = L_n(\pi),$$

and can be described as follows:

$L_{4k}(\pi)$	nonsingular symmetric quadratic forms over the group ring $\mathbb{Z}\pi$
$L_{4k+2}(\pi)$	nonsingular anti-symmetric quadratic forms over $\mathbb{Z}\pi$
$L_{2k+1}(\pi)$	Grothendieck group of automorphisms of the trivial class in $L_{2k}(\pi)$

[5] Naturally, other decorations than $Diff$ can be defined and have been investigated. Besides working with smooth manifolds, one is also interested in $\mathcal{S}^{PL}(X)$ (piecewise linear manifolds) and $\mathcal{S}^{Top}(X)$ (topological manifolds).

For the case of the trivial group $\pi = 1$, one obtains

$$L_n(1) = \begin{cases} \mathbb{Z}, & n = 4k, \\ \mathbb{Z}/2, & n = 4k+2, \\ 0, & n \text{ odd}. \end{cases} \tag{5.2}$$

Theorem 5.5.9 (*The* (*geometric*) *structure sequence*) *Let X be a Poincaré complex of dimension $n \geq 5$, $\pi = \pi_1(X)$. There is an exact sequence of sets*[6]*:*

$$L_{n+1}(\pi) \overset{\omega}{\dashrightarrow} \mathcal{S}^{Diff}(X) \overset{\eta}{\longrightarrow} \mathcal{N}(X) \overset{\sigma}{\longrightarrow} L_n(\pi).$$

Remark 5.5.10 *In this sequence, the term $\mathcal{N}(X)$ is regarded as quite computable. This is due to the fact that we have several different interpretations of $\mathcal{N}(X)$: Let G/O be the homotopy fiber of $BO \to BG$. Then, provided $\mathcal{N}(X) \neq \varnothing$, $\mathcal{N}(X) = [X, G/O]$. Since BG is a "small" space, G/O is quite close to BO, e.g. $(G/O) \otimes \mathbb{Q} \simeq BO \otimes \mathbb{Q}$. In particular,*

$$H^*(G/O; \mathbb{Q}) \cong H^*(BO; \mathbb{Q}) \cong \mathbb{Q}[\textit{Pontrjagin classes}].$$

Even more interesting from a computational standpoint is the description of $\mathcal{N}(X)$ as an exotic homology theory. Many of the constructions in surgery do not really depend so much on the geometry of manifolds, but rather on the algebra of their underlying chain complexes together with the Poincaré duality map. For an n-dimensional topological manifold M ($n \geq 5$), the algebraic theory of surgery identifies the geometric structure sequence

$$L_{n+1}(\pi_1(M)) \dashrightarrow \mathcal{S}^{Top}(M) \longrightarrow [M, G/TOP] \longrightarrow L_n(\pi_1(M))$$

with (a portion of) the algebraic surgery exact sequence

$$L_{n+1}(\pi_1(M)) \longrightarrow \mathbb{S}_{n+1}(M) \longrightarrow H_n(M; \mathbb{L}_\bullet) \overset{A}{\longrightarrow} L_n(\pi_1(M)),$$

see [Ran92, Theorem 18.5]. In particular, $\mathcal{N}(M) = [M, G/TOP] \cong H_n(M; \mathbb{L}_\bullet)$. Here $H_n(M; \mathbb{L}_\bullet)$ denotes the generalized homology groups of M with coefficients in the 1-connective simply-connected surgery spectrum $\mathbb{L}_\bullet$. The map $A : H_n(M; \mathbb{L}_\bullet) \to L_n(\pi_1(M))$ is called the assembly map. *Rationally, the identification $\mathcal{N}(M) \cong H_n(M; \mathbb{L}_\bullet)$ is given by sending a normal map $f : N \to M$ to the Poincaré duals of the L-genera*

$$f_*(L(N) \cap [N]_{\mathbb{Q}}) - L(M) \cap [M]_{\mathbb{Q}} \in H_n(M; \mathbb{L}_\bullet) \otimes \mathbb{Q} \cong H_{n-4*}(M; \mathbb{Q}).$$

We shall discuss the maps in the geometric structure sequence of Theorem 5.5.9:

- The map η has already been explained: Given a homotopy equivalence $f : M \to X$, let $g : X \to M$ be its homotopy inverse and set $\xi^k = g^*\nu_M$. The image of f under η is f, a degree one map, together with the bundle isomorphism $f^*\xi \cong f^*g^*\nu_M \cong \nu_M$. Equivalent homotopy equivalences lead to normally bordant normal maps.

[6] We suppress the orientation character.

- We explain $\sigma : \mathcal{N}(X) \to L_n(\pi)$, the *surgery obstruction*, only for $\pi = \pi_1(X) = 1$. According to (5.2), σ is nonzero only if $n = \dim X$ is even. Assume $n \equiv 0(4)$. Then $L_n(1) \cong \mathbb{Z}$ and the value of σ on a normal map $f : M \to X$ is defined to be
$$\sigma(f) = \frac{1}{8}(\sigma(M) - \sigma(X)).$$
Note that this definition uses implicitly the result:

Theorem 5.5.11 *If $f : M \to X$ is a normal map, then the difference $\sigma(M) - \sigma(X)$ is divisible by* 8.

This explains how the signature invariant fits into, and plays a key role in, the classification theory of high-dimensional manifolds. If $n \equiv 2(4)$, then $\sigma(f) \in \mathbb{Z}/2 \cong L_n(1)$ is the *Arf invariant*: One defines a certain quadratic form q over $\mathbb{Z}/2$ on the kernel $K(f; \mathbb{Z}/2)$ of the normal map, $q : K(f; \mathbb{Z}/2) \to \mathbb{Z}/2$. Now $\mathbb{Z}/2$-quadratic forms have an Arf-invariant, which is characterized nicely by a "majority-vote":

Theorem 5.5.12 *If q is a nonsingular quadratic form over $\mathbb{Z}/2$, then $\sigma(q) = 1$ if and only if q sends a majority of elements to $1 \in \mathbb{Z}/2$.*

If f and g are normally bordant, then indeed $\sigma(f) = \sigma(g)$.
- The broken arrow $\omega : L_{n+1}(\pi) \dashrightarrow \mathcal{S}^{Diff}(X)$ was chosen to indicate that ω is not actually a map, but rather an action $L_{n+1}(\pi) \times \mathcal{S}(X) \to \mathcal{S}(X)$. If ω were simply a map, we could conclude that always $\mathcal{S}^{Diff}(X) \neq \emptyset$, which is incorrect. The construction of the action rests on the following realization result:

Theorem 5.5.13 (*Plumbing Theorem*) *If $m = 2k > 4$, then there exists a manifold $(M, \partial M)$ and a normal map $g : (M, \partial M) \to (D^m, S^{m-1})$ such that $g|_{\partial M}$ is a homotopy equivalence and $\sigma(g)$ takes on any desired value.*

Let $[M^n \xrightarrow{f} X^n] \in \mathcal{S}(X)$ be a homotopy equivalence and $x \in L_{n+1}(\pi)$. We invoke the plumbing theorem to get a normal map $g : (V^{n+1}, \partial V) \to (D^{n+1}, S^n)$, $g|_{\partial V}$ a homotopy equivalence (i.e. ∂V is a homotopy sphere), and $\sigma(g) = x$. Consider the cylinder $M \times I$ and take the connected sum along the boundary component $M \times \{0\}$ with V. We obtain a $(n+1)$-manifold with boundary $M \times \{1\} \sqcup (M \# \partial V)$ (where # denotes connected sum). Sending $\partial V - D^n$ to a point in M defines a collapse map $h : M \# \partial V \to M$, which is a homotopy equivalence since ∂V is a homotopy sphere. Define $\omega(x, f)$ to be the composition
$$[M \# \partial V \xrightarrow[\text{he}]{h} M \xrightarrow[\text{he}]{f} X] \in \mathcal{S}(X).$$

Exactness of the structure sequence at $\mathcal{S}(X)$ is to be interpreted as follows: Let $f, g \in \mathcal{S}(X)$. Then $\eta(f) = \eta(g)$ iff $g = \omega(x, f)$ for some $x \in L_{n+1}$ (that is, f and g are in the same orbit under the action). Exactness at $\mathcal{N}(X)$ is an elegant reformulation of the *fundamental surgery theorem*:

Theorem 5.5.14 *Let X be a simply connected Poincaré complex, $n = \dim X \geq 5$, and let $f : M \to X$ be a normal map. Then*

if n odd: f is normally bordant to a homotopy equivalence $f' : M' \to X$;
if n even: f is normally bordant to a homotopy equivalence $f' : M' \to X$ iff $\sigma(f) = 0$.

Example 5.5.15 *William Browder originally investigated H-spaces, i.e. spaces X that are equipped with a continuous map $\mu : X \times X \to X$ and a distinguished point $e \in X$ such that $\mu(e, -)$ and $\mu(-, e)$ are homotopic to id_X (a topological group would be an example). He discovered:*

Theorem 5.5.16 *(Browder) H-spaces satisfy Poincaré duality over $\mathbb{Z}$.*

Thus he was lead to ask: Is an H-space homotopy equivalent to a manifold? We now carry out the steps outlined in this section. It turns out that the Spivak normal space of an H-space is stably trivial (compare this to the fact that the tangent bundle of a Lie group is trivial). Therefore, $X \to BG$ has a lift $X \to BO$ so that ν_X is a (trivial) vector bundle. As described above, transversality gives us a normal map $f : M \to X$. Now use surgery to make M look like X homotopically. According to the surgery exact sequence of Theorem 5.5.9, *this works right away in odd dimensions. This proves:*

Theorem 5.5.17 *A finite-dimensional simply connected H-space of odd dimension is homotopy equivalent to a manifold.*

In the even-dimensional case, we have to contend with the surgery obstruction.

Example 5.5.18 *By considering $X = S^n$, one recovers the work of Kervaire and Milnor on classifying homotopy spheres.*[7] *Using $\mathcal{N}(X) \cong [X, G/O]$, the structure sequence becomes*

$$L_{n+1}(1) \dashrightarrow \mathcal{S}^{Diff}(S^n) \longrightarrow \pi_n(G/O) \longrightarrow L_n(1), \tag{5.3}$$

$n \geq 5$. Now $\pi_n(G/O) = \operatorname{coker} J \oplus A_n$, with

$$A_n = \begin{cases} \mathbb{Z}, & n \equiv 0(4), \\ 0, & n \not\equiv 0(4) \end{cases}$$

and J is the J-homomorphism. The Kervaire–Arf invariant defines a homomorphism $\operatorname{coker} J \to \mathbb{Z}/2$. Further work shows that with $C_n = \ker(\operatorname{coker} J \to \mathbb{Z}/2)$, the sequence (5.3) *breaks down into a split short exact sequence*

$$0 \to bP_{n+1} \longrightarrow \mathcal{S}(S^n) \longrightarrow C_n \to 0,$$

where bP_{n+1} is always finite cyclic, and in fact 0 for n even. The homomorphism $\operatorname{coker} J \to \mathbb{Z}/2$ is trivial for $n \neq 2^i - 2$, but known to be nontrivial for $n = 6, 14, 30, 62$. Thus, we obtain for example,

$$\mathcal{S}^{Diff}(S^n) = bP_{n+1} \oplus \operatorname{coker} J, \quad \text{if } n \neq 2^i - 2.$$

[7] Note that by the high-dimensional Poincaré conjecture, $\mathcal{S}^{Top}(S^n) = 0, n \geq 5$.

5.6 The L-Class of a Manifold: Approach via Tangential Geometry

We define certain characteristic classes, the Hirzebruch L-classes, of a smooth manifold, which are closely linked to the signature of a manifold via the Hirzebruch signature theorem (Theorem 5.6.9 below) and whose importance has already been demonstrated at various points in this book (e.g. the Browder–Novikov Theorem 5.1.3). Moreover, we will define the L-genus—a homomorphism over $\mathbb{Q}$ from smooth oriented bordism to rational numbers. The L-classes are polynomials in the Pontrjagin classes discussed in Sect. 5.3. Let us define these polynomials.

Let $Q(x) = 1 + a_2x^2 + a_4x^4 + \cdots$ be an even power series, say with rational coefficients a_{2i}. If $x_1, \dots, x_n$ are indeterminates of weight 2, then the product $Q(x_1)\cdots Q(x_n)$ is symmetric in the x_i^2:

$$Q(x_1)\cdots Q(x_n) = 1 + a_2\sum_{i=1}^{n} x_i^2 + \cdots .$$

With σ_j the j-th elementary symmetric function as in Sect. 5.4.1, set $p_j = \sigma_j(x_1^2, \dots, x_n^2)$. Then we can express the term of weight $4r$ in $Q(x_1)\cdots Q(x_n)$ as a homogeneous polynomial $K_r(p_1, \dots, p_r)$, i.e.

$$\begin{aligned} Q(x_1)\cdots Q(x_n) = 1 &+ K_1(p_1) + K_2(p_1, p_2) + \cdots + K_n(p_1, \dots, p_n) \\ &+ K_{n+1}(p_1, \dots, p_n, 0) + \cdots . \end{aligned}$$

The sequence of polynomials $\{K_r\}$ is called the *multiplicative sequence of polynomials associated to $Q(x)$*.

Example 5.6.1 *If*

$$Q(x) = \frac{x}{\tanh(x)} = 1 + \frac{1}{3}x^2 - \frac{1}{45}x^4 \pm \cdots + (-1)^{k-1}\frac{2^{2k}B_k}{(2k)!}x^{2k} \pm \cdots$$

then, with $p_1 = \sigma_1(x_1^2, x_2^2) = x_1^2 + x_2^2$ and $p_2 = \sigma_2(x_1^2, x_2^2) = x_1^2x_2^2$, we obtain

$$\begin{aligned} Q(x_1)Q(x_2) &= \left(1 + \frac{1}{3}x_1^2 - \frac{1}{45}x_1^4 \pm \ldots\right)\left(1 + \frac{1}{3}x_2^2 - \frac{1}{45}x_2^4 \pm \cdots\right) \\ &= 1 + \frac{1}{3}(x_1^2 + x_2^2) + \left(-\frac{1}{45}x_1^4 + \frac{1}{9}x_1^2x_2^2 - \frac{1}{45}x_2^4\right) + \cdots \\ &= 1 + \frac{1}{3}p_1 + \left(\frac{1}{9}p_2 - \frac{1}{45}(x_1^4 + x_2^4)\right) + \cdots \\ &= 1 + \frac{1}{3}p_1 + \left(\frac{1}{9}p_2 - \frac{1}{45}((x_1^2 + x_2^2)^2 - 2x_1^2x_2^2)\right) + \cdots \\ &= 1 + \frac{1}{3}p_1 + \left(\frac{1}{9}p_2 - \frac{1}{45}p_1^2 + \frac{2}{45}p_2\right) + \cdots \\ &= 1 + \frac{1}{3}p_1 + \frac{1}{45}(7p_2 - p_1^2) + \cdots , \end{aligned}$$

whence $K_1(p_1) = \frac{1}{3}p_1$ and $K_2(p_1, p_2) = \frac{1}{45}(7p_2 - p_1^2)$.

Definition 5.6.2 *The* L-polynomials $L_i(p_1, \ldots, p_i)$ *are the multiplicative sequence associated to* $Q(x) = x/\tanh(x)$.

Given a multiplicative sequence, we define characteristic numbers as follows:

Definition 5.6.3 *Let* M^{4n} *be a closed oriented smooth manifold and* $\{K_i\}$ *a multiplicative sequence. Put*

$$K[M] = \langle K_n(p_1(TM), \ldots, p_n(TM)), [M]\rangle \in \mathbb{Q}.$$

Definition 5.6.4 *A* genus *is a ring homomorphism* $\Omega_*^{SO} \otimes \mathbb{Q} \to \mathbb{Q}$. *(More generally, one may define genera over any ring.)*

Example 5.6.5 *The signature* $\sigma \otimes \mathbb{Q}$ *is a genus.*

Multiplicative sequences provide us with a rich class of genera:

Theorem 5.6.6 $K[-]$ *is a genus.*

Proof. The number $K[M]$ is bordism invariant: If $M^{4n} = \partial W^{4n+1}$, either use the argument from Sect. 5.4.1 which established Pontrjagin's theorem, or observe that $K[M]$ is a linear combination of Pontrjagin numbers. Additivity is clear. An important point is multiplicativity: The sequence $\{K_i\}$ has the following multiplicative property (hence the name). If

$$1 + p_1 + p_2 + \cdots = (1 + p_1' + p_2' + \cdots)(1 + p_1'' + p_2'' + \cdots) \tag{5.4}$$

then

$$\sum_{n\geq 0} K_n(p_1, \ldots, p_n) = \sum_{n\geq 0} K_n(p_1', \ldots, p_n') \cdot \sum_{n\geq 0} K_n(p_1'', \ldots, p_n'').$$

But (5.4) holds for the Pontrjagin classes of a cross product of two manifolds. □

Set $f(x) = \frac{x}{Q(x)}$. Note f is an odd power series beginning with x. Put $y = f(x)$ and let $g = f^{-1}$ be the formal inverse $g(y) = x$.

Lemma 5.6.7 $g'(y) = \sum_{n=0}^{\infty} K[\mathbb{C}P^n]y^n$.

Proof. Let $x \in H^2(\mathbb{C}P^n; \mathbb{Z})$ be the standard generator so that $p(\mathbb{C}P^n) = (1 + x^2)^{n+1}$. Consequently, $p_1(\mathbb{C}P^n) = \sigma_1(x^2, \ldots, x^2)$, $p_2(\mathbb{C}P^n) = \sigma_2(x^2, \ldots, x^2)$, etc., and

$$\begin{aligned}
&1 + K_1(p_1\mathbb{C}P^n) + K_2(p_1\mathbb{C}P^n, p_2\mathbb{C}P^n) + \cdots \\
&= 1 + K_1(\sigma_1(x^2, \ldots, x^2)) + K_2(\sigma_1(x^2, \ldots, x^2), \sigma_2(x^2, \ldots, x^2)) + \cdots \\
&= Q(x) \cdots Q(x) \\
&= Q(x)^{n+1}.
\end{aligned}$$

Then for n even,

$$\begin{aligned} K[\mathbb{C}P^n] &= \langle K_{n/2}(p_1\mathbb{C}P^n, \ldots, p_{n/2}\mathbb{C}P^n), [\mathbb{C}P^n]\rangle \\ &= \text{coefficient } A_n \text{ of } x^n = (x^2)^{n/2} \text{ in } Q(x)^{n+1}. \end{aligned}$$

Now dividing $Q(x)^{n+1} = 1 + A_2x^2 + \cdots + A_nx^n + \cdots$ by x^{n+1} we get

$$\frac{Q(x)^{n+1}}{x^{n+1}} = \frac{1}{x^{n+1}} + \cdots + \frac{A_n}{x} + A_{n+1} + \cdots$$

and so

$$\begin{aligned} A_n &= \frac{1}{2\pi i}\oint_C \left(\frac{Q(x)}{x}\right)^{n+1} dx \\ &= \frac{1}{2\pi i}\oint_C \frac{1}{f(x)^{n+1}}dx \\ &= \frac{1}{2\pi i}\oint_{f(C)} \frac{1}{y^{n+1}}g'(y)dy \\ &= \text{coefficient of } y^n \text{ in } g'(y), \end{aligned}$$

where C is a simple closed curve around the origin. □

Example 5.6.8 *The genus $L[M]$ given by $Q(x) = \frac{x}{\tanh(x)}$ is the* L-genus. *In this case $f(x) = \tanh(x)$, which satisfies the differential equation $f'(x) = 1 - f(x)^2$. Hence*

$$g'(y) = \frac{1}{f'(g(y))} = \frac{1}{1-y^2} = 1 + y^2 + y^4 + y^6 + \cdots.$$

We conclude from the above lemma that $L[\mathbb{C}P^{2n}] = 1$ for all n.

The following result is an astonishing relation between the signature, a number defined only using the homological intersection structure of the manifold, and the L-classes, which are defined using the tangential geometry of the manifold.

Theorem 5.6.9 (*Hirzebruch Signature Theorem*) *If M^{4n} is a closed smooth oriented manifold, then*

$$\sigma(M) = L[M].$$

Proof. The signature is a genus $\sigma \otimes \mathbb{Q} : \Omega_*^{SO} \otimes \mathbb{Q} \to \mathbb{Q}$, and so is $L[M]$. Now by our bordism calculation, Theorem 5.4.8, $\Omega_*^{SO} \otimes \mathbb{Q} \cong \mathbb{Q}[\mathbb{C}P^2, \mathbb{C}P^4, \ldots]$ and $\sigma(\mathbb{C}P^{2k}) = 1 = L[\mathbb{C}P^{2k}]$. □

Corollary 5.6.10 *$L[M]$ is an integer.*

Corollary 5.6.11 *$L[M]$ is an oriented homotopy invariant of M.*

5.7 The L-Class of a Manifold: Approach via Maps to Spheres

Let us describe an alternative way of constructing the L-class of a manifold. Suppose we wish to define the L-class of a piecewise linear manifold for which no smooth tangent bundle is directly available. Then Thom observed that, given the Hirzebruch signature theorem (Theorem 5.6.9), one can still build up an L-class, provided one has the signature defined on "enough" submanifolds. This is yet another application of the powerful principle of transversality which we have already used twice before: In Sect. 5.4.2 to establish an upper bound on the rank of the smooth oriented bordism groups Ω_i^{SO}, and in Sect. 5.5 to construct a normal map starting from a Spivak normal fiber space, provided a reduction to a linear bundle exists.

Let M^n be a smooth closed oriented manifold. Given a smooth map $f : M \to S^{n-4i}$, the theorem of Brown and Sard tells us that the set of regular values of f is dense in S^{n-4i}. Pick any regular value $p \in S^{n-4i}$. Then the inverse image $f^{-1}(p)$ is a smooth closed submanifold of M. Its normal bundle is trivial, since it is induced from the normal bundle of the point $\{p\}$ in S^{n-4i}. Consider

$$\sigma(f^{-1}(p)) \in \mathbb{Z}.$$

This number depends only on the homotopy class of f: If $f \simeq g$, then we can construct a smooth homotopy $H : M \times I \to S^{n-4i}$ having p as a regular value. The compact manifold-with-boundary $H^{-1}(p)$ is a bordism from $f^{-1}(p)$ to $g^{-1}(p)$, whence $\sigma(f^{-1}(p)) = \sigma(g^{-1}(p))$, using bordism invariance of the signature (Theorem 6.1.2). Keeping in mind that by the smooth approximation theorem every continuous map is homotopic to a smooth one, we have thus constructed a map

$$\pi^{n-4i}(M) \xrightarrow{\sigma} \mathbb{Z}.$$

Here, $\pi^{n-4i}(M)$ denotes the Borsuk–Spanier cohomotopy set $[M, S^{n-4i}]$, which is a group when $4i < \frac{n-1}{2}$. In that range our map is indeed a group homomorphism. Over the rationals, we have $\sigma \otimes \mathbb{Q} : \pi^{n-4i}(M) \otimes \mathbb{Q} \to \mathbb{Q}$. Recall the Hurewicz map

$$\begin{aligned} \pi^k(M^n) &\longrightarrow H^k(M; \mathbb{Z}) \\ f &\mapsto f^*(u), \end{aligned}$$

with $u \in H^k(S^k; \mathbb{Z})$ the generator such that $\langle u, [S^k] \rangle = +1$. In this situation, Serre's theorem (cf. Theorem 5.4.7 for the homological version) asserts that the rational Hurewicz map

$$\pi^k(M^n) \otimes \mathbb{Q} \longrightarrow H^k(M; \mathbb{Q})$$

is an isomorphism for $n < 2k - 1$. For $k = n - 4i$, this range translates to $4i < \frac{n-1}{2}$, i.e. coincides with the range for which the cohomotopy sets are groups. Let us now assume that i is in that range. We obtain a linear map

$$\sigma \otimes \mathbb{Q} : H^{n-4i}(M; \mathbb{Q}) \longrightarrow \mathbb{Q},$$

which by the universal coefficient theorem (non-degeneracy of the Kronecker pairing) defines a homology class $\sigma \otimes \mathbb{Q} \in H_{n-4i}(M;\mathbb{Q})$. We define the *homology L-classes* of M to be

$$\ell_{n-4i}(M) = \sigma \otimes \mathbb{Q} \in H_{n-4i}(M;\mathbb{Q}).$$

This very method will be used (suitably adapted and with the relevant transversality tools in hand) later on (Sect. 6.3) in constructing the Goresky–MacPherson L-class of a Whitney stratified space, and, more generally, the L-class of any Verdier self-dual complex of sheaves (Sect. 8.2). This explains already that the L-class of a singular space will in general only be a homology class (which need not lift to cohomology under cap product with the fundamental class). In the present discussion, M was assumed to be smooth, so that we also have the Hirzebruch L-classes $L_i(M) \in H^{4i}(M;\mathbb{Q})$ and their Poincaré duals $L_i(M) \cap [M] \in H_{n-4i}(M)$. We claim that

$$L_i(M) \cap [M] = \ell_{n-4i}(M).$$

In the proof, we will use the following lemma.

Lemma 5.7.1 *Let $f : M^n \to S^{n-4i}$ be a smooth map with regular value $p \in S^{n-4i}$. Put $N = f^{-1}(p)$ and let $j : N \hookrightarrow M$ be the inclusion. Then*

$$f^*(u) \cap [M] = j_*[N]$$

in $H_{4i}(M)$.

Proof. Let ν_N be the (trivial) normal bundle of $N \hookrightarrow M$ and let $D\nu$, $S\nu$ denote the associated disk- and sphere-bundle, respectively. The Thom-class $t(\nu_N)$ is a generator $t(\nu_N) \in H^{n-4i}(D\nu, S\nu) \cong H^{n-4i}(M, M-N)$. Write $S = S^{n-4i}$ and let r^* denote the restriction isomorphism $r^* : H^{n-4i}(S, S-p) \xrightarrow{\cong} H^{n-4i}(S)$. Let $\alpha \in H^{n-4i}(S, S-p)$ be the generator such that $r^*\alpha = u$. The map f induces $f|^* : H^{n-4i}(S, S-p) \to H^{n-4i}(M, M-N)$ and we note that $f|^*(\alpha) = t(\nu_M)$ (we may think of α as the Thom-class of $\{p\} \hookrightarrow S$). The commutative diagram

$$\begin{array}{ccc} H^{n-4i}(S, S-p) & \xrightarrow{r^*} & H^{n-4i}(S) \\ \downarrow{\scriptstyle f|^*} & & \downarrow{\scriptstyle f^*} \\ H^{n-4i}(M, M-N) & \xrightarrow{r^*} & H^{n-4i}(M) \end{array}$$

shows that $f^*u = f^*r^*\alpha = r^*f|^*\alpha = r^*t(\nu_N)$. Using the commutative diagram

$$\begin{array}{ccccc} H^{n-4i}(M,M-N)\otimes H_n(M) & \longrightarrow & H^{n-4i}(M,M-N)\otimes H_n(M,M-N) & \cong & H^{n-4i}(D\nu,S\nu)\otimes H_n(D\nu,S\nu) \\ \downarrow{\scriptstyle r^*\otimes 1} & & \downarrow{\scriptstyle \cap} & & \downarrow{\scriptstyle \cap} \\ H^{n-4i}(M)\otimes H_n(M) & \xrightarrow{\cap} & H_{4i}(M) & \xleftarrow{j_*} & H_{4i}(N) \end{array}$$

we calculate

$$\begin{aligned} f^*u \cap [M] &= r^*t(\nu_N) \cap [M] = j_*(t(\nu_N) \cap [D\nu, S\nu]) \\ &= j_*[\text{zero section}] = j_*[N]. \end{aligned}$$ □

Proposition 5.7.2

$$L_i(M) \cap [M] = \ell_{n-4i}(M).$$

Proof. By nondegeneracy of the Kronecker product it suffices to prove

$$\langle x, L_i \cap [M] \rangle = \langle x, \ell_{n-4i} \rangle$$

for all $x \in H^{n-4i}(M)$. By Serre's theorem we can represent such an x as $x = f^*(u)$. Now $\langle f^*u, \ell_{n-4i} \rangle = \sigma(f^{-1}(p))$ by definition. We thus have to demonstrate

$$\langle f^*(u), L_i(M) \cap [M] \rangle = \sigma(f^{-1}(p)).$$

By the Hirzebruch signature theorem (Theorem 5.6.9) the right-hand side is $\sigma(f^{-1}(p)) = \langle L_i(f^{-1}(p)), [f^{-1}(p)] \rangle$. Set $N = f^{-1}(p)$ and let $j : N \hookrightarrow M$ be the inclusion. The Hirzebruch L-class of N is in fact the restriction of $L_i(M)$,

$$\begin{aligned} j^*L_i(M) &= j^*L_i(TM) = L_i(j^*TM) \\ &= L_i(TN \oplus \nu_N) = L_i(TN \oplus \epsilon) = L_i(TN) \\ &= L_i(N) \end{aligned}$$

(using naturality and stability of the Pontrjagin classes; also recall the normal bundle of j is trivial). By Lemma 5.7.1, $f^*(u) \cap [M] = j_*[N]$. Thus

$$\begin{aligned} \langle f^*u, L_i(M) \cap [M] \rangle &= \langle L_i(M), (f^*u) \cap [M] \rangle \\ &= \langle L_i(M), j_*[N] \rangle \\ &= \langle j^*L_i(M), [N] \rangle \\ &= \langle L_i(N), [N] \rangle. \end{aligned}$$ □

To complete the definition of the class $\ell_*(M)$, we have to define $\ell_{n-4i}(M)$ when $4i \geq \frac{n-1}{2}$. This is done by crossing with a high-dimensional sphere. The details are as follows: Let $\tilde{M} = M^n \times S^m$, with m large and write $\tilde{n} = \dim \tilde{M} = n + m$. For m large enough, $4i < \frac{n+m-1}{2} = \frac{\tilde{n}-1}{2}$, and so $\ell_{\tilde{n}-4i}(\tilde{M})$ is defined. We have

$$\begin{aligned} \ell_{n+m-4i}(\tilde{M}) &\in H_{n+m-4i}(M^n \times S^m) \\ &= H_{n+m-4i}(M^n) \otimes H_0(S^m) \oplus H_{n-4i}(M) \otimes H_m(S^m) \\ &\cong H_{n-4i}(M). \end{aligned}$$

Let $\phi : H_{n+m-4i}(\tilde{M}) \xrightarrow{\cong} H_{n-4i}(M)$ denote this isomorphism. We define

$$\ell_{n-4i}(M) = \phi(\ell_{n+m-4i}(\tilde{M})).$$

6

Invariants of Witt Spaces

6.1 The Signature of Spaces with only Even-Codimensional Strata

Let X^n be a topological stratified pseudomanifold which has only strata of even codimension:

$$X_n \supset X_{n-2} \supset X_{n-4} \supset X_{n-6} \supset \cdots .$$

A wide class of examples is given by complex algebraic varieties. In this case, the intersection pairing (4.15) of Sect. 4.4 will allow us to define a signature $\sigma(X)$ by using the two middle perversities. Assume that $n = 2k$. We recall that the two middle perversities $\bar{m}, \bar{n}$ have the following values:

c	2	3	4	5	6	7	8	9	...
$\bar{m}(c)$	0	0	1	1	2	2	3	3	...
$\bar{n}(c)$	0	1	1	2	2	3	3	4	...

The intersection pairing on the middle dimension between the middle perversity groups is

$$IH_k^{\bar{m}}(X) \otimes IH_k^{\bar{n}}(X) \longrightarrow \mathbb{R}. \tag{6.1}$$

Since $\bar{m}(c) = \bar{n}(c)$ for even values of c, and only these values are relevant for our present X, we have $\mathbf{IC}^\bullet_{\bar{m}}(X) = \mathbf{IC}^\bullet_{\bar{n}}(X)$ and thus $IH_k^{\bar{m}}(X) = IH_k^{\bar{n}}(X)$. Therefore, the pairing (6.1) becomes

$$IH_k^{\bar{m}}(X) \otimes IH_k^{\bar{m}}(X) \longrightarrow \mathbb{R}$$

(symmetric if k even), i.e. defines a quadratic form on the vector space $IH_k^{\bar{m}}(X)$. Let $\sigma(X)$ be the signature of this quadratic form.

Remark 6.1.1 *As* $\mathbf{IC}^\bullet_{\bar{m}}(X) = \mathbf{IC}^\bullet_{\bar{n}}(X)$, *we have on the sheaf level* $\mathcal{D}\mathbf{IC}^\bullet_{\bar{m}}(X)[n] \cong \mathbf{IC}^\bullet_{\bar{m}}(X)$. *Thus* $\mathbf{IC}^\bullet_{\bar{m}}(X)$ *for* X *with only even-codimensional strata is our first example of a* self-dual *sheaf. Self-dual sheaves will be discussed in generality later in this book.*

The most important property of the signature is its bordism invariance, which we shall now discuss in detail. We review first the classical statement for nonsingular spaces (Thom's theorem), followed by a sheaf-theoretic proof of the corresponding statement for singular spaces with only even codimensional strata.

1. For manifolds: We digress briefly to explain some linear algebra that we will need: Given linear maps $f : A \to B$ and $g : C \to D$ between k-vector spaces and a commutative diagram

$$\begin{array}{ccc} A & \xrightarrow{f} & B \\ \cong \downarrow \phi & & \psi \downarrow \cong \\ D^* & \xrightarrow{g^*} & C^* \end{array}$$

where we use V^* to denote the linear dual $V^* = \mathrm{Hom}(V, k)$ and $g^* = \mathrm{Hom}(g, 1)$. Then the factorization of g as a surjection followed by an injection,

$$C \xrightarrow{\bar{g}} \operatorname{im} g \hookrightarrow D,$$

induces a dual factorization of g^* as a surjection followed by an injection,

$$D^* \longrightarrow (\operatorname{im} g)^* \overset{\bar{g}^*}{\hookrightarrow} C^*,$$

and thus

$$(\operatorname{im} g)^* \cong \bar{g}^*((\operatorname{im} g)^*) = g^*(D^*) = g^*\phi(A) = \psi(\operatorname{im} f) \cong \operatorname{im} f,$$

whence

$$A/\ker f \cong (\operatorname{im} g)^*. \tag{6.2}$$

Theorem 6.1.2 (*Thom*) *If the closed oriented manifold M^{4k} is an oriented boundary $M = \partial W^{4k+1}$ of a compact oriented manifold W^{4k+1}, then $\sigma(M) = 0$.*

Proof. Consider the commutative diagram with exact rows

$$\begin{array}{ccccc} H^{2k}(W) & \xrightarrow{j^*} & H^{2k}(M) & \xrightarrow{\delta^*} & H^{2k+1}(W, M) \\ PD \downarrow \cong & & PD \downarrow \cong & & PD \downarrow \cong \\ H_{2k+1}(W, M) & \xrightarrow{\partial_*} & H_{2k}(M) & \xrightarrow{j_*} & H_{2k}(W) \end{array}$$

where PD denotes the Poincaré duality isomorphism. Let $L = \operatorname{im} j^* \subset H^{2k}(M)$ and $K = \ker j_* \subset H_{2k}(M)$. As an input for the digression preceding the theorem, take $f = j_*$, $g = j^*$ and as ϕ, ψ universal coefficient isomorphisms as in the following commutative diagram:

$$\begin{array}{ccc} H_{2k}(M) & \xrightarrow{j_*} & H_{2k}(W) \\ \cong \downarrow \phi & & \psi \downarrow \cong \\ H^{2k}(M)^* & \xrightarrow{(j^*)^*} & H^{2k}(W)^* \end{array}$$

Then (6.2) shows

$$H_{2k}(M)/K \cong L^*.$$

Now $\operatorname{im} j^* = \ker \delta^* \overset{PD}{\cong} \ker j_*$, i.e. $L \cong K$. Consequently, $\operatorname{rk} K = \operatorname{rk} L = \operatorname{rk} H_{2k}(M) - \operatorname{rk} K$ and we conclude that

$$\operatorname{rk} L = \tfrac{1}{2} \operatorname{rk} H_{2k}(M).$$

If $x, y \in L$ then

$$\langle x \cup y, [M] \rangle = \langle j^* x' \cup j^* y', [M] \rangle = \langle x' \cup y', j_*[M] \rangle = 0,$$

since $j_*[M] = 0$ as M bounds. Thus L is a *Lagrangian* (maximally self-annihilating) subspace for the intersection form. The existence of a Lagrangian subspace implies that the signature of the given form vanishes, $\sigma(M) = 0$. □

2. For singular spaces:

Definition 6.1.3 *An n-dimensional* pseudomanifold with boundary *is a pair (X^n, A) such that*

1. *A is an $(n-1)$-dimensional pseudomanifold with singular set Σ_A.*
2. *There exists a closed subset $\Sigma_X \subset X$, $\dim \Sigma_X \leq n-2$ such that $X - (\Sigma_X \cup A)$ is an oriented n-manifold, dense in X.*
3. *A is collared in X: There exists a closed neighborhood N of A in X and an orientation- and stratum-preserving isomorphism $A \times [0,1] \xrightarrow{\simeq} N$.*
4. *X has a filtration by closed subsets*

$$X = X_n \supset X_{n-2} = \Sigma_X \supset X_{n-3} \supset \cdots \supset X_0$$

such that the sets $A_{j-1} = X_j \cap A$ stratify A and the $X_j - A_{j-1}$ stratify $X - A$.

The following result is due to Goresky and MacPherson; see [GM80], where it is proven for PL spaces. The proof we present here is entirely sheaf-theoretic and thus works for any topological pseudomanifold.

Theorem 6.1.4 *Let $(X^{4k+1}, \partial X = A)$ be a compact oriented pseudomanifold with boundary. Assume that X has only strata of even codimension. Then*

$$\sigma(A) = 0.$$

Proof. Let Y^{4k+1} be the closed pseudomanifold obtained from X by coning off the boundary:

$$Y = X \cup_A cA.$$

Let $\mathbf{A}^\bullet$ be the algebraic mapping cone of the canonical morphism (4.14) (Sect. 4.2), $\mathbf{IC}^\bullet_{\bar{m}}(Y) \to \mathbf{IC}^\bullet_{\bar{n}}(Y)$, so that we have a distinguished triangle

$$\begin{array}{ccc} \mathbf{IC}^\bullet_{\bar{m}}(Y) & \longrightarrow & \mathbf{IC}^\bullet_{\bar{n}}(Y) \\ & {\scriptstyle [1]}\nwarrow \quad \swarrow & \\ & \mathbf{A}^\bullet & \end{array} \tag{6.3}$$

We observe that away from the cone-point, the canonical morphism is an isomorphism

$$\mathbf{IC}^\bullet_{\bar{m}}(Y)|_{Y-\{c\}} \xrightarrow{\simeq} \mathbf{IC}^\bullet_{\bar{n}}(Y)|_{Y-\{c\}}$$

(since $Y - \{c\}$ has only strata of even codimension) so $\mathbf{A}^\bullet|_{Y-\{c\}} \cong 0$ and $\mathbf{A}^\bullet$ is supported over the cone-point. Moreover, $\mathbf{A}^\bullet_c$ is concentrated in degree $s = \bar{n}(4k+1) - (4k+1)$. As $\bar{n}(4k+1) = 2k$, that degree is $s = -2k-1$. For hypercohomology with compact supports we have the long exact sequence (cf. e.g. [Ive86])

$$\begin{array}{ccccccc} \mathcal{H}^s_c(Y-\{c\};\mathbf{A}^\bullet) & \longrightarrow & \mathcal{H}^s_c(Y;\mathbf{A}^\bullet) & \longrightarrow & \mathcal{H}^s_c(\{c\};\mathbf{A}^\bullet) & \longrightarrow & \mathcal{H}^{s+1}_c(Y-\{c\};\mathbf{A}^\bullet) \\ \| & & \| \wr & & \| & & \| \\ 0 & & \mathcal{H}^s(Y;\mathbf{A}^\bullet) & & \mathcal{H}^s(\{c\};\mathbf{A}^\bullet) & & 0 \end{array}$$

This allows us to identify the hypercohomology of $\mathbf{A}^\bullet$ as the intersection homology of the boundary A:

$$\begin{aligned} \mathcal{H}^s(Y;\mathbf{A}^\bullet) &\cong \mathcal{H}^s(\{c\};\mathbf{A}^\bullet) \\ &\cong \mathbf{H}^s(\mathbf{A}^\bullet)_c \\ &\cong \mathbf{H}^s(\mathbf{IC}^\bullet_{\bar{n}}(Y))_c \quad \text{(using the triangle (6.3))} \\ &\cong \mathcal{H}^s(Link(c);\mathbf{IC}^\bullet_{\bar{n}}(Y)) \\ &\cong \mathcal{H}^{-2k}(A;\mathbf{IC}^\bullet_{\bar{n}}(A)) \\ &= IH^{\bar{m}}_{2k}(A). \end{aligned}$$

The triangle (6.3) induces on hypercohomology an exact sequence

$$\mathcal{H}^s(Y;\mathbf{IC}^\bullet_{\bar{m}}(Y)) \longrightarrow \mathcal{H}^s(Y;\mathbf{IC}^\bullet_{\bar{n}}(Y)) \longrightarrow \mathcal{H}^s(Y;\mathbf{A}^\bullet) \longrightarrow \mathcal{H}^{s+1}(Y;\mathbf{IC}^\bullet_{\bar{m}}(Y)),$$

that is,

$$IH^{\bar{m}}_{2k+1}(Y) \longrightarrow IH^{\bar{n}}_{2k+1}(Y) \xrightarrow{j^*} IH^{\bar{m}}_{2k}(A) \longrightarrow IH^{\bar{m}}_{2k}(Y).$$

Let $L = \operatorname{im} j^* \subset IH^{\bar{m}}_{2k}(A)$. Dualizing triangle (6.3) yields a distinguished triangle

$$\mathcal{D}\mathbf{A}^\bullet \longrightarrow \mathcal{D}\mathbf{IC}^\bullet_{\bar{n}} \longrightarrow \mathcal{D}\mathbf{IC}^\bullet_{\bar{m}} \longrightarrow \mathcal{D}\mathbf{A}^\bullet[1].$$

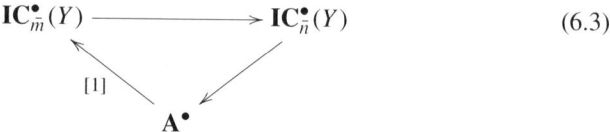

We turn this once to obtain

$$\mathcal{D}\mathbf{IC}^\bullet_{\bar{n}} \longrightarrow \mathcal{D}\mathbf{IC}^\bullet_{\bar{m}} \longrightarrow \mathcal{D}\mathbf{A}^\bullet[1] \longrightarrow \mathcal{D}\mathbf{IC}^\bullet_{\bar{n}}[1],$$

and then shift by $[n]$:

$$\mathcal{D}\mathbf{IC}^\bullet_{\bar{n}}[n] \longrightarrow \mathcal{D}\mathbf{IC}^\bullet_{\bar{m}}[n] \longrightarrow \mathcal{D}\mathbf{A}^\bullet[n+1] \longrightarrow \mathcal{D}\mathbf{IC}^\bullet_{\bar{n}}[n+1].$$

The commutative square

$$\begin{array}{ccc} \mathcal{D}\mathbf{IC}^\bullet_{\bar{n}}[n] & \longrightarrow & \mathcal{D}\mathbf{IC}^\bullet_{\bar{m}}[n] \\ \| \wr & & \| \wr \\ \mathbf{IC}^\bullet_{\bar{m}} & \longrightarrow & \mathbf{IC}^\bullet_{\bar{n}} \end{array}$$

can be completed to an isomorphism of triangles

$$\begin{array}{ccccccc} \mathcal{D}\mathbf{IC}^\bullet_{\bar{n}}[n] & \longrightarrow & \mathcal{D}\mathbf{IC}^\bullet_{\bar{m}}[n] & \longrightarrow & \mathcal{D}\mathbf{A}^\bullet[n+1] & \longrightarrow & \mathcal{D}\mathbf{IC}^\bullet_{\bar{n}}[n+1] \\ \| \wr & & \| \wr & & \downarrow \cong & & \| \wr \\ \mathbf{IC}^\bullet_{\bar{m}} & \longrightarrow & \mathbf{IC}^\bullet_{\bar{n}} & \longrightarrow & \mathbf{A}^\bullet & \longrightarrow & \mathbf{IC}^\bullet_{\bar{m}}[1] \end{array}$$

so that $\mathbf{A}^\bullet$ is a self-dual complex, $\mathcal{D}\mathbf{A}^\bullet[n+1] \cong \mathbf{A}^\bullet$ (but note that the duality shift is off by one). The self-duality isomorphism induces a nonsingular pairing

$$\mathbf{H}^s(\mathbf{A}^\bullet)_c \otimes \mathbf{H}^s(\mathbf{A}^\bullet)_c \longrightarrow \mathbb{R}$$

which is the intersection pairing

$$IH^{\bar{m}}_{2k}(A) \otimes IH^{\bar{m}}_{2k}(A) \longrightarrow \mathbb{R}.$$

We obtain thus a commutative diagram of paired sequences

$$\begin{array}{ccccc} IH^{\bar{n}}_{2k+1}(Y) & \xrightarrow{j^*} & IH^{\bar{m}}_{2k}(A) & \longrightarrow & IH^{\bar{m}}_{2k}(Y) \\ \otimes & & \otimes & & \otimes \\ IH^{\bar{m}}_{2k}(Y) & \longleftarrow & IH^{\bar{m}}_{2k}(A) & \xleftarrow{j^*} & IH^{\bar{n}}_{2k+1}(Y) \\ \downarrow & & \downarrow & & \downarrow \\ \mathbb{R} & & \mathbb{R} & & \mathbb{R} \end{array}$$

which implies as in the proof of Theorem 6.1.2 that L is a Lagrangian subspace and $\sigma(A) = 0$. □

6.2 Introduction to Whitney Stratifications

We have previously defined topological and piecewise linear stratified spaces (Sect. 4.1.2). To be able to apply methods from differential topology to stratified spaces we need a suitable notion of "smoothly stratified" spaces. One such notion is provided by the theory of Whitney stratifications. Within the world of Whitney stratified spaces we will have powerful geometric tools available, such as certain transversality techniques, tube neighborhoods of strata, stratified submersion and Thom's first isotopy lemma, and the important concept of a stratified map.

Definition 6.2.1 *Let $\mathcal{S}$ be a partially ordered set with order relation $<$. An* $\mathcal{S}$-decomposition *of a topological space Z is a locally finite collection of disjoint locally closed subsets* (*called* pieces) $S_\alpha \subset Z$ $(\alpha \in \mathcal{S})$ *such that*

1. $Z = \bigcup_{\alpha \in \mathcal{S}} S_\alpha$,
2. $S_\alpha \cap \bar{S}_\beta \neq \varnothing \Leftrightarrow S_\alpha \subset \bar{S}_\beta \Leftrightarrow \alpha = \beta$ *or* $\alpha < \beta$ (*we will write* $S_\alpha < S_\beta$).

A Whitney stratified space is a decomposed space with each piece a smooth manifold, and pieces fit together via Whitney's conditions A and B.

Definition 6.2.2 *Let M be a smooth manifold and $Z \subset M$ a closed subset. An $\mathcal{S}$-decomposition $Z = \bigcup_{\alpha \in \mathcal{S}} S_\alpha$ is a* Whitney stratification *of Z if:*

1. *Each piece S_α is a locally closed smooth submanifold of M.*
2. *Every pair $S_\alpha < S_\beta$ satisfies* Whitney's conditions A and B, *see Fig. 6.1: Suppose*
   - $(x_i) \subset S_\beta$ *is a sequence of points converging to a point $y \in S_\alpha$.*
   - *The tangent planes $T_{x_i} S_\beta$ converge (in local coordinates of M near y) to a limiting plane τ.*

   Then, whenever $(y_i) \subset S_\alpha$ is a sequence $y_i \to y$ such that the secant lines $l_i = \overline{x_i y_i}$ converge to a limiting line l, we require that

   $$l \subset \tau.$$

   This condition is called condition B. It is not hard to verify that it implies condition A: $T_y S_\alpha \subset \tau$.

Figures 6.2–6.4 illustrate conditions A and B with algebraic surfaces in $M = \mathbb{R}^3$. Figure 6.2 shows $Z = \{x^4 + y^4 = xyz\}$ with $\mathcal{S}$-decomposition $S_1 = z$-axis, $S_2 = Z - S_1$, $S_1 < S_2$. Taking $(x_i) \subset S_2$ to be $x_i = (\frac{1}{i}, \frac{1}{i}, \frac{2}{i^2})$, we have $y = 0 \in \mathbb{R}^3$, $T_y S_1 = z$-axis, $\tau = xy$-plane, and $T_y S_1 \not\subset \tau$, so that condition A is violated. In a similar fashion, the reader may investigate the example of Fig. 6.3 and of Fig. 6.4.

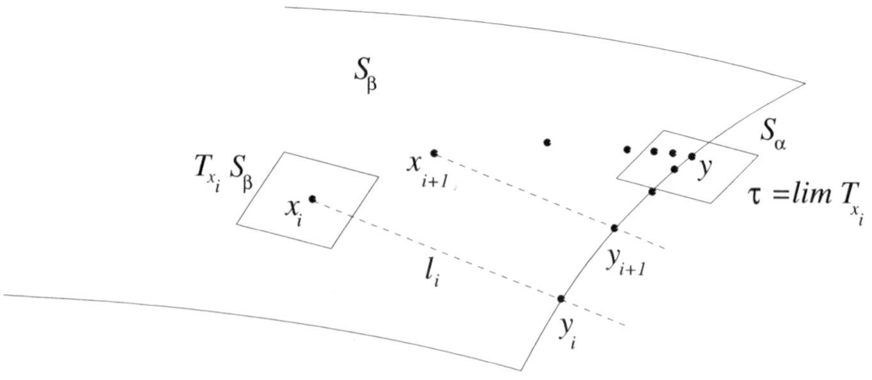

Fig. 6.1. Whitney's conditions A and B.

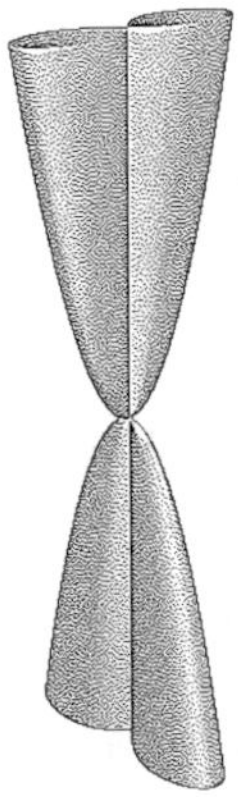

Fig. 6.2. $x^4 + y^4 = xyz$: Neither A nor B hold.

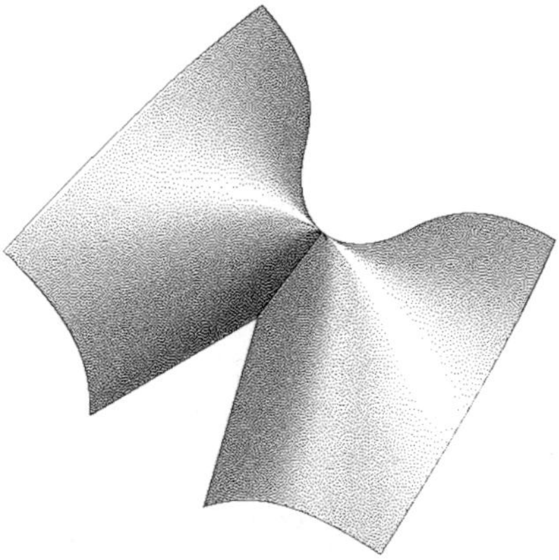

Fig. 6.3. $x^2 = zy^2$: Neither A nor B hold.

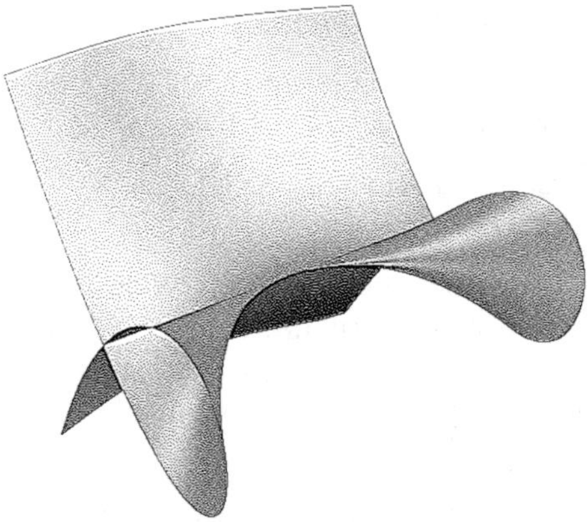

Fig. 6.4. $y^2 = z^2x^2 - x^3$: A holds, but not B.

Any compact complex analytic variety has a canonical Whitney stratification with only even codimensional strata. The Whitney conditions imply the following properties:

- Whitney stratifications are locally topologically trivial along the strata.
- Whitney stratified spaces can be triangulated (compare J. H. C. Whitehead's theorem for smooth manifolds).
- The transverse intersection of two Whitney stratified spaces is again Whitney stratified by the intersections of the strata of the two spaces.

Definition 6.2.3 *Let $Z_1 \subset M_1$ and $Z_2 \subset M_2$ be Whitney subsets of smooth manifolds and let $f : M_1 \to M_2$ be a smooth map. We say that the restriction $f|_{Z_1} : Z_1 \to M_2$ is* transverse *to Z_2 if, for each stratum $A \subset Z_1$ and for each stratum $B \subset Z_2$, the map $f|_A : A \to M_2$ is transverse to B (i.e. for each $x \in f^{-1}(B)$, $df(x)(T_x A) + T_{f(x)} B = T_{f(x)} M_2$).*

If $f|_{Z_1}$ is transverse to Z_2, it follows that $(f|_{Z_1})^{-1}(Z_2)$ is Whitney stratified by strata of the form $A \cap f^{-1}(B)$. For the special case of an inclusion map, we obtain the notion of a *transverse intersection* of two Whitney stratified sets, $(f|_{Z_1})^{-1}(Z_2) = Z_1 \cap Z_2$.

6.2.1 Local Structure

Choose a Riemannian metric on M and let $Z \subset M$ be Whitney stratified. Fix a point $p \in Z$; let S be the connected component containing p of the stratum containing p. Locally at p we can always find a smooth submanifold N' of M, transverse to each stratum of Z, intersecting S only at p, and satisfying $\dim S + \dim N' = \dim M$. Let

$$B_\delta(p) = \{z \in M | \operatorname{dist}(z, p) \leq \delta\}$$

be the closed ball of radius δ around p with boundary

$$\partial B_\delta(p) = \{z \in M | \operatorname{dist}(z, p) = \delta\}.$$

Whitney's condition B implies that if δ is sufficiently small, then $\partial B_\delta(p)$ is transverse to each stratum of Z and also transverse to each stratum of $Z \cap N'$. Fix such $\delta > 0$.

Definition 6.2.4 *The* normal slice *to S at p is $N(p) = N' \cap Z \cap B_\delta(p)$. The* link *of S at p is $L(p) = N' \cap Z \cap \partial B_\delta(p)$.*

The normal slice and the link are canonically Whitney stratified, since they are transverse intersections of Whitney stratified spaces. For δ sufficiently small, the homeomorphism type of $(N(p), L(p))$ is independent of the choice of δ, the Riemannian metric, N', and $p \in S$. There is a homeomorphism $N(p) \cong cL(p)$. In fact, one can show:

Proposition 6.2.5 *S has a closed neighborhood T_S ("tube") in Z and a locally trivial projection $\pi : T_S \to S$ such that the fiber $\pi^{-1}(p)$ is homeomorphic to the cone on the link of S.*

6.2.2 Stratified Submersions

Recall that for manifolds, a smooth map $f : \bar{M}^{n+k} \to M^n$ is a *submersion*, if f is surjective and for all $x \in \bar{M}$, $df(x) : T_x\bar{M} \to T_{f(x)}M$ has rank n. (It follows that the fiber $f^{-1}(p)$ is a submanifold of $\bar{M}$.)

Definition 6.2.6 *Let $Z \subset M$ be Whitney stratified and N be a smooth manifold. A* stratified submersion *$f : M \to N$ is a smooth map such that*

1. *$f|_Z$ is proper.*
2. *For each stratum A of Z, $f|_A : A \to N$ is a submersion.*

If $f : Z \to \mathbb{R}^n$ is a stratified submersion, then for all $t \in \mathbb{R}^n$, $Z \cap f^{-1}(t)$ is Whitney stratified by its intersection with the strata of Z.

Theorem 6.2.7 (*Thom's first isotopy lemma*) *Given a stratified submersion $f : Z \to \mathbb{R}^n$, there exists a stratum preserving homeomorphism*

$$Z \xrightarrow{\cong} \mathbb{R}^n \times (f^{-1}(0) \cap Z),$$

which is smooth on every stratum and commutes with the projection to $\mathbb{R}^n$.

6.2.3 Stratified Maps

The concept of a stratified map between Whitney stratified spaces generalizes the concept of a fiber bundle between smooth manifolds.

Definition 6.2.8 *Let $Z_1 \subset M_1$, $Z_2 \subset M_2$ be Whitney stratified spaces, and let $f : M_1 \to M_2$ be a smooth map between the ambient manifolds such that $f|_{Z_1}$ is proper and $f(Z_1) \subset Z_2$. The restriction $f|_{Z_1} : Z_1 \to Z_2$ is called a* stratified map, *if*

1. *for each stratum $A_2 \subset Z_2$, $f^{-1}(A_2)$ is a union of connected components of strata of Z_1, and*
2. *f takes each of these components submersively to A_2.*

If B is a connected component of a stratum of Z_2 then $f|_{f^{-1}(B)}$ is a stratified submersion. Thom's first isotopy lemma implies that $f|_{f^{-1}(B)}$ is a locally trivial fiber bundle in a stratum preserving way. In particular, the fibers over any two points in B are homeomorphic. Also, the homeomorphism type of $f^{-1}N(b)$ is independent of $b \in B$, where $N(b)$ is the normal slice to B at b.

6.2.4 Normally Nonsingular Maps

Definition 6.2.9 *An inclusion $i : X \hookrightarrow Y$ of locally compact Hausdorff spaces is* normally nonsingular *if there exists a vector bundle $E \to X$, an open neighborhood $U \subset E$ of the zero section, and a homeomorphism $j : U \xrightarrow{\cong} j(U) \subset Y$ such that $j(U)$ is open in Y and the composition $X \to U \to Y$ is i.*

Transverse intersections give rise to normally nonsingular inclusions:

Proposition 6.2.10 *Let $Y \subset M$ be Whitney stratified and suppose $N \subset M$ is a smooth submanifold transverse to Y. Then $X = N \cap Y$ is Whitney stratified and*

$$X \hookrightarrow Y$$

is normally nonsingular.

Normally nonsingular inclusions play an important role in the theory of self-dual sheaves, since self-duality is preserved under restriction to a normally nonsingular subvariety.[1] This observation will be used e.g. in Sect. 8.2 to construct the L-class of an arbitrary self-dual sheaf.

6.3 The Goresky–MacPherson L-Class

Let X^n be a Whitney stratified, closed, oriented pseudomanifold which has *only even-codimensional strata*. We will show how to adapt the Thom–Pontrjagin type approach of Sect. 5.7 to construct the Goresky–MacPherson L-class $L(X)$ of X. Let $M \supset X$ denote the ambient smooth manifold. In the context of the present section, let us adopt the following language:

Definition 6.3.1 *A continuous map $f : X \to S^k$ is* transverse, *if*

1. *f is the restriction of a smooth map $\tilde{f} : M \to S^k$,*
2. *the north pole $N \in S^k$ is a regular value of $\tilde{f}$, and*
3. *$\tilde{f}^{-1}(N)$ is transverse to each stratum of X (that is, the submanifold $\tilde{f}^{-1}(N)$ is transverse to X in the sense of Whitney stratified sets).*

Let $f : X \to S^k$ be transverse. Then $f^{-1}(N) = \tilde{f}^{-1}(N) \cap X$ is Whitney stratified with strata $\tilde{f}^{-1}(N) \cap A$, for each stratum A of X. Note that all of these strata $\tilde{f}^{-1}(N) \cap A$ are of even codimension in $f^{-1}(N)$. Therefore, the signature $\sigma(f^{-1}(N))$ is well-defined using middle perversity intersection homology as described in Sect. 6.1.

Lemma 6.3.2 *The map*

$$\begin{array}{rcl} \sigma : \pi^k(X) & \longrightarrow & \mathbb{Z} \\ {[f]} & \mapsto & \sigma(f^{-1}(N)) \end{array}$$

is a well-defined homomorphism $(2k - 1 > n)$.

[1] Self-duality is also preserved by restriction to open subsets, but it is not preserved under general restrictions. For example, the inclusion of a stratum into the whole space is usually not normally nonsingular, and the restriction of a self-dual sheaf to a lower-dimensional stratum is typically not self-dual.

Proof. Any continuous $f : X \to S^k$ is homotopic to a transverse map. Suppose f and g are homotopic transverse maps. There exists a transverse homotopy $H : X \times I \to S^k$ from f to g. Then $H^{-1}(N)$ is a bordism with only even codimensional strata between $f^{-1}(N)$ and $g^{-1}(N)$. By bordism invariance, Theorem 6.1.4, $\sigma(f^{-1}(N)) = \sigma(g^{-1}(N))$. □

By Serre's theorem, the Hurewicz map $\pi^k(X) \otimes \mathbb{Q} \to H^k(X; \mathbb{Q})$ is an isomorphism for $2k - 1 > n$. Thus, we may view the homomorphism of Lemma 6.3.2 as a linear functional $\sigma \otimes \mathbb{Q} \in \mathrm{Hom}(H^k(X), \mathbb{Q}) \cong H_k(X)$. We define

$$L_k(X) = \sigma \otimes \mathbb{Q} \in H_k(X; \mathbb{Q}).$$

- The restriction $2k-1 > n$ is removed by crossing with a high-dimensional sphere as shown at the end of Sect. 5.7.
- In general, $L(X)$ is not in the image of the duality map $H^*(X) \xrightarrow{\cap[X]} H_*(X)$. Thus the Goresky–MacPherson L-class is in general only a homology characteristic class.
- Using results of Sullivan, one can find an element in $KO(X) \otimes \mathbb{Z}[\frac{1}{2}]$ whose Pontrjagin character is $L(X)$.

6.4 Witt Spaces

Let us start out with some remarks on singular bordism theories. Naively, we could consider bordism based on all (say topological, or PL) closed pseudomanifolds,

$$\Omega_*^{\text{all pseudomfds}}(Y) = \{[X \xrightarrow{f} Y] \mid X \text{ a pseudomanifold}\},$$

where the admissible bordisms are compact pseudomanifolds with boundary, without further restrictions. Now it is immediately clear that the associated coefficient groups vanish, $\Omega_*^{\text{all pseudomfds}}(pt) = 0$, since any pseudomanifold X is the boundary of the cone cX, which is an admissible bordism. Moreover, under such a notion of bordism, the signature is not a bordism invariant—complex projective space $\mathbb{C}P^2$ with $\sigma(\mathbb{C}P^2) = 1$ is the boundary of the cone $c\mathbb{C}P^2$, whence we are supposed to have $\sigma(\mathbb{C}P^2) = 0$. Thus this naive definition is not useful, and we conclude that a subclass of pseudomanifolds has to be selected to define useful singular bordism theories. Given the results on middle perversity intersection homology presented so far, our next approach would be to select the class of all closed pseudomanifolds with only even-codimensional strata,

$$\Omega_*^{\text{ev}}(Y) = \{[X \xrightarrow{f} Y] \mid X \text{ has only even codim strata}\}$$

(and the same condition is imposed on all admissible bordisms). While we do know that the signature is well-defined on $\Omega_*^{\text{ev}}(pt)$, this is however still not a good theory as this definition leads to a large number of geometrically insignificant generators.

For instance, X may not even be bordant to a space X^{ref} obtained by merely refining the stratification of X. This demonstrates that we need to allow some strata of odd codimension, but so as not to destroy Poincaré duality. This problem was solved by Paul Siegel in [Sie83] by introducing a class of stratified spaces called Witt spaces (because of a relation to Witt groups, as will be explained in the next section).

Let X^n be a stratified oriented pseudomanifold without boundary. We ask the following question: Under precisely which conditions on X is $\mathbf{IC}^\bullet_{\bar{m}}(X)$ a self-dual sheaf, $\mathcal{D}\mathbf{IC}^\bullet_{\bar{m}}(X)[n] \cong \mathbf{IC}^\bullet_{\bar{m}}(X)$? Since the lower middle perversity intersection sheaves are always dual to the upper middle perversity intersection sheaves, $\mathbf{IC}^\bullet_{\bar{m}}(X)$ is self-dual if and only if the canonical morphism $\mathbf{IC}^\bullet_{\bar{m}}(X) \to \mathbf{IC}^\bullet_{\bar{n}}(X)$ is an isomorphism (in the derived category). This can be analyzed by studying the mapping cone:

$$\begin{array}{ccc} \mathbf{IC}^\bullet_{\bar{m}} & \longrightarrow & \mathbf{IC}^\bullet_{\bar{n}} \\ & {}_{[1]}\nwarrow \quad \swarrow & \\ & \mathbf{S}^\bullet & \end{array}$$

We make use of the following lemma.

Lemma 6.4.1 *Suppose $\bar{p}(c) = \bar{q}(c)$ for all $c \neq k$ and $\bar{q}(k) = \bar{p}(k) + 1$. If*

$$\begin{array}{ccc} \mathbf{IC}^\bullet_{\bar{p}} & \longrightarrow & \mathbf{IC}^\bullet_{\bar{q}} \\ & {}_{[1]}\nwarrow \quad \swarrow & \\ & \mathbf{S}^\bullet & \end{array}$$

is a distinguished triangle on the canonical morphism, then

1. $\operatorname{supp} \mathbf{H}^i(\mathbf{S}^\bullet) = \operatorname{cl}(\operatorname{supp} \mathbf{H}^i(\mathbf{S}^\bullet) \cap (X_{n-k} - X_{n-k-1}))$.
2. *For $x \in X_{n-k} - X_{n-k-1}$:*

$$\mathbf{H}^i(\mathbf{S}^\bullet)_x = \begin{cases} 0, & i \neq \bar{q}(k) - n, \\ \mathbf{H}^i(\mathbf{IC}^\bullet_{\bar{q}})_x, & i = \bar{q}(k) - n. \end{cases}$$

Factoring $\mathbf{IC}^\bullet_{\bar{m}} \to \mathbf{IC}^\bullet_{\bar{n}}$ through several perversities and applying the lemma to each one-step increase, we see that $\mathbf{IC}^\bullet_{\bar{m}} \to \mathbf{IC}^\bullet_{\bar{n}}$ is an isomorphism iff

$$\mathbf{H}^i(\mathbf{IC}^\bullet_{\bar{n}})_x = 0, \quad i = \bar{n}(k) - n, k \geq 3 \text{ odd}.$$

For $k = 2l + 1$,

$$\mathbf{H}^{\bar{n}(2l+1)-n}(\mathbf{IC}^\bullet_{\bar{n}})_x \cong IH_l(Link(x)).$$

Thus, self-duality of $\mathbf{IC}^\bullet_{\bar{m}}(X)$ is characterized by the vanishing of the middle-perversity middle-dimensional intersection homology of the links of odd-codimensional strata.

Definition 6.4.2 *X^n is a* Witt space, *if $IH_l^{\bar{m}}(Link(x)) = 0$ for all $x \in X_{n-2l-1} - X_{n-2l-2}$ and all $l \geq 1$.*

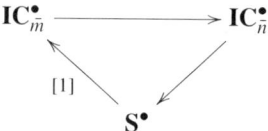

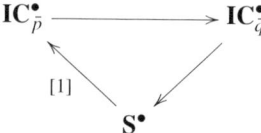

Example 6.4.3 *The suspension $X^7 = \Sigma\mathbb{C}P^3$ has two singular points which form a stratum of odd codimension 7. The link is $\mathbb{C}P^3$ with middle homology $H_3(\mathbb{C}P^3) = 0$. Hence X^7 is a Witt space. The suspension $X^3 = \Sigma T^2$ has two singular points which form a stratum of odd codimension 3. The space X^3 is not Witt, since the middle Betti number of the link T^2 is 2.*

We summarize:

Theorem 6.4.4 (*Siegel*) *X is Witt iff $\mathbf{IC}^\bullet_{\bar{m}}(X) \to \mathbf{IC}^\bullet_{\bar{n}}(X)$ is an isomorphism. On a Witt space X, $\mathbf{IC}^\bullet_{\bar{m}}(X)$ is self-dual, and if X is compact, we have a nonsingular pairing $IH_i^{\bar{m}}(X) \otimes IH_{n-i}^{\bar{m}}(X) \to \mathbb{R}$.*

Let Ω_i^{Witt} denote the bordism group based on i-dimensional Witt spaces, i.e.

$$\Omega_i^{Witt} = \{[X^i] \mid X^i \text{ PL closed oriented Witt}\},$$

where two spaces X and X' represent the same class, $[X] = [X']$, if there exists an $(i+1)$-dimensional oriented compact PL stratified pseudomanifold-with-boundary Y^{i+1} such that each stratum of odd codimension in $Y - \partial Y$ satisfies the Witt condition and $\partial Y \cong X \sqcup -X'$ under an orientation-preserving PL-isomorphism.

When is a Witt space X^i a boundary? Suppose i is odd. Then $X = \partial Y$ with $Y = cX$, the cone on X. The cone Y is a Witt space, since the cone-point is a stratum of even codimension in Y (we stratify the cone of a stratified pseudomanifold as in Example 4.1.15). This shows that $\Omega_{2k+1}^{Witt} = 0$ for all k. If $i = 2k$ is even, then $Y = cX$ is Witt only if $IH_k(X) = 0$. Thus we have $[X] = 0 \in \Omega_{2k}^{Witt}$ provided the middle-dimensional middle-perversity intersection homology of X vanishes.

6.5 Siegel's Calculation of Witt Bordism

We give a complete calculation of the Witt bordism coefficient groups Ω_*^{Witt}, due to P. Siegel. Our exposition is guided by M. Goresky's report [B⁺84, VII] on [Sie83]. Let us first recall definition and structure of the *Witt group* $W(\mathbb{Q})$ of the ring $\mathbb{Q}$ of rational numbers. Define $W(\mathbb{Q})$ to be the free abelian group generated by $\mathbb{Q}$-vector spaces V equipped with a symmetric bilinear nondegenerate pairing $\beta : V \times V \to \mathbb{Q}$, modulo the relations $(V_1, \beta_1) + (V_2, \beta_2) = (V_1 \oplus V_2, \beta_1 \oplus \beta_2)$ (orthogonal sum), and $(V, \beta) = 0$ if V contains a self-annihilating subspace W such that $\dim W = \frac{1}{2} \dim V$. The structure of $W(\mathbb{Q})$ is known:

$$W(\mathbb{Q}) \cong W(\mathbb{Z}) \oplus \bigoplus_{p \text{ prime}} W(\mathbb{Z}/_p),$$

where $W(\mathbb{Z}) \cong \mathbb{Z}$ via the signature, $W(\mathbb{Z}/_2) \cong \mathbb{Z}/_2$, and for $p \neq 2$, $W(\mathbb{Z}/_p) \cong \mathbb{Z}/_4$ or $\mathbb{Z}/_2 \oplus \mathbb{Z}/_2$.

Let X^{4k} be a (PL) Witt space. Define an invariant $w(X)$ as follows:

$$w(X) = (IH_{2k}^{\bar{m}}(X; \mathbb{Q}), \text{intersection pairing}) \in W(\mathbb{Q}).$$

Theorem 6.5.1 *(Siegel) The assignment $X \mapsto w(X)$ induces an isomorphism*

$$w : \Omega_{4k}^{Witt} \xrightarrow{\cong} W(\mathbb{Q})$$

for $k \geq 1$ and $\Omega_0^{Witt} \cong \mathbb{Z}$. If $i \not\equiv 0(4)$, then $\Omega_i^{Witt} = 0$.

Proof. 1. w is well-defined: If $X = \partial Y$, then the image of $IH_{2k+1}^{\bar{n}}(Y \cup_X cX) \to IH_{2k}^{\bar{m}}(X)$ is a self-annihilating subspace of half the dimension (cf. the proof of Theorem 6.1.4). Thus $w(X) = 0$ in the Witt ring.

2. w is surjective: This is a geometric realization problem, which is solved by the following plumbing result:

Theorem 6.5.2 *Let A be a symmetric $n \times n$-matrix with integer entries and only even entries on the diagonal. For $k \geq 1$, there exists a manifold $(M^{4k}, \partial M)$ such that the matrix of intersections $H_{2k}(M; \mathbb{Z}) \otimes H_{2k}(M; \mathbb{Z}) \to \mathbb{Z}$ is given by A.*[2]

Given $(V, \beta) \in W(\mathbb{Q})$, we can find a basis of V with respect to which the intersection matrix $A = A(\beta)$ satisfies the conditions of the theorem. Then $H_{2k}(M; \mathbb{Q})$ represents (V, β) and

$$IH_{2k}^{\bar{m}}(M \cup_{\partial M} c\partial M; \mathbb{Q}) \cong H_{2k}(M; \mathbb{Q}).$$

The space $X = M \cup_{\partial M} c\partial M$ is Witt, since the cone vertex is of even codimension, and we have $w(X) = (V, \beta)$.

3. w is injective: Suppose that X^{4k} is a connected Witt space such that $w(X) = 0$. This means there exists a maximally self-annihilating subspace in $IH_{2k}^{\bar{m}}(X)$. It suffices to find a bordism $X \sim X'$ such that $IH_{2k}^{\bar{m}}(X') = 0$, for then $X' = \partial cX'$ and the cone cX' is Witt. There exists a symplectic basis (see e.g. [MH73]) for $IH_{2k}^{\bar{m}}(X)$, i.e. a basis in which the intersection matrix takes the form

$$\begin{pmatrix} 0 & I \\ I & 0 \end{pmatrix}.$$

Siegel shows that every basis element can be represented by an irreducible[3] PL-cycle in X. Let ξ be such a cycle, let $N(\xi)$ be a regular neighborhood of $|\xi|$ with boundary $\partial N(\xi)$. We will construct the required X' in stages, where each stage is an elementary singular surgery as follows:

$$\begin{aligned} &\textit{Claim: } X' = (X - \operatorname{int} N(\xi)) \cup c\partial N(\xi) \text{ is Witt-bordant to } X, \\ &\operatorname{rk} IH_{2k}^{\bar{m}}(X') = \operatorname{rk} IH_{2k}^{\bar{m}}(X) - 2. \end{aligned}$$

We supply some of the arguments involved in proving the claim:

---

[2] In fact, we can find $(M, \partial M)$ with M $(2k-1)$-connected and ∂M a homology sphere. Then $H_{2k}(M)$ is free abelian and $i_* : H_{2k}(M) \to H_{2k}(M, \partial M)$ is an isomorphism. Hence the pairing $H_{2k}(M) \otimes H_{2k}(M) \to \mathbb{Z}$ can be identified with the Poincaré duality pairing $H_{2k}(M) \otimes H_{2k}(M, \partial M) \to \mathbb{Z}$ via i_*.

[3] Definition: A representative cycle z for $\alpha \in IH_j^{\bar{p}}(X; \mathbb{Q})$ is *irreducible* if (1) $H_j(|z|; \mathbb{Z}) = \mathbb{Z}$, and (2) the generator of $H_j(|z|; \mathbb{Z})$ has coefficient ± 1 on every j-simplex of $|z|$.

- Thinking of X' as obtained from X by collapsing $N(\xi)$ to a point, it is clear that the class $[\xi]$ is killed in X'.
- Let $[\xi^*]$ denote the class dual to $[\xi]$ in the symplectic basis, that is, we have the intersection $[\xi] \cap [\xi^*] = 1$. Then $[\xi^*]$ is also killed by the surgery: Since $|\xi^*|$ intersects $|\xi|$ in X, it passes through the cone-point c in X'. But if ξ^* were an allowable chain in $IH_{2k}^{\bar{m}}(X')$, it would have to satisfy

$$\dim(|\xi^*| \cap \{c\}) \leq 2k - 4k + (2k - 1) = -1.$$

  Thus ξ^* is not $\bar{m}$-allowable in X'.
- The bordism between X and X' is given by the "trace of the surgery": Let $coll : X \to X'$ be the collapse map, and let $cyl(coll)$ be the mapping cylinder. Define Y^{4k+1} to be $cyl(coll)$ with a collar $X' \times I$ attached on X',

$$Y^{4k+1} = cyl(coll) \cup_{X'} X' \times I.$$

  Then $\partial Y = X \sqcup -X'$.
- Y is Witt: No problems arise from the stratum $\{c\} \times (0, 1)$ in the collar, since it is of even codimension in Y. The stratum $\{c\} \times \{0\}$ in the interior of Y is of odd codimension, so that we need to check the Witt condition there. The link of $\{c\} \times \{0\}$ in Y is

$$L^{4k} = Lk(\{c\} \times \{0\}) = N(\xi) \cup_{\partial N(\xi)} c\partial N(\xi).$$

  We have to show that $IH_{2k}^{\bar{m}}(L) = 0$. Any $2k$-cycle in L must miss the cone-point (again by $\bar{m}$-allowability) and can therefore be deformed into $N(\xi)$ and then further deformed onto $|\xi|$. Now ξ was chosen irreducible. We conclude that

$$\operatorname{rk} IH_{2k}^{\bar{m}}(L) \leq 1,$$

  generated by $[\xi]$. As L is a Witt space, $IH_{2k}^{\bar{m}}(L) \times IH_{2k}^{\bar{m}}(L) \to \mathbb{Q}$ is nondegenerate. But the original basis was symplectic. Thus the self-intersection number $[\xi] \cap [\xi] = 0$ in (X and hence in) L. It follows that $IH_{2k}^{\bar{m}}(L) = 0$. □

6.6 Application of Witt Bordism: Novikov Additivity

Definition 6.6.1 *If $(X^{4k}, \partial X)$ is a Witt space with boundary, let*

$$w(X) = w(X \cup_{\partial X} c(\partial X)).$$

Theorem 6.6.2 (*Additivity*) *Let Y^{4k} be an oriented closed Witt space and $Z, X_1, X_2 \subset Y$ such that*

1. *$Y = X_1 \cup X_2$,*
2. *$X_1 \cap X_2 = Z$,*
3. *Z is bicollared in Y,*
4. *(X_1, Z_1) and $(X_2, -Z)$ are oriented $4k$-dimensional Witt spaces with boundary.*

Then

$$w(Y) = w(X_1) + w(X_2).$$

Remark 6.6.3 *Composing with the signature* $W(\mathbb{Q}) \to \mathbb{Z}$, *it follows that* $\sigma(Y) = \sigma(X_1) + \sigma(X_2)$. *In the manifold situation, this property is referred to as* Novikov additivity. *In general, this formula does not suggest that we can calculate the signature of a space by decomposing it into small pieces, since "corners" will develop and the formula does not remain applicable. There is, however, a refinement of the formula due to C. T. C. Wall involving Maslov indices.*

Proof. (of Theorem 6.6.2) The following geometric argument rests on the pinch bordism, see Fig. 6.5, and is due to Siegel. Let

$$Y' = Y/Z \cong (X_1 \cup cZ) \cup_c (X_2 \cup cZ),$$

where c denotes the common cone-point. We define the pinch bordism P to be the mapping cylinder of the collapse $Y \to Y'$ with a collar $Y' \times I$ attached. The main point is to verify that P is a Witt space. One needs to check that

$$IH_{2k}^{\bar{m}}(Lk(\{c\} \times \{0\}, P)) = 0.$$

Now the link is

$$\begin{aligned} Lk &= Z \times [-1, 1] \cup c\partial(Z \times [-1, 1]) \\ &\cong \text{suspension}(Z)/(\text{cone-points identified}). \end{aligned}$$

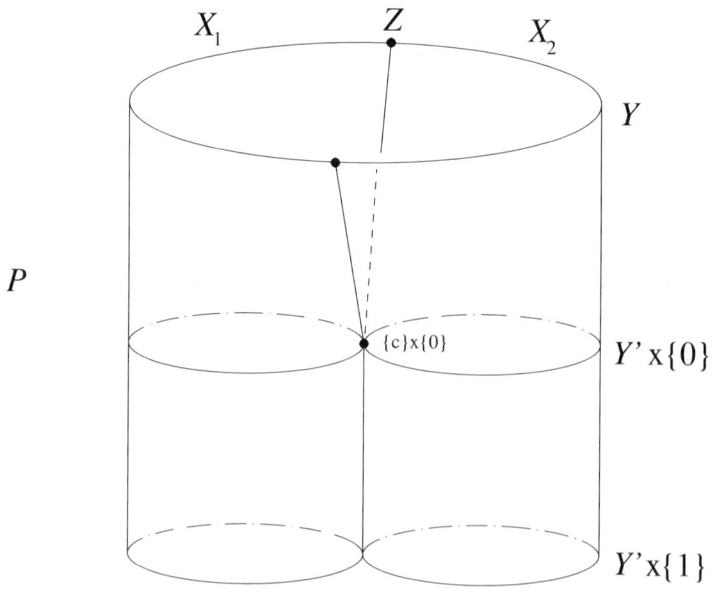

Fig. 6.5. The pinch bordism P.

As the intersection homology of a space does not change under normalization, see Remark 6.6.4 below, we have $IH_{2k}^{\bar{m}}(Lk) \cong IH_{2k}^{\bar{m}}(\Sigma(X))$, where Σ denotes suspension. We claim that the latter group vanishes. If $\xi \in IC_{2k}^{\bar{m}}(\Sigma Z)$ is an intersection cycle, then by allowability $\xi \cap \{c_\pm\} = \varnothing$ (with $c_\pm$ the two suspension points). Thus we can find a representative $[\xi'] = [\xi]$, $|\xi'| \subset Z$. But $\xi' = \partial(c_+\xi')$, and $c_+\xi'$ is $\bar{m}$-allowable. Therefore $[\xi'] = 0$.

Knowing that P is a Witt bordism, we can now invoke bordism invariance to conclude

$$w(Y) = w(Y').$$

By Definition 6.6.1, $w(Y') = w(X_1) + w(X_2)$. □

Remark 6.6.4 *We call a pseudomanifold X* normal, *if it has connected links.*

Definition 6.6.5 *A* normalization *of an n-dimensional pseudomanifold X is a normal pseudomanifold $\tilde{X}$ together with a finite-to-one projection $\pi : \tilde{X} \to X$ such that for all $p \in X$*

$$\pi_* : \bigoplus_{q \in \pi^{-1}(p)} H_n(\tilde{X}, \tilde{X} - q) \longrightarrow H_n(X, X - p)$$

is an isomorphism.

Every PL pseudomanifold can be normalized by the following procedure: Given a triangulation T of X, we construct $\tilde{X}$. Let

$$Y = \bigsqcup_{\substack{\sigma_n \in T \\ n-\text{simplex}}} \sigma.$$

If we let $f : Y \to X$ be the characteristic map and put

$$\tilde{X} = Y/\sim,$$

where $\sim$ means identify two $(n-1)$-simplices τ and τ' iff $f(\tau) = f(\tau')$, then $\tilde{X}$ is a simplicial complex such that the link of every simplex σ_i, $i \leq n-2$, is connected.

Theorem 6.6.6 *If X is a pseudomanifold with normalization $\tilde{X} \xrightarrow{\pi} X$, then $\pi_* : C_*(\tilde{X}) \to C_*(X)$ induces isomorphisms $IH_*^{\bar{p}}(\tilde{X}) \cong IH_*^{\bar{p}}(X)$ for any $\bar{p}$.*

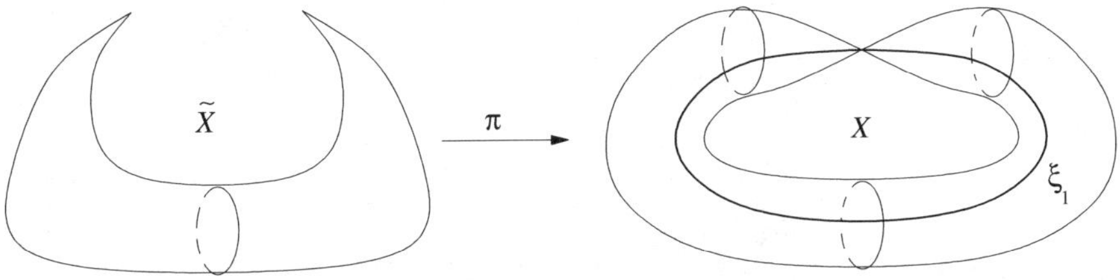

Fig. 6.6. The normalization of a pinched torus.

Proof. Check $IC_i^{\bar{p}}(\tilde{X}) \cong IC_i^{\bar{p}}(X)$. □

Example 6.6.7 *Let X be the pinched torus, Fig. 6.6. Its normalization is the 2-sphere, $\tilde{X} = S^2$. The 1-cycle ξ_1 in X (which disappears in $\tilde{X}$) is not $\bar{p}$-allowable for any $\bar{p}$, since the condition is*

$$\dim(|\xi_1| \cap pt) \leq 1 - 2 + \bar{p}(2) = -1.$$

7

T-Structures

7.1 Basic Definitions and Properties

In discussing the proof of the Kazhdan–Lusztig conjecture, Beilinson, Bernstein and Deligne discovered that the essential image of the category of regular holonomic $\mathcal{D}$-modules under the Riemann–Hilbert correspondence gives a natural abelian subcategory of the nonabelian bounded constructible derived category on a smooth complex algebraic variety. An intrinsic characterization of this abelian subcategory was obtained by Deligne (based on discussions with Beilinson, Bernstein, and MacPherson), and independently by Kashiwara. It was then realized that one still gets an abelian subcategory if the axioms of the characterization are modified to accommodate an arbitrary perversity function, with the original axioms corresponding to the middle perversity. The objects of these abelian subcategories were termed *perverse sheaves* (they are actually still complexes of sheaves, but behave in many ways like sheaves). The middle perversity intersection chain sheaves $\mathbf{IC}^{\bullet}_{\bar{m}}(X)$ are examples of perverse sheaves. For a detailed account of the history of this subject we refer the reader to [Kle89].

Our presentation of the theory is based on the authoritative monograph [BBD82], as well as [KS90]. First we develop the general theory of *t-structures* on any triangulated category. We define the notion of the *heart* of a t-structure and prove that the heart is always an abelian category (Theorem 7.1.11). In Sect. 7.2, we describe a procedure which takes as its input a t-structure on an open subset and a t-structure on the closed complement, and produces as an output a t-structure on the union. This is the important *gluing Theorem* 7.2.2. Starting with natural t-structures shifted by the values of a chosen perversity function, one applies the gluing theorem inductively to the various strata of a filtered space X to obtain the *perverse t-structure* on X. We define the category of perverse sheaves on X as the heart of the perverse t-structure. In the process, we shall also obtain numerous functors operating on t-structures and perverse sheaves.

Let D be a triangulated category. We will write the translation functor as $[1]$.

Definition 7.1.1 *Let $D^{\leq 0}$ and $D^{\geq 0}$ be strictly*[1] *full subcategories of D. Put $D^{\leq n} = D^{\leq 0}[-n]$, $D^{\geq n} = D^{\geq 0}[-n]$. The pair $(D^{\leq 0}, D^{\geq 0})$ is a* t-structure *on D if*

1. $D^{\leq -1} \subset D^{\leq 0}$, $D^{\geq 1} \subset D^{\geq 0}$,
2. $\mathrm{Hom}_D(X, Y) = 0$ *for* $X \in D^{\leq 0}$, $Y \in D^{\geq 1}$,
3. *For every $X \in D$ there exists a distinguished triangle*

$$X_0 \longrightarrow X \longrightarrow X_1 \xrightarrow{+1}$$

in D with $X_0 \in D^{\leq 0}$, $X_1 \in D^{\geq 1}$.

Remark 7.1.2 *If $(D^{\leq 0}, D^{\geq 0})$ is a t-structure, then so is $(D^{\leq n}, D^{\geq n})$ ("shifted t-structure").*

Example 7.1.3 *The* natural t-structure: *Let A be an abelian category and $D = D(A)$ its derived category. We claim that setting*

$$\begin{aligned} D^{\leq 0}(A) &= \textit{full subcategory of } D \textit{ of objects } X^\bullet \textit{ such that} \\ &\qquad H^j(X^\bullet) = 0 \textit{ for } j > 0, \\ D^{\geq 0}(A) &= \textit{full subcategory of } D \textit{ of objects } X^\bullet \textit{ such that} \\ &\qquad H^j(X^\bullet) = 0 \textit{ for } j < 0, \end{aligned}$$

yields a t-structure $(D^{\leq 0}(A), D^{\geq 0}(A))$. Requirement (1) *in Definition* 7.1.1 *is obvious. We prove* (2): *Let $X^\bullet \in D^{\leq 0}(A)$, $Y^\bullet \in D^{\geq 1}(A)$. In $D(A)$, the natural morphism $Y^\bullet \to \tau_{\geq 1} Y^\bullet$ is an isomorphism and thus*

$$\mathrm{Hom}_D(X^\bullet, Y^\bullet) \simeq \mathrm{Hom}_D(X^\bullet, \tau_{\geq 1} Y^\bullet).$$

Let

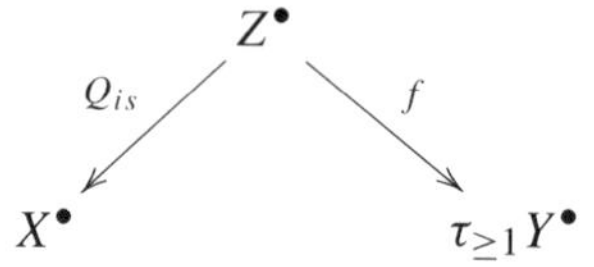

be a morphism in $\mathrm{Hom}_D(X^\bullet, \tau_{\geq 1} Y^\bullet)$ *and consider the diagram*

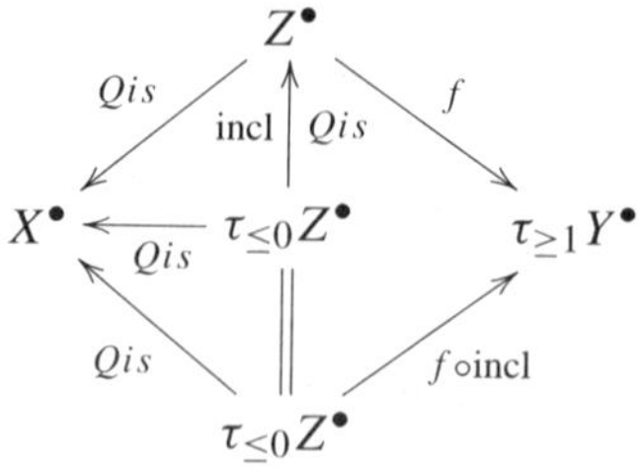

[1] *Strictly full* means that $X \in D$, $X' \in D^{\leq 0}$, and $X \cong X'$ in D implies $X \in D^{\leq 0}$.

But $f \circ \mathrm{incl} : \tau_{\leq 0} Z^\bullet \to \tau_{\geq 1} Y^\bullet$, *a morphism in the homotopy category* $K(A)$, *can only be represented by the zero-morphism in* $C(A)$. *Therefore,* $\mathrm{Hom}_D(X^\bullet, \tau_{\geq 1} Y^\bullet) = 0$. *To satisfy requirement* (3), *take the distinguished triangle*

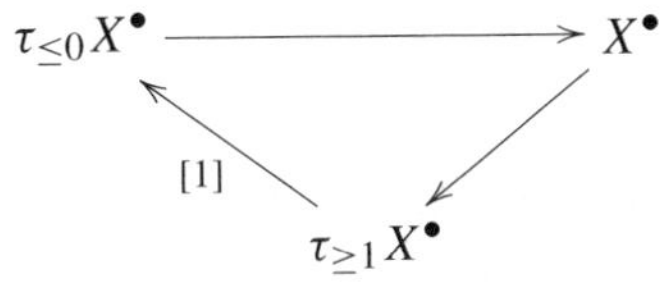

from Example 2.4.15.

Definition 7.1.4 *The* heart (*or* core) *of a t-structure is the full subcategory* $D^{\leq 0} \cap D^{\geq 0}$.

Example 7.1.5 *The heart of the natural t-structure* $(D^{\leq 0}(A), D^{\geq 0}(A))$, $D^{\leq 0}(A) \cap D^{\geq 0}(A)$, *is equivalent to* A.

Proposition 7.1.6

1. *The inclusion* $D^{\leq n} \to D$ (*resp.* $D^{\geq n} \to D$) *has a right adjoint functor* $\tau^{\leq n} : D \to D^{\leq n}$ (*resp. left adjoint* $\tau^{\geq n} : D \to D^{\geq n}$), *i.e. there exist natural morphisms* $\tau^{\leq n} X \to X$ (*resp.* $X \to \tau^{\geq n} X$) *such that the induced map*

$$\mathrm{Hom}_{D^{\leq n}}(X, \tau^{\leq n} Y) \xrightarrow{\simeq} \mathrm{Hom}_D(X, Y)$$

is an isomorphism for all $X \in D^{\leq n}$, $Y \in D$ (*resp.*

$$\mathrm{Hom}_{D^{\geq n}}(\tau^{\geq n} X, Y) \xrightarrow{\simeq} \mathrm{Hom}_D(X, Y)$$

for all $X \in D$, $Y \in D^{\geq n}$).

2. *There exists a unique morphism*

$$d : \tau^{\geq n+1}(X) \to (\tau^{\leq n}(X))[1]$$

such that the triangle

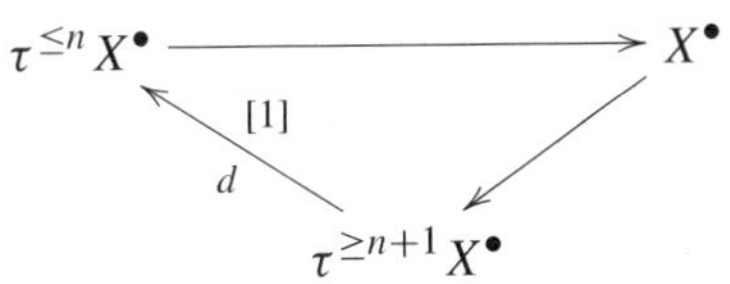

is distinguished. Moreover, d *is functorial.*

Proof. We will assume $n = 0$. For every $X \in D$ choose a distinguished triangle

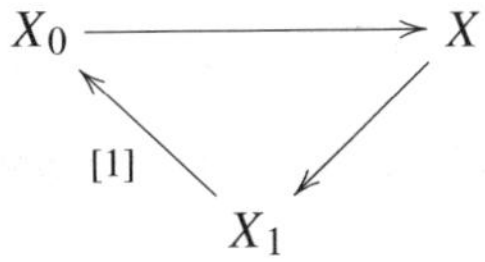

with $X_0 \in D^{\leq 0}$, $X_1 \in D^{\geq 1}$. This is possible by axiom (3). Put

$$\tau^{\leq 0} X = X_0,$$
$$\tau^{\geq 1} X = X_1.$$

Let Y be any object in $D^{\leq 0}$. In the long exact sequence

$$\mathrm{Hom}(Y, X_1[-1]) \longrightarrow \mathrm{Hom}_D(Y, X_0) \longrightarrow \mathrm{Hom}_D(Y, X) \longrightarrow \mathrm{Hom}(Y, X_1)$$

the term $\mathrm{Hom}(Y, X_1) = 0$ by (2) since $Y \in D^{\leq 0}$, and $X_1 \in D^{\geq 1}$. Similarly, $\mathrm{Hom}(Y, X_1[-1]) = 0$ since $Y \in D^{\leq 0}$, and $X_1[-1] \in D^{\geq 2}$ so that

$$\mathrm{Hom}(Y, X_0) \xrightarrow{\simeq} \mathrm{Hom}(Y, X).$$

We take $\tau^{\leq 0} X \to X$ to be $X_0 \to X$ from the triangle. Next, we define $\tau^{\leq 0}$ on morphisms. If we are given a morphism $f : X \to Y$, let $\tau^{\leq 0} f : \tau^{\leq 0} X \to \tau^{\leq 0} Y$ be the unique preimage of the composition $(\tau^{\leq 0} X \to X \xrightarrow{f} Y) \in \mathrm{Hom}(\tau^{\leq 0} X, Y)$ under the isomorphism $\mathrm{Hom}(\tau^{\leq 0} X, \tau^{\leq 0} Y) \cong \mathrm{Hom}(\tau^{\leq 0} X, Y)$. We verify

$$\tau^{\leq 0}(gf) = \tau^{\leq 0} g \circ \tau^{\leq 0} f$$

for $X \xrightarrow{f} Y \xrightarrow{g} Z$, leaving $\tau^{\leq 0} 1_X = 1_{\tau^{\leq 0} X}$ to the reader. Let

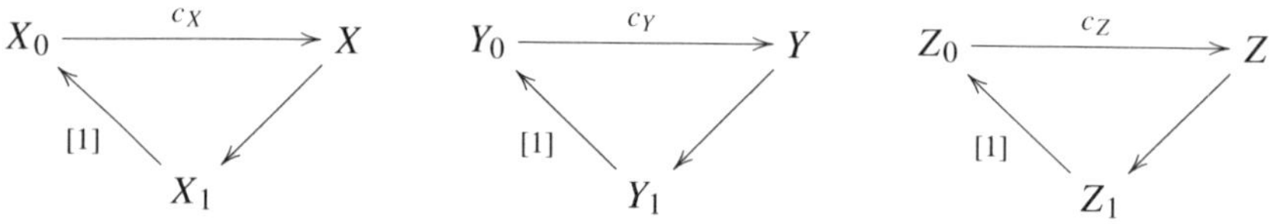

be the triangles associated to X, Y, Z. Now $\tau^{\leq 0} f$ is defined by the equation

$$c_Y \circ \tau^{\leq 0} f = f \circ c_X$$

and $\tau^{\leq 0} g$ is defined by the equation

$$c_Z \circ \tau^{\leq 0} g = g \circ c_Y.$$

Then

$$\begin{aligned} c_Z \circ (\tau^{\leq 0} g \circ \tau^{\leq 0} f) &= g \circ c_Y \circ \tau^{\leq 0} f \\ &= g \circ f \circ c_X \\ &= c_Z \circ (\tau^{\leq 0}(gf)) \end{aligned}$$

shows that $\tau^{\leq 0}(gf) = \tau^{\leq 0} g \circ \tau^{\leq 0} f$. Analogous considerations apply to $\tau^{\geq 0}$. Take

$$d : \tau^{\geq 1} X \longrightarrow \tau^{\leq 0} X[1]$$

to be $X_1 \to X_0[1]$ from the distinguished triangle. Its uniqueness follows from Lemma 2.2.6 as indeed $\mathrm{Hom}(X_0[1], X_1) = 0$ $(X_0[1] \in D^{\leq -1}, X_1 \in D^{\geq 1})$. □

Note that

$$\tau^{\leq n}(X[m]) \cong (\tau^{\leq n+m}(X))[m],$$
$$\tau^{\geq n}(X[m]) \cong (\tau^{\geq n+m}(X))[m].$$

It is frequently convenient to write $\tau^{<n} = \tau^{\leq n-1}$ and $\tau^{>n} = \tau^{\geq n+1}$ (similarly $D^{<n}, D^{>n}$).

Proposition 7.1.7

1. *If $X \in D^{\leq n}$, then $\tau^{\leq n}X \to X$ is an isomorphism.*
2. *Let $X \in D$. Then $X \in D^{\leq n}$ iff $\tau^{>n}X \cong 0$. Similar statements hold for $D^{\geq n}$.*

Proof. Use the morphism of triangles

$$\begin{array}{ccccccc} X & = & X & \longrightarrow & 0 & \xrightarrow{+1} & \\ \uparrow & & \| & & \uparrow & & \\ \tau^{\leq n}X & \longrightarrow & X & \longrightarrow & \tau^{>n}X & \xrightarrow{+1} & \end{array}$$

□

Proposition 7.1.8 *Let*

$$\begin{array}{ccc} X' & \longrightarrow & X \\ & {}_{[1]}\nwarrow \quad \swarrow & \\ & X'' & \end{array}$$

be distinguished in D. If $X', X'' \in D^{\geq 0}$, then $X \in D^{\geq 0}$ (similarly for $D^{\leq 0}$).

Proof. The assumption $X', X'' \in D^{\geq 0}$ implies

$$\mathrm{Hom}(\tau^{<0}X, X') = 0 = \mathrm{Hom}(\tau^{<0}X, X'')$$

by axiom (2). Looking at the long exact sequence obtained from applying $\mathrm{Hom}(\tau^{<0}X, -)$ to the triangle, we conclude $\mathrm{Hom}(\tau^{<0}X, X) = 0$. By Proposition 7.1.6, this latter group is isomorphic to $\mathrm{Hom}(\tau^{<0}X, \tau^{<0}X)$. Hence $1_{\tau^{<0}X} = 0$ and so $\tau^{<0}X \cong 0$. This means that $X \in D^{\geq 0}$ using Proposition 7.1.7. □

Proposition 7.1.9

1. $b \geq a \Rightarrow \tau^{\geq b} \circ \tau^{\geq a} \cong \tau^{\geq a} \circ \tau^{\geq b} \cong \tau^{\geq b}$,
   $\tau^{\leq b} \circ \tau^{\leq a} \cong \tau^{\leq a} \circ \tau^{\leq b} \cong \tau^{\leq a}$.
2. $a > b \Rightarrow \tau^{\leq b} \circ \tau^{\geq a} \cong \tau^{\geq a} \circ \tau^{\leq b} \cong 0$.
3. *For $X \in D$, there exists an isomorphism $\tau^{\geq a} \circ \tau^{\leq b}X \xrightarrow{\simeq} \tau^{\leq b} \circ \tau^{\geq a}X$.*

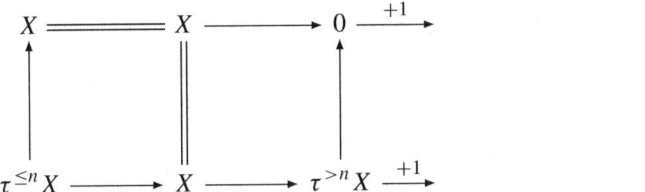

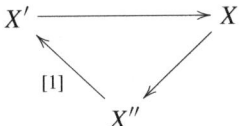

Proof. 1. Since $\tau^{\geq b}X \in D^{\geq b} \subset D^{\geq a}$, the canonical morphism $\tau^{\geq b}X \to \tau^{\geq a}(\tau^{\geq b}X)$ is an isomorphism by Proposition 7.1.7, (1).

2. Follows from Proposition 7.1.7, (2).

3. The nontrivial case is $b \geq a$. The distinguished triangles

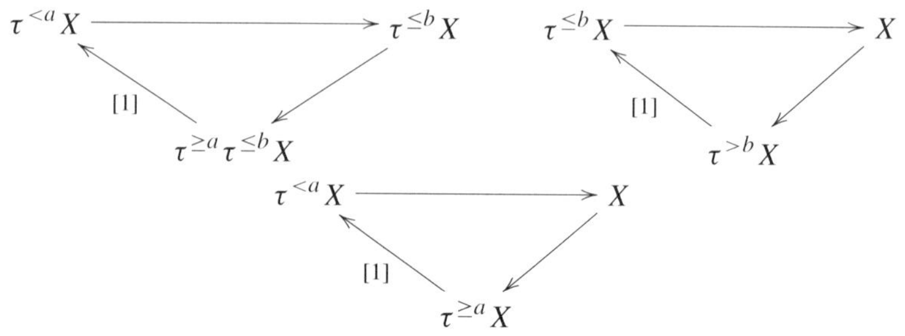

can be placed into an octahedron

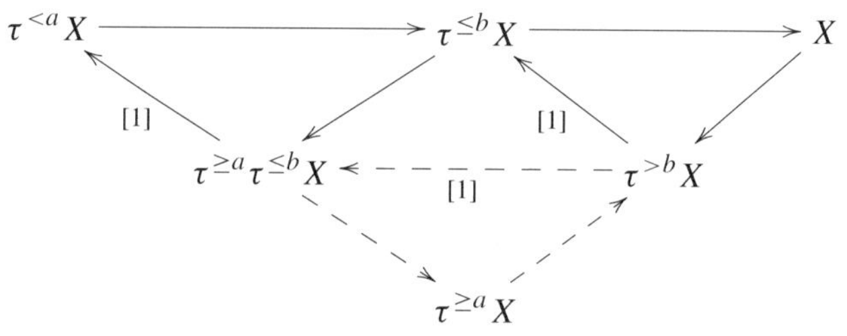

yielding the distinguished triangle

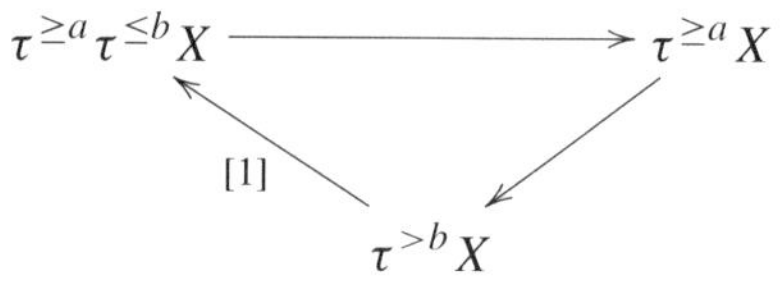

$$\tau^{\geq a}\tau^{\leq b}X \longrightarrow \tau^{\geq a}X \longrightarrow \tau^{>b}X \xrightarrow{[1]} \quad (7.1)$$

Comparing this triangle to the canonical

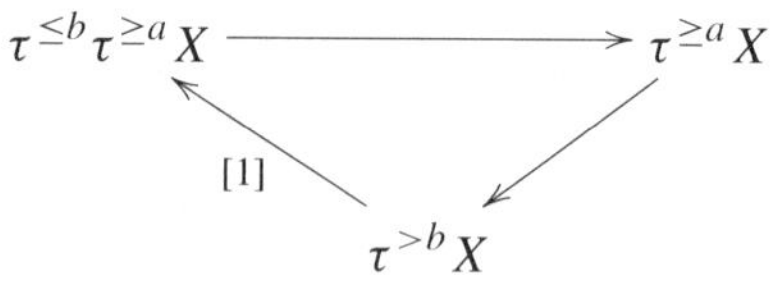

induces the sought isomorphism. □

Let $C = D^{\leq 0} \cap D^{\geq 0}$ be the heart in D. By the last proposition, $\tau^{\leq 0}\tau^{\geq 0} \cong \tau^{\geq 0}\tau^{\leq 0}$. Define a functor

$$H^0 : D \longrightarrow C$$

by

$$H^0(X) = \tau^{\leq 0}\tau^{\geq 0}X \cong \tau^{\geq 0}\tau^{\leq 0}X.$$

Also set $H^n(X) = H^0(X[n])$.

Proposition 7.1.10 *Suppose* $X \in D^{\geq a}$ *for some* a. *Then*:

$$X \in D^{\geq 0} \;\Leftrightarrow\; H^n(X) = 0 \text{ for } n < 0.$$

Proof. This is clear from the axioms if $a \geq 0$. Let $a < 0$. The triangle (7.1)

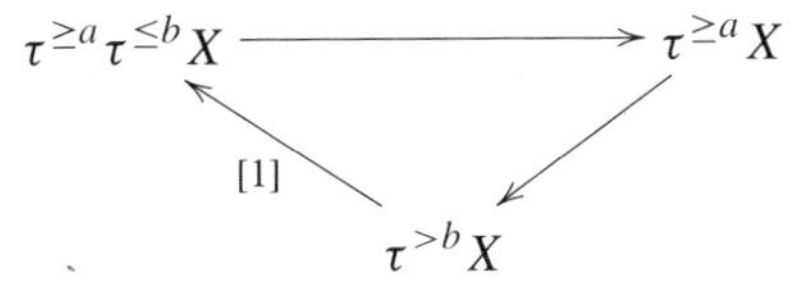

obtained from the octahedron above becomes

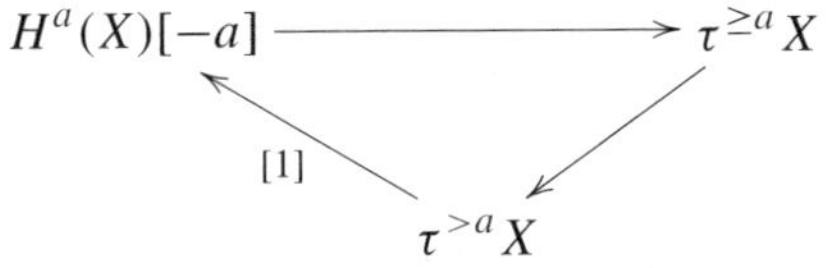

after taking $b = a$. Thus, if $X \xrightarrow{\simeq} \tau^{\geq a}X$ is an isomorphism and $H^a(X) = 0$, then $X \xrightarrow{\simeq} \tau^{>a}X$ is an isomorphism. □

Theorem 7.1.11 *The heart* $C = D^{\leq 0} \cap D^{\geq 0}$ *is an abelian category.*

Proof.

- C is an additive category: In any triangulated category, the following criterion holds:

  Let $X_i \to Y_i \to Z_i \overset{+1}{\to}$, $i = 1, 2$, *be two triangles. They are distinguished iff* $X_1 \oplus X_2 \to Y_1 \oplus Y_2 \to Z_1 \oplus Z_2 \overset{+1}{\to}$ *is distinguished.*

  Given $X, Y \in C$, we apply the criterion to $X \overset{1}{\to} X \to 0 \overset{+1}{\to}$ and $0 \to Y \overset{1}{\to} Y \overset{+1}{\to}$ to see that

  $$X \longrightarrow X \oplus Y \longrightarrow Y \overset{+1}{\longrightarrow}$$

  is distinguished. By Proposition 7.1.8, $X \oplus Y \in C$.
- Construction of cokernels: Before discussing the construction, we briefly review the universal property characterizing the cokernel. Let $f : X \to Y$ be a morphism, say, in the category of modules over a ring. The module $\operatorname{coker} f = Y/\operatorname{im} f$ satisfies the following universal property: If $g : Y \to W$ is any homomorphism such that $gf = 0$, then there exists a unique homomorphism $\bar{g} : \operatorname{coker} f \to W$ such that the diagram

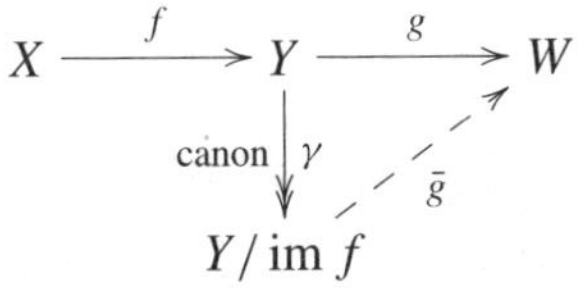

commutes. More generally, in any additive category C an object $\operatorname{coker} f$ is called a *cokernel for* f, if there exists a morphism $\gamma : Y \to \operatorname{coker} f$ solving the universal problem

$$\begin{array}{ccccc} X & \xrightarrow{f} & Y & \xrightarrow{g} & W \\ & & \downarrow{\scriptstyle\gamma} & \nearrow{\scriptstyle \exists!\text{ if } gf=0} & \\ & & \operatorname{coker} f & & \end{array}$$

Now let $f : X \to Y$ be a given morphism in the heart C. Embed f in a distinguished triangle

$$\begin{array}{ccc} X & \xrightarrow{f} & Y \\ & {\scriptstyle [1]}\nwarrow \quad \swarrow & \\ & Z & \end{array}$$

Using Proposition 7.1.8, the shifted triangle $Y \to Z \to X[1] \xrightarrow{+1}$ shows that

$$Z \in D^{\leq 0} \cap D^{\geq -1}$$

since $Y \in D^{\leq 0} \cap D^{\geq 0} \subset D^{\leq 0} \cap D^{\geq -1}$ and $X[1] \in D^{\leq -1} \cap D^{\geq -1} \subset D^{\leq 0} \cap D^{\geq -1}$. We claim

$$\boxed{H^0(Z) \cong \tau^{\geq 0} Z \cong \operatorname{coker} f.}$$

Define γ to be the composition $Y \longrightarrow Z \overset{\text{canon}}{\longrightarrow} \tau^{\geq 0} Z$. Let $W \in C$. Studying the long exact sequence

$$\operatorname{Hom}(X[1], W) \longrightarrow \operatorname{Hom}(Z, W) \longrightarrow \operatorname{Hom}(Y, W) \overset{-\circ f}{\longrightarrow} \operatorname{Hom}(X, W),$$

we observe that $\operatorname{Hom}(X[1], W) = 0$ as $X[1] \in D^{\leq -1}$, $W \in D^{\geq 0}$. Suppose $g \in \operatorname{Hom}(Y, W)$ such that $gf = 0$. Since $\tau^{\geq 0}$ is an adjoint for the inclusion, the sequence can be rewritten as

$$\begin{array}{ccccccc} 0 \longrightarrow \operatorname{Hom}(\tau^{\geq 0} Z, W) & \overset{-\circ\gamma}{\longrightarrow} & \operatorname{Hom}(Y, W) & \overset{-\circ f}{\longrightarrow} & \operatorname{Hom}(X, W), \\ & & g & \mapsto & 0 = gf. \end{array}$$

By exactness, there exists a unique $\bar{g} \in \operatorname{Hom}(\tau^{\geq 0} Z, W)$ such that $\bar{g}\gamma = g$; the claim is established.

- Construction of kernels: We claim

$$\boxed{H^0(Z[-1]) \cong \tau^{\leq 0}(Z[-1]) \cong \ker f.}$$

We take $\ker f \to X$ to be the composition $\tau^{\leq 0}(Z[-1]) \overset{\text{canon}}{\longrightarrow} Z[-1] \longrightarrow X$, and consider the exact sequence

$$\operatorname{Hom}(W, Y[-1]) \longrightarrow \operatorname{Hom}(W, Z[-1]) \longrightarrow \operatorname{Hom}(W, X) \overset{f\circ -}{\longrightarrow} \operatorname{Hom}(W, Y).$$

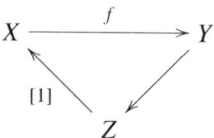

After observing $\mathrm{Hom}(W, Y[-1]) = 0$ and

$$\mathrm{Hom}(W, Z[-1]) \cong \mathrm{Hom}(W, \tau^{\leq 0}(Z[-1])),$$

the claim is easily established as above.

- $\mathrm{Coim}\, f \cong \mathrm{im}\, f$: Coimage and image are defined as

$$\mathrm{Coim}\, f = \mathrm{coker}(\ker f \to X), \qquad \mathrm{im}\, f = \ker(Y \to \mathrm{coker}\, f).$$

To construct these objects in C, embed $Y \to \tau^{\geq 0} Z$ in a distinguished triangle

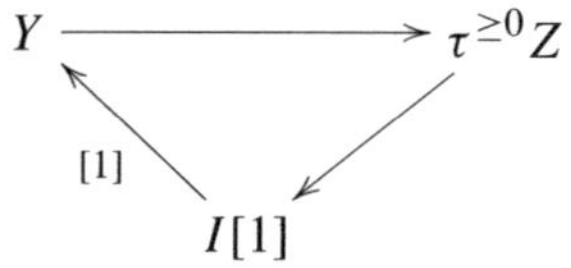

defining I (up to isomorphism). We prove first that $I \in C$. Since $\tau^{\geq 0} Z \in D^{\geq 0} \subset D^{\geq -1}$ and $Y[1] \in D^{\geq -1}$, we conclude from the shifted triangle $\tau^{\geq 0} Z \to I[1] \to Y[1] \overset{+1}{\to}$ that $I[1] \in D^{\geq -1}$ and thus $I \in D^{\geq 0}$. Use the triangles

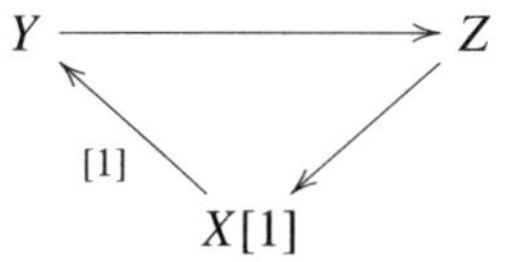

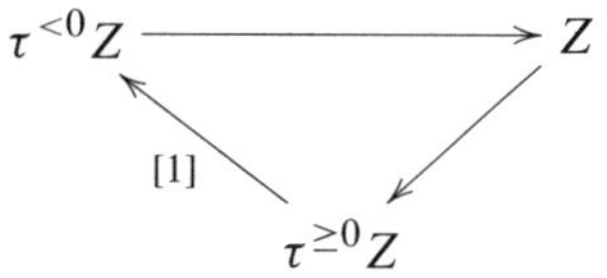

to build an octahedron

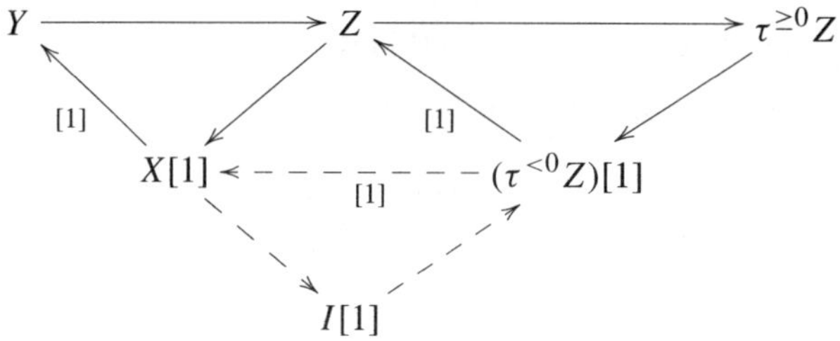

which produces a distinguished triangle $X \to I \to (\tau^{<0} Z)[1] \overset{+1}{\to}$ (with $X, (\tau^{<0} Z)[1] \in D^{\leq 0}$), i.e.

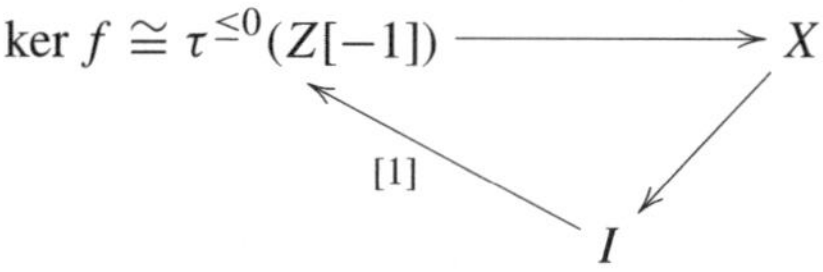

(7.2)

It follows that $I \in D^{\leq 0}$, and thus $I \in C$. Triangle (7.2) shows that

$$\operatorname{Coim} f = \operatorname{coker}(\ker f \to X) = \tau^{\geq 0} I \cong I.$$

On the other hand, the triangle

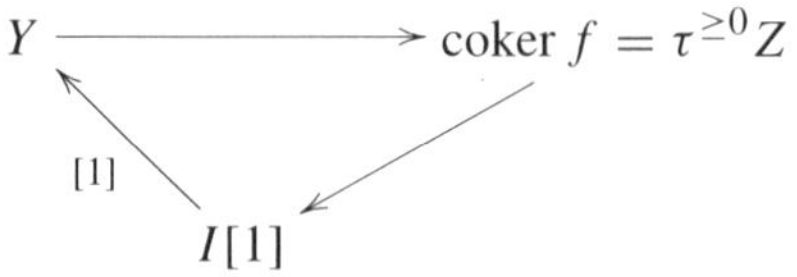

implies that

$$\operatorname{im} f = \ker(Y \to \operatorname{coker} f) = \tau^{\leq 0}(I[1][-1]) \cong I.$$

Thus $\operatorname{Coim} f \cong I \cong \operatorname{im} f$. □

Proposition 7.1.12 $H^0 : D \to C$ *is a cohomological functor.*

Proof. (Sketch.) Given a distinguished triangle

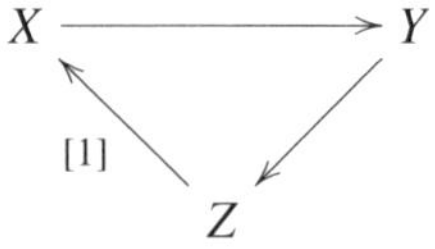

in D, one has to show that

$$H^0(X) \longrightarrow H^0(Y) \longrightarrow H^0(Z)$$

is exact in C. The argument can be divided into four steps:

1. We prove that $X, Y, Z \in D^{\geq 0}$ implies that

$$0 \to H^0(X) \longrightarrow H^0(Y) \longrightarrow H^0(Z)$$

is exact: If $W \in C$, then

$$\operatorname{Hom}_C(W, H^0(X)) \cong \operatorname{Hom}_D(W, \tau^{\leq 0}\tau^{\geq 0} X) \cong \operatorname{Hom}_D(W, \tau^{\geq 0} X)$$

as $\tau^{\leq 0}$ is a right adjoint for the inclusion, and

$$\operatorname{Hom}_D(W, \tau^{\geq 0} X) \cong \operatorname{Hom}_D(W, X)$$

since $X \in D^{\geq 0}$. We note that $\operatorname{Hom}_D(W, Z[-1]) = 0$ follows from $W \in D^{\leq 0}$ and $Z[-1] \in D^{\geq 1}$. Thus, the long exact sequence

$$\operatorname{Hom}_D(W, Z[-1]) \to \operatorname{Hom}_D(W, X) \to \operatorname{Hom}_D(W, Y) \to \operatorname{Hom}_D(W, Z)$$

induced by the triangle is isomorphic to

$$0 \to \mathrm{Hom}_C(W, H^0(X)) \to \mathrm{Hom}_C(W, H^0(Y)) \to \mathrm{Hom}_C(W, H^0(Z)).$$

2. Only assume $Z \in D^{\geq 0}$ and reach the same conclusion as in the previous step (to carry this out, use an appropriate octahedron to reduce to the previous step).
3. Similarly, show the dual statement: If $X \in D^{\leq 0}$, then

$$H^0(X) \longrightarrow H^0(Y) \longrightarrow H^0(Z) \to 0$$

is exact.
4. General case: Consider the octahedron based on the composition

$$\tau^{\leq 0}X \longrightarrow X \longrightarrow Y.$$

□

Remark 7.1.13 *If* $0 \to X \to Y \to Z \to 0$ *is an exact sequence in* C*, then there exists a unique morphism* $h : Z \to X[1]$ *such that*

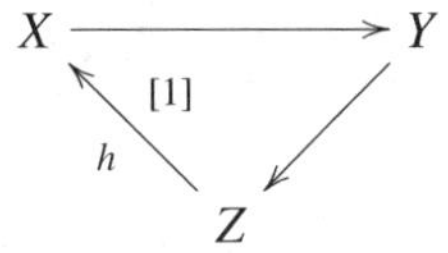

is a distinguished triangle in D.

Definition 7.1.14 *Let* D_i*,* $i = 1, 2$*, be two triangulated categories with t-structures* $(D_i^{\leq 0}, D_i^{\geq 0})$*, let* C_i *be the heart of* D_i *and* $\epsilon_i : C_i \to D_i$ *the inclusion functor. A functor* $F : D_1 \to D_2$ *of triangulated categories is called* left t-exact *if* $F(D_1^{\geq 0}) \subset D_2^{\geq 0}$*, and* right t-exact *if* $F(D_1^{\leq 0}) \subset D_2^{\leq 0}$*.* F *is* t-exact *if it is both left and right t-exact. Furthermore, we put*

$${}^pF = H^0 \circ F \circ \epsilon_1 : C_1 \to C_2.$$

Proposition 7.1.15 *If* $F : D_1 \to D_2$ *is left t-exact, then*

1. $X \in D_1^{\geq 0} \Rightarrow H^0(F(X)) \cong {}^pF(H^0(X))$.
2. ${}^pF : C_1 \to C_2$ *is left exact.*

(Analogously: F *right t-exact,* $X \in D_1^{\leq 0}$*,* ${}^pF$ *right exact.)*

Remark 7.1.16 *If* $F : D_1 \to D_2$ *is t-exact, then* F *sends* C_1 *to* C_2 *and* $F|_{C_1}$ *is exact. Moreover,* $F|_{C_1} \cong {}^pF$ *and* $F(H^n(X)) \cong H^n(F(X))$ *for all* $X \in D_1$.

Example 7.1.17 *Let* C_1, C_2 *any two abelian categories and* $F : C_1 \to C_2$ *an additive, left exact functor. Assume that* C_1 *has enough injectives. Then the derived functor*

$$RF : D^+(C_1) \longrightarrow D^+(C_2)$$

is left t-exact with respect to the natural t-structures.

7.2 Gluing of t-Structures

7.2.1 Gluing Data

Let X be a topological space, $U \subset X$ an open subset, and $F = X - U \subset X$ the closed complement. We denote the inclusions by

$$U \overset{i}{\hookrightarrow} X \overset{j}{\hookleftarrow} F.$$

The following discussion will involve the derived categories $D^+(U)$, $D^+(F)$, and $D^+(X)$. We wish to describe a construction, which, given a t-structure on $D^+(U)$ and a t-structure on $D^+(F)$, produces a t-structure on $D^+(X)$. The following table summarizes the various sheaf theoretic functors associated to the two inclusions, as well as their exactness properties:

$i_!$:	$Sh(U) \to Sh(X)$	extension by 0	exact
$i^! = i^*$:	$Sh(X) \to Sh(U)$	restriction	exact
i_* :	$Sh(U) \to Sh(X)$	direct image	left exact
j^* :	$Sh(X) \to Sh(F)$	restriction	exact
$j_! = j_*$:	$Sh(F) \to Sh(X)$	extension by 0	exact
$j^!$:	$Sh(X) \to Sh(F)$	sections supported in F	left exact

On the corresponding derived categories, these induce the derived functors $i_!, i^*$, Ri_*, j^*, j_*, $Rj^!$. In our notation we will henceforth drop the R—the appropriate derived functor will be understood. It is convenient to set $D_U = D^+(U)$, $D_F = D^+(F)$, $D = D^+(X)$. Thus, we have three triangulated categories D_U, D, D_F which are related by two functors of triangulated categories

$$D_F \xrightarrow{j_*} D \xrightarrow{i^*} D_U$$

satisfying the following properties:

(G1) j_* has a left adjoint j^*, and a right adjoint $j^!$. The functors j^* and $j^!$ are functors of triangulated categories.

(G2) i^* has a left adjoint $i_!$, and a right adjoint i_*. The functors $i_!$ and i_* are functors of triangulated categories.

(G3) Since $j_* = j_!$ is extension by 0, we have $i^* j_* = 0$. By adjunction, we also have $j^* i_! = 0$:

$$\begin{aligned} \operatorname{Hom}(j^* i_! \mathbf{A}^\bullet, \mathbf{B}^\bullet) &= \operatorname{Hom}(i_! \mathbf{A}^\bullet, j_* \mathbf{B}^\bullet) \\ &= \operatorname{Hom}(\mathbf{A}^\bullet, i^* j_* \mathbf{B}^\bullet) \\ &= \operatorname{Hom}(\mathbf{A}^\bullet, 0) \\ &= 0. \end{aligned}$$

Similarly, $j^! i_* = 0$.

(G4) In $Sh(-)$, the adjunction morphisms give rise to exact sequences

$$0 \longrightarrow i_! i^* \mathbf{A} \longrightarrow \mathbf{A} \longrightarrow j_* j^* \mathbf{A} \longrightarrow 0,$$
$$0 \longrightarrow j_* j^! \mathbf{A} \longrightarrow \mathbf{A} \longrightarrow i_* i^* \mathbf{A} \longrightarrow 0$$

(for the surjectivity of $\mathbf{A} \to i_* i^* \mathbf{A}$, assume $\mathbf{A}$ injective). In $D^+(-)$, these give rise to distinguished triangles

$$\begin{array}{ccc} i_! i^* \mathbf{A}^\bullet & \longrightarrow & \mathbf{A}^\bullet \\ & {\scriptstyle [1]}\nwarrow \quad \swarrow & \\ & j_* j^* \mathbf{A}^\bullet & \end{array} \qquad \text{and} \qquad \begin{array}{ccc} j_* j^! \mathbf{A}^\bullet & \longrightarrow & \mathbf{A}^\bullet \\ & {\scriptstyle [1]}\nwarrow \quad \swarrow & \\ & i_* i^* \mathbf{A}^\bullet & \end{array}$$

(G5) The following adjunction morphisms are isomorphisms:

$$j^* j_* \xrightarrow{\simeq} 1 \xrightarrow{\simeq} j^! j_*, \qquad i^* i_* \xrightarrow{\simeq} 1 \xrightarrow{\simeq} i^* i_!.$$

The following definition emphasizes that we may as well forget that D_U, D, D_F are the derived categories on various spaces and that j_*, i^* are induced by inclusions, to arrive at an abstract version of a notion of gluing data.

Definition 7.2.1 *A triple of triangulated categories D_U, D, D_F together with functors $D_F \xrightarrow{j_*} D \xrightarrow{i^*} D_U$ satisfying* (G1)–(G5) *is called* gluing data.

7.2.2 The Gluing Theorem

Let D_U, D, D_F be a triple of triangulated categories, for example $D_U = D^+(U)$, $D = D^+(X)$, $D_F = D^+(F)$. Suppose gluing data $D_F \xrightarrow{j_*} D \xrightarrow{i^*} D_U$ are given. Let $(D_F^{\leq 0}, D_F^{\geq 0})$ be a t-structure on D_F, and $(D_U^{\leq 0}, D_U^{\geq 0})$ a t-structure on D_U. The gluing of the two t-structures is achieved by defining

$$\boxed{\begin{aligned} D^{\leq 0} &= \{X \in D \mid i^* X \in D_U^{\leq 0},\ j^* X \in D_F^{\leq 0}\}, \\ D^{\geq 0} &= \{X \in D \mid i^* X \in D_U^{\geq 0},\ j^! X \in D_F^{\geq 0}\}. \end{aligned}}$$

Theorem 7.2.2 *$(D^{\leq 0}, D^{\geq 0})$ is a t-structure on D.*

Proof. We have to verify axioms (1)–(3) of Definition 7.1.1. Axiom (1) requires $D^{\leq -1} \subset D^{\leq 0}$ and $D^{\geq 1} \subset D^{\geq 0}$. This follows immediately from axiom (1) for D_U and D_F.

Axiom (2): Let $X \in D^{\leq 0}$, $Y \in D^{\geq 1}$. The triangle

$$i_! i^* X \longrightarrow X \longrightarrow j_* j^* X \xrightarrow{+1}$$

induces an exact sequence

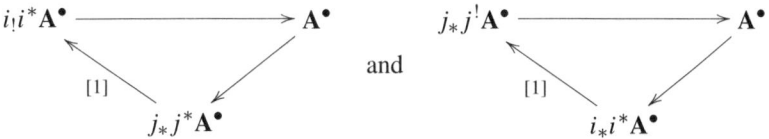

$$\mathrm{Hom}(j_* j^* X, Y) \longrightarrow \mathrm{Hom}(X, Y) \longrightarrow \mathrm{Hom}(i_! i^* X, Y).$$

Since $j^!$ is a right adjoint for j_*, we have $\mathrm{Hom}(j_* j^* X, Y) = \mathrm{Hom}(j^* X, j^! Y)$. But as $j^* X \in D_F^{\leq 0}$ and $j^! Y \in D_F^{\geq 1}$, the group $\mathrm{Hom}(j^* X, j^! Y) = 0$ by axiom (2) for D_F. Similarly, $\mathrm{Hom}(i_! i^* X, Y) = \mathrm{Hom}(i^* X, i^* Y) = 0$ as $i^* X \in D_U^{\leq 0}$, $i^* Y \in D_U^{\geq 1}$, and using axiom (2) for D_U. Thus $\mathrm{Hom}(X, Y) = 0$.

Axiom (3): Let $X \in D$. We need to construct a distinguished triangle

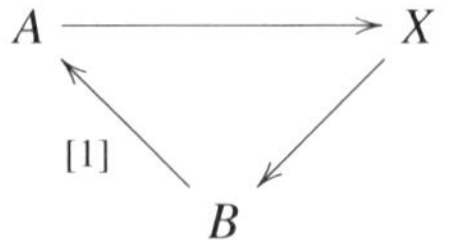

with $A \in D^{\leq 0}$ and $B \in D^{\geq 1}$. In the following discussion, the symbols $\tau^{\leq 0}$, $\tau^{\geq 0}$, etc. denote t-structure truncation in D_U when applied to an object in D_U (such as e.g. $i^* X$), and they denote t-structure truncation in D_F when applied to an object in D_F (such as e.g. $j^* X$). The canonical morphism $i^* X \to \tau^{>0} i^* X$ induces a morphism

$$i_* i^* X \longrightarrow i_* \tau^{>0} i^* X.$$

Composing with the adjunction $X \to i_* i^* X$ yields a morphism

$$X \longrightarrow i_* \tau^{>0} i^* X.$$

Embed this morphism in a distinguished triangle

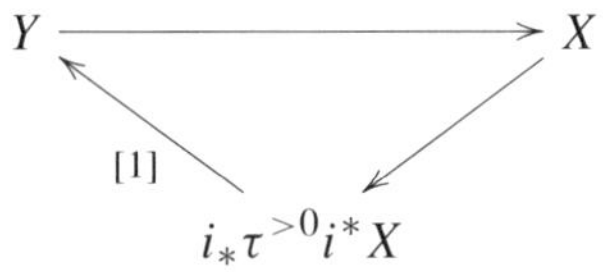

to define (up to isomorphism) the object $Y \in D$. Embedding the morphism

$$Y \longrightarrow j_* \tau^{>0} j^* Y$$

in a distinguished triangle

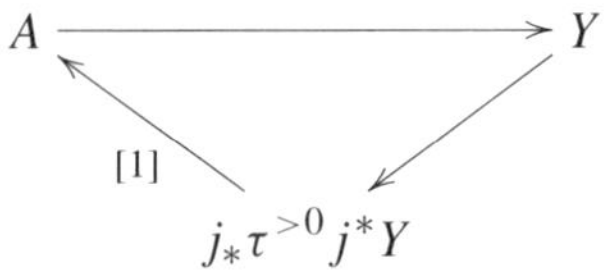

defines $A \in D$. Consider the octahedron built on the two triangles:

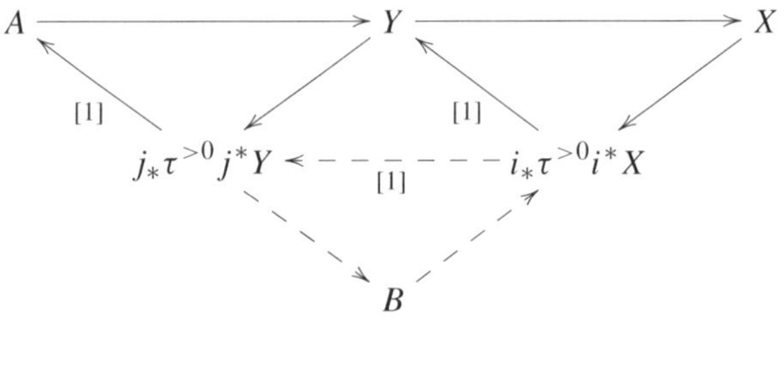

In particular, this structure defines $B \in D$ and gives a distinguished triangle

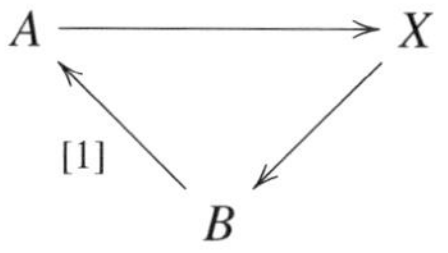

To prove that $A \in D^{\leq 0}$ and $B \in D^{\geq 1}$, we apply the functors i^*, j^*, $j^!$ to various triangles in the octahedron. Applying i^*, we obtain

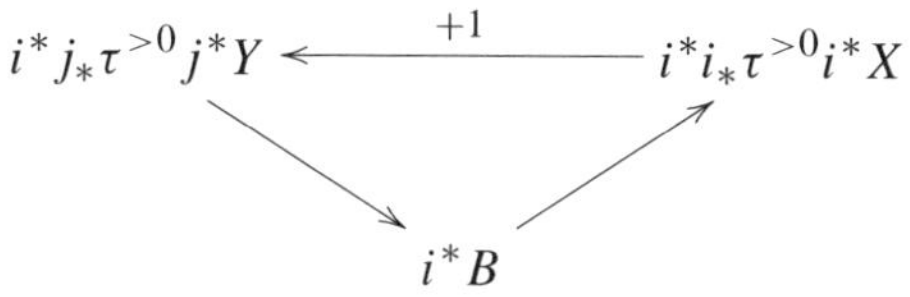

which is

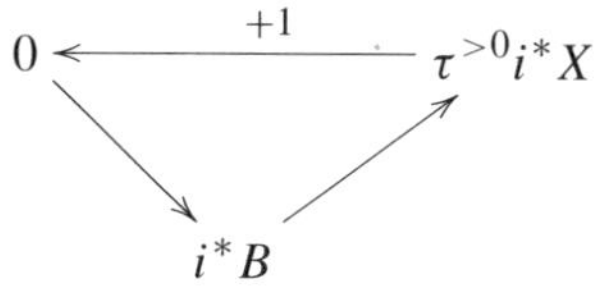

so that

$$i^* B \cong \tau^{>0} i^* X. \tag{7.3}$$

Thus the triangle

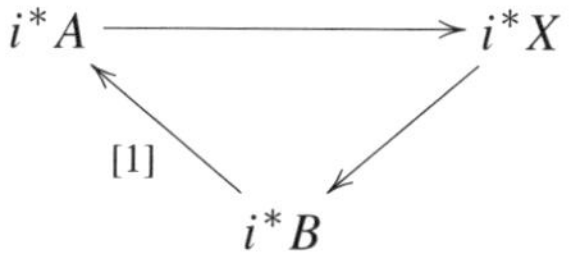

is

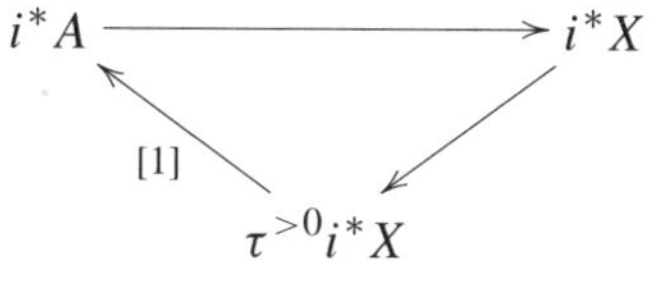

which implies the identification

$$i^* A \cong \tau^{\leq 0} i^* X. \tag{7.4}$$

Applying j^* yields

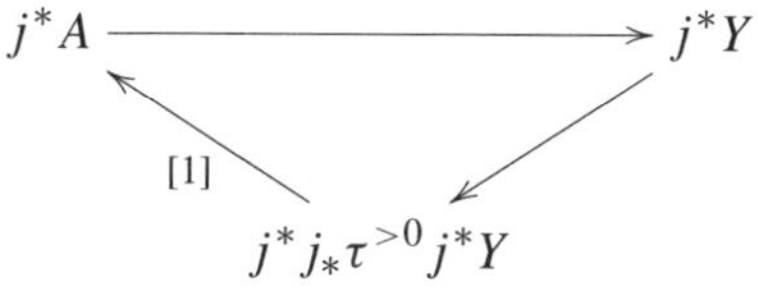

which is

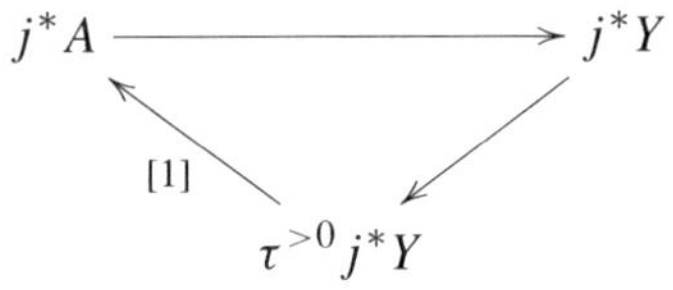

Consequently,

$$j^*A \cong \tau^{\leq 0} j^*Y. \tag{7.5}$$

Finally, we apply $j^!$ to get a triangle

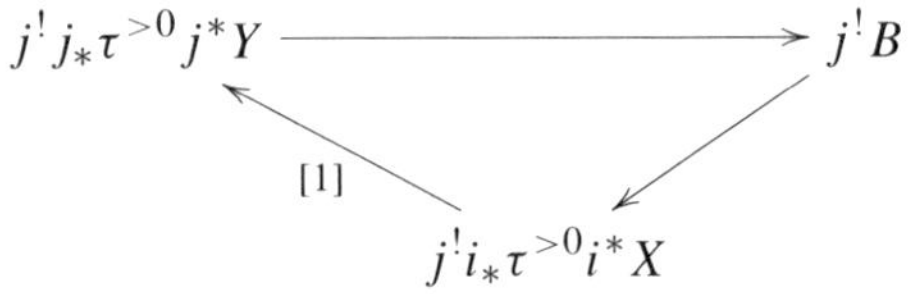

isomorphic to

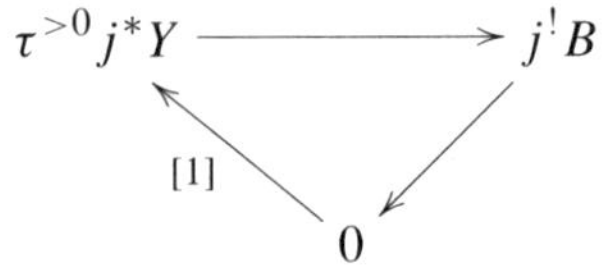

This proves that

$$j^!B \cong \tau^{>0} j^*Y. \tag{7.6}$$

The identifications (7.3)–(7.6) place A in $D^{\leq 0}$ and B in $D^{\geq 1}$. □

If, as in Sect. 7.2.1, X is a topological space, $U \subset X$ an open subset, $F = X - U$ with inclusions

$$U \xhookrightarrow{i} X \xhookleftarrow{j} F,$$

and $D_U = D^+(U)$, $D_F = D^+(F)$, $D = D^+(X)$, then let C_U, C_F, C be the hearts of D_U, D_F, D, respectively, and let ϵ denote any of the inclusion functors

$$C_U \hookrightarrow D_U, \qquad C_F \hookrightarrow D_F, \qquad C \hookrightarrow D.$$

For $G \in \{i_!, i^*, i_*, j^*, j_*, j^!\}$, we define functors

$${}^pG = H^0 \circ G \circ \epsilon$$

(cf. Definition 7.1.14). For example $i_! : D_U \to D_X$ gives rise to a functor ${}^pi_! : C_U \to C$, and j^* gives rise to ${}^pj^* : C \to C_F$, etc. The definition of the t-structure $(D^{\leq 0}, D^{\geq 0})$ implies the following t-exactness properties:

i^*	t-exact
j^*	right t-exact
$j^!$	left t-exact

By adjointness,

$i_!$	right t-exact
j_*	t-exact
i_*	left t-exact

Thus, by Proposition 7.1.15:

${}^p i^*, {}^p j_*$	exact
${}^p j^*, {}^p i_!$	right exact
${}^p j^!, {}^p i_*$	left exact

Without proof, we record the following facts:

Proposition 7.2.3

- ${}^p i^* \circ {}^p j_* = 0$; *hence also* ${}^p j^* \circ {}^p i_! = 0 = {}^p j^! \circ {}^p i_*$.
- *There are exact sequences* $(A \in C)$

$$0 \to {}^p j_* H^{-1} j^* A \longrightarrow {}^p i_! {}^p i^* A \longrightarrow A \longrightarrow {}^p j_* {}^p j^* A \to 0$$

and

$$0 \to {}^p j_* {}^p j^! A \longrightarrow A \longrightarrow {}^p i_* {}^p i^* A \longrightarrow {}^p j_* H^1 j^! A \to 0.$$

- ${}^p j^* {}^p j_* \xrightarrow{\simeq} 1 \xrightarrow{\simeq} {}^p j^! {}^p j_*$, ${}^p i^* {}^p i_* \xrightarrow{\simeq} 1 \xrightarrow{\simeq} {}^p i^* {}^p i_!$.

We have adjunction morphisms ${}^p j_* {}^p j^! \to 1 \to {}^p j_* {}^p j^*$. Applying ${}^p j^*$, one obtains a morphism of functors

$${}^p j^! \longrightarrow {}^p j^*.$$

Similarly, an application of ${}^p i^*$ to ${}^p i_! {}^p i^* \to 1 \to {}^p i_* {}^p i^*$ produces a morphism of functors

$${}^p i_! \longrightarrow {}^p i_*.$$

The functor

$$i_{!*} : C_U \longrightarrow C$$

defined by

$$i_{!*}(X) = \operatorname{im}({}^p i_! X \longrightarrow {}^p i_* X)$$

is characterized by the following property:

Proposition 7.2.4 *For $X \in C_U$, $i_{!*} X$ is the unique extension Y of X in D (i.e. $i^* Y = X$), such that*

$$j^* Y \in D_F^{\leq -1} \quad \textit{and} \quad j^! Y \in D_F^{\geq 1}.$$

7.3 The Perverse t-Structure

Let $\bar{p}$ be any function $\bar{p} : \{U, F\} \to \mathbb{Z}$ (a "perversity").[2] We think of X as a two-strata space $X \supset F \supset \varnothing$, so that U is the "top-stratum." Let $(D^{\leq 0}(U), D^{\geq 0}(U))$

[2] In this context, we do not place any additional conditions (as in Definition 4.1.7) on the function $\bar{p}$.

and $(D^{\leq 0}(F), D^{\geq 0}(F))$ be the natural t-structures on the derived categories $D^+(U)$ and $D^+(F)$, respectively (Example 7.1.3). Then $(D^{\leq \bar{p}(U)}(U), D^{\geq \bar{p}(U)}(U))$ and $(D^{\leq \bar{p}(F)}(F), D^{\geq \bar{p}(F)}(F))$ are t-structures as well (the shifted t-structures of Remark 7.1.2). We will denote these shifted t-structures by $({}^{\bar{p}}D_{\bar{U}}^{\leq 0}, {}^{\bar{p}}D_{\bar{U}}^{\geq 0})$ and $({}^{\bar{p}}D_{\bar{F}}^{\leq 0}, {}^{\bar{p}}D_{\bar{F}}^{\geq 0})$, resp. Thus, more explicitly,

$$
\begin{aligned}
{}^{\bar{p}}D_{\bar{U}}^{\leq 0} &= \{\mathbf{A}^\bullet \in D^+(U) \mid \mathbf{H}^n(\mathbf{A}^\bullet) = 0,\ n > \bar{p}(U)\}, \\
{}^{\bar{p}}D_{\bar{U}}^{\geq 0} &= \{\mathbf{A}^\bullet \in D^+(U) \mid \mathbf{H}^n(\mathbf{A}^\bullet) = 0,\ n < \bar{p}(U)\}, \\
{}^{\bar{p}}D_{\bar{F}}^{\leq 0} &= \{\mathbf{B}^\bullet \in D^+(F) \mid \mathbf{H}^n(\mathbf{B}^\bullet) = 0,\ n > \bar{p}(F)\}, \\
{}^{\bar{p}}D_{\bar{F}}^{\geq 0} &= \{\mathbf{B}^\bullet \in D^+(F) \mid \mathbf{H}^n(\mathbf{B}^\bullet) = 0,\ n < \bar{p}(F)\}.
\end{aligned}
$$

Definition 7.3.1 *The* perverse t-structure $({}^{\bar{p}}D^{\leq 0}, {}^{\bar{p}}D^{\geq 0})$ *(for perversity $\bar{p}$) on $D = D^+(X)$ is the t-structure obtained by gluing the t-structures $({}^{\bar{p}}D_{\bar{U}}^{\leq 0}, {}^{\bar{p}}D_{\bar{U}}^{\geq 0})$ on $D_U = D^+(U)$ and $({}^{\bar{p}}D_{\bar{F}}^{\leq 0}, {}^{\bar{p}}D_{\bar{F}}^{\geq 0})$ on $D_F = D^+(F)$.*

Again, more concretely we have

$$
\begin{aligned}
{}^{\bar{p}}D^{\leq 0} &= \{\mathbf{X}^\bullet \in D^+(X) \mid i^*\mathbf{X}^\bullet \in {}^{\bar{p}}D_{\bar{U}}^{\leq 0},\ j^*\mathbf{X}^\bullet \in {}^{\bar{p}}D_{\bar{F}}^{\leq 0}\} \\
&= \{\mathbf{X}^\bullet \in D^+(X) \mid \mathbf{H}^n(i^*\mathbf{X}^\bullet) = 0,\ n > \bar{p}(U), \\
&\qquad\qquad \mathbf{H}^n(j^*\mathbf{X}^\bullet) = 0,\ n > \bar{p}(F)\}, \\
{}^{\bar{p}}D^{\geq 0} &= \{\mathbf{X}^\bullet \in D^+(X) \mid i^*\mathbf{X}^\bullet \in {}^{\bar{p}}D_{\bar{U}}^{\geq 0},\ j^!\mathbf{X}^\bullet \in {}^{\bar{p}}D_{\bar{F}}^{\geq 0}\} \\
&= \{\mathbf{X}^\bullet \in D^+(X) \mid \mathbf{H}^n(i^*\mathbf{X}^\bullet) = 0,\ n < \bar{p}(U), \\
&\qquad\qquad \mathbf{H}^n(j^!\mathbf{X}^\bullet) = 0,\ n < \bar{p}(F)\}
\end{aligned}
$$

(recall $i^! = i^*$).

Definition 7.3.2 *We call $\mathbf{X}^\bullet \in D^+(X)$ ($\bar{p}$-)*perverse, *if $\mathbf{X}^\bullet$ is in the heart of the $\bar{p}$-perverse t-structure, that is, if $\mathbf{X}^\bullet \in {}^{\bar{p}}D^{\leq 0} \cap {}^{\bar{p}}D^{\geq 0}$.*

By applying the gluing Theorem 7.2.2 inductively, the notion of a $\bar{p}$-perverse t-structure quickly generalizes to spaces with more than two strata: Let X be filtered by closed subsets, let $\{S\}$ be the set of components of pure strata, let $\bar{p} : \{S\} \to \mathbb{Z}$ be any function and $j_S : S \hookrightarrow X$ the inclusion. Then the perverse t-structure on X is given by

$$
\begin{aligned}
{}^{\bar{p}}D^{\leq 0} &= \{\mathbf{X}^\bullet \in D^+(X) \mid \mathbf{H}^n(j_S^*\mathbf{X}^\bullet) = 0,\ n > \bar{p}(S),\ \text{all } S\}, \\
{}^{\bar{p}}D^{\geq 0} &= \{\mathbf{X}^\bullet \in D^+(X) \mid \mathbf{H}^n(j_S^!\mathbf{X}^\bullet) = 0,\ n < \bar{p}(S),\ \text{all } S\}.
\end{aligned}
$$

Definition 7.3.3 *The* category ${}^{\bar{p}}P(X)$ of $\bar{p}$-perverse sheaves on X *is the heart*

$$
{}^{\bar{p}}P(X) = {}^{\bar{p}}D^{\leq 0} \cap {}^{\bar{p}}D^{\geq 0}.
$$

By Theorem 7.1.11, the category of perverse sheaves is an abelian subcategory of the nonabelian derived category $D^+(X)$. The general theory of t-structures developed above automatically provides us with perverse truncation functors

$$
\begin{aligned}
{}^{\bar{p}}\tau^{\leq 0} &: D^+(X) \longrightarrow {}^{\bar{p}}D^{\leq 0}, \\
{}^{\bar{p}}\tau^{\geq 0} &: D^+(X) \longrightarrow {}^{\bar{p}}D^{\geq 0}, \\
{}^{\bar{p}}H^0 &: D^+(X) \longrightarrow {}^{\bar{p}}P(X)
\end{aligned}
$$

(${}^{\bar{p}}H^0 = {}^{\bar{p}}\tau^{\geq 0} \circ {}^{\bar{p}}\tau^{\leq 0}$). If U is an open union of strata with inclusion $i : U \hookrightarrow X$, then we have e.g.

$$
{}^{\bar{p}}i_*, {}^{\bar{p}}i_! : {}^{\bar{p}}P(U) \longrightarrow {}^{\bar{p}}P(X),
$$

etc.

8

Methods of Computation

8.1 Stratified Maps and Topological Invariants

8.1.1 Introduction

Given spaces X and Y and a continuous map $f : X \to Y$, it is a fundamental problem in topology to relate the invariants of X to the invariants of Y via f.

Example 8.1.1 *All spaces in this example are assumed to be finite CW complexes. Consider the Euler characteristic*

$$\chi(X) = \sum_i (-1)^i \operatorname{rk} H_i(X) = \sum_i (-1)^i (\textit{number of } i\textit{-cells of } X).$$

A well-known result asserts that χ is multiplicative for fiber bundles:

Theorem 8.1.2 *If $F \longrightarrow E \xrightarrow{\pi} B$ is a fiber bundle, then the Euler characteristic of the total space is related to the Euler characteristics of base and fiber by the formula*

$$\chi(E) = \chi(B)\chi(F).$$

Proof. 1. *If A and B are identified along a subspace C, then $\chi(A \cup_C B) = \chi(A) + \chi(B) - \chi(C)$. Obtain this either by counting the numbers of cells, or use the Mayer–Vietoris sequence*

$$H_i(C) \to H_i(A) \oplus H_i(B) \to H_i(A \cup_C B) \stackrel{\partial}{\to} H_{i-1}(C)$$

and the fact that exactness of a long sequence $\cdots \to V_i \to V_{i-1} \to \cdots$ implies $\sum(-1)^i \operatorname{rk} V_i = 0$.

2. *$\chi(A \times B) = \chi(A)\chi(B)$ follows from the Künneth theorem.*

3. *Consider the fiber bundle $F \longrightarrow E \xrightarrow{\pi} B$. We can assume that the bundle is trivial when restricted to a cell. We proceed by induction on the number of cells. Let $B' \subset B$ be a subcomplex where the result has been established. Then, using step 1, the effect of attaching a k-cell e^k is*

$$\chi(\pi^{-1}(B' \cup_{S^{k-1}} e^k)) = \chi(\pi^{-1}(B')) + \chi(\pi^{-1}(e^k)) - \chi(\pi^{-1}(S^{k-1})).$$

The induction hypothesis is $\chi(\pi^{-1}(B')) = \chi(B')\chi(F)$, *and on* $\pi^{-1}(e^k)$ *and* $\pi^{-1}(S^{k-1})$ *we use step 2, since these spaces are products.* □

Example 8.1.3 *The signature of fiber bundles: Let* $F \longrightarrow E \xrightarrow{\pi} B$ *be a fiber bundle of oriented closed manifolds. Then the theorem of Chern, Hirzebruch and Serre [CHS57] states that under the assumption that the base* B *is simply connected (or, more generally, that the action of the fundamental group* $\pi_1(B)$ *on the middle cohomology of the fiber is trivial), the signatures are multiplicative,*

$$\sigma(E) = \sigma(B)\sigma(F).$$

The proof uses the Serre spectral sequence.

8.1.2 Behavior of Invariants under Stratified Maps

The Chern–Hirzebruch–Serre formula assumes all spaces to be nonsingular, and it assumes the map $E \to B$ to be a locally trivial bundle projection. One may thus raise the questions: How does the signature behave if

- the involved spaces are singular?
- the fiber of the map is allowed to change from point to point?

For target spaces with only even-codimensional strata and stratified maps, these questions have been answered by Cappell and Shaneson [CS91]. We will first describe their results, and subsequently discuss the proofs in some detail. We will make use of the material on t-structures and perverse sheaves as developed in Sects. 7.1–7.3. Let X and Y be closed Whitney stratified spaces with only even-codimensional strata. All strata and links are assumed to be compatibly oriented. Let $f : X \longrightarrow Y$ be a stratified map (cf. Definition 6.2.8) of even relative dimension. If $\mathcal{V}$ denotes the set of components of pure strata of Y, then the theory of Whitney stratifications provides us, for every $y \in V \in \mathcal{V}$, with a normal slice $N(y)$ and a link $L(y) = \partial N(y)$ of V at y ($N(y) \cong cL(y)$). Define

$$E_y = f^{-1}N(y) \cup_{f^{-1}L(y)} cf^{-1}L(y),$$

that is, E_y is obtained from the preimage of the normal slice at y by coning off the boundary. Hence E_y is a closed stratified pseudomanifold. If y is a point in the top stratum, we set $E_y = f^{-1}(y)$. Since the dimension of E_y is the sum of the codimension of V and the relative dimension of f, the cone-point is a stratum of even codimension in E_y. Consequently, the lower middle perversity chain sheaf $\mathbf{IC}^\bullet_{\bar{m}}(E_y)$ is self-dual and the signature $\sigma(E_y) = \sigma(\mathbf{IC}^\bullet_{\bar{m}}(E_y))$ is defined, see Sect. 6.1. For $V \in \mathcal{V}$, let $j : \overline{V} \hookrightarrow Y$ denote the inclusion of the closure $\overline{V}$ of V in Y. On rational homology, j induces $j_* : H_*(\overline{V}) \to H_*(Y)$. As X has only strata of even codimension, the Goresky–MacPherson L-classes $L_i(X) \in H_i(X; \mathbb{Q})$ are defined (Sect. 6.3). Similarly, every $\overline{V}$ has L-classes $L_i(\overline{V})$.

Theorem 8.1.4 (*Cappell–Shaneson*) *Assume each $V \in \mathcal{V}$ is simply connected. Choose a base point $y_V \in V$ for every $V \in \mathcal{V}$. Then*

$$f_* L_i(X) = \sum_{V \in \mathcal{V}} \sigma(E_{y_V}) j_* L_i(\overline{V}). \tag{8.1}$$

In particular, for the signature $L_0(X) = \sigma(X)$:

Corollary 8.1.5

$$\sigma(X) = \sum_{V \in \mathcal{V}} \sigma(E_{y_V}) \sigma(\overline{V}). \tag{8.2}$$

If V is a component of the top stratum, then the term

$$\sigma(E_{y_V})\sigma(\overline{V}) = \sigma(f^{-1}(y_V))\sigma(\overline{V})$$

corresponds to the term $\sigma(F)\sigma(B)$ in the formula of Chern, Hirzebruch and Serre, Example 8.1.3. Thus, formulae (8.1) and (8.2) show that in general the singularities of Y and of f may contribute additional terms. It is also interesting to note that only the bottom component $L_0(E_{y_V})$ of the L-class in the fiber-direction enters into formula (8.1).

The above formulae follow rather rapidly from a sheaf-theoretic orthogonal decomposition theorem, which comprises the core of the proof: Suppose $\mathbf{S}^\bullet \in D_c^b(X)$ is self-dual, $\mathcal{D}\mathbf{S}^\bullet[\dim X] \cong \mathbf{S}^\bullet$. Let $f : X^p \to Y^n$ be a stratified map as above, except that for the following it is only necessary to assume that Y has no strata of odd-codimension. Put $t = \frac{1}{2}(p-n) \in \mathbb{Z}$. For $y \in V \in \mathcal{V}$ we have the inclusions

$$E_y \xleftarrow{i_y} f^{-1}\mathring{N}(y) \xhookrightarrow{\rho_y} X.$$

On the category $D_c^b(E_y)$, let $\tau_{\le}^{\{cone\}}$ denote truncation over the distinguished cone-point of E_y, i.e. the point with link $f^{-1}L(y)$. Define a sheaf $\mathbf{S}^\bullet(y)$ on E_y by

$$\mathbf{S}^\bullet(y) = \tau_{\le -c-t-1}^{\{cone\}} Ri_{y*}\rho_y^! \mathbf{S}^\bullet,$$

where $c = \frac{1}{2}\operatorname{codim} V$. Let us explain why $\mathbf{S}^\bullet(y)$ is again self-dual.

Lemma 8.1.6 *If $\alpha : R^r \hookrightarrow S^s$ is a normally nonsingular inclusion (Sect. 6.2.4), and $\mathbf{T}^\bullet \in D_c^b(S)$ a self-dual sheaf, then $\alpha^!\mathbf{T}^\bullet \in D_c^b(R)$ is again self-dual.*

Proof. The point is that if R sits in a vector-bundle neighborhood, then the restriction with compact supports $\alpha^!$ is merely a degree-shift by the negative of the codimension: $\alpha^! \cong \alpha^*[r-s]$. Therefore, an isomorphism $\mathcal{D}_S\mathbf{T}^\bullet[s] \cong \mathbf{T}^\bullet$ induces

$$\mathcal{D}_R(\alpha^!\mathbf{T}^\bullet)[r] \cong \alpha^*\mathcal{D}_S\mathbf{T}^\bullet[r] \cong \alpha^*\mathbf{T}^\bullet[r-s] \cong \alpha^!\mathbf{T}^\bullet. \qquad \square$$

The normal slice $N(y)$ is clearly transverse to the strata of Y. Since f is a stratified map, $f^{-1}N(y)$ is transverse to the strata of X. By Proposition 6.2.10, ρ_y is a normally nonsingular inclusion, so that $\rho_y^!\mathbf{S}^\bullet$ is self-dual by the lemma. The functor

$\tau^{\{cone\}}_{\leq -c-t-1} Ri_{y*}$ is a one-step Deligne extension of the type used in formula (4.13) of Sect. 4.2 to construct an incarnation of the intersection chain complex. Since the truncation-value $-c-t-1$ is precisely the middle perversity, $\mathbf{S}^\bullet(y)$ is self-dual. Note that we are only truncating over the cone-point since we do not assume any stalk conditions on the complement of that point, and thus a global truncation may destroy self-duality. Let $\mathfrak{H}_V(\mathbf{S}^\bullet)$ be the local coefficient system over V with stalks

$$\mathfrak{H}_V(\mathbf{S}^\bullet)_y = \mathcal{H}^{-c-t}(E_y; \mathbf{S}^\bullet(y)).$$

This local system comes equipped with a nonsingular bilinear pairing on each stalk which is induced by the self-duality of $\mathbf{S}^\bullet(y)$. Alternatively, we may regard $\mathfrak{H}_V(\mathbf{S}^\bullet)$ as a Verdier self-dual locally constant sheaf on V. The Deligne extension process defines the intersection chain sheaf $\mathbf{IC}^\bullet_{\bar{m}}(\overline{V}; \mathfrak{H}_V(\mathbf{S}^\bullet))$ on $\overline{V}$ with coefficients $\mathfrak{H}_V(\mathbf{S}^\bullet)$. This sheaf is self-dual on $\overline{V}$ (since $\overline{V}$ has only strata of even codimension), and the self-duality isomorphism restricts on V to the isomorphism induced by the bilinear pairing on $\mathfrak{H}_V(\mathbf{S}^\bullet)$. In the following theorem, $\sim$ denotes the equivalence relation of algebraic bordism of self-dual sheaves, a concept which will be defined precisely later on in the present section. Suffice it to say at the moment that the invariants associated to self-dual sheaves are preserved under algebraic bordisms.

Theorem 8.1.7 (*Cappell–Shaneson* [CS91])

$$Rf_*\mathbf{S}^\bullet[-t] \sim \bigoplus_{V \in \mathcal{V}} j_*\mathbf{IC}^\bullet_{\bar{m}}(\overline{V}; \mathfrak{H}_V(\mathbf{S}^\bullet))[c].$$

Some preparatory homological algebra in the derived category is required.

Proposition 8.1.8 *Let $f : \mathbf{A}^\bullet \to \mathbf{B}^\bullet$ be a morphism in $D^b(X)$ between complexes satisfying $\mathbf{H}^i(\mathbf{A}^\bullet) = 0$ for $i > p$ and $\mathbf{H}^i(\mathbf{B}^\bullet) = 0$ for $i < p$. Then taking cohomology induces an isomorphism*

$$\mathrm{Hom}_{D^b(X)}(\mathbf{A}^\bullet, \mathbf{B}^\bullet) \xrightarrow{\simeq} \mathrm{Hom}_{Sh(X)}(\mathbf{H}^p(\mathbf{A}^\bullet), \mathbf{H}^p(\mathbf{B}^\bullet)).$$

Proof. Replace $\mathbf{B}^\bullet$ by an injective resolution $\mathbf{I}^\bullet$ with $\mathbf{I}^i = 0$ for $i < p$. Then f is represented by an actual morphism (in $C(X)$) of complexes $\mathbf{A}^\bullet \to \mathbf{I}^\bullet$. □

Proposition 8.1.9 *Let $\mathbf{A}^\bullet, \mathbf{B}^\bullet, \mathbf{C}^\bullet \in D^b(X)$. If $\mathbf{H}^i(\mathbf{A}^\bullet) = 0$ for $i > p$ and $\psi : \mathbf{B}^\bullet \to \mathbf{C}^\bullet$ is a morphism such that $\tau_{\leq p}\psi$ is an isomorphism, then*

$$\mathrm{Hom}_{D^b(X)}(\mathbf{A}^\bullet, \mathbf{B}^\bullet) \xrightarrow{\psi \circ -} \mathrm{Hom}_{D^b(X)}(\mathbf{A}^\bullet, \mathbf{C}^\bullet)$$

is an isomorphism.

Proof. Embed ψ into a distinguished triangle

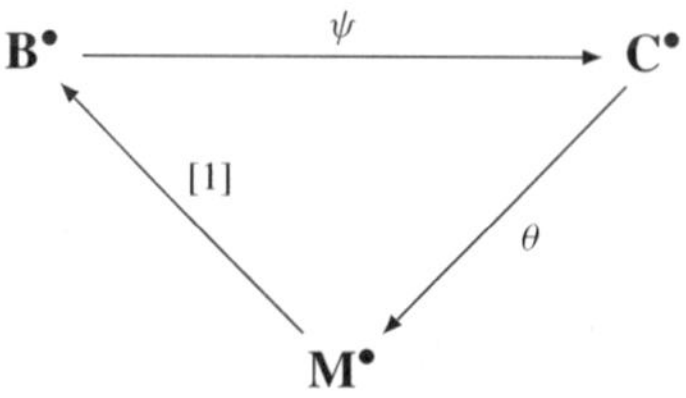

and observe that $\mathbf{H}^i(\mathbf{M}^\bullet) = 0$ for $i < p$. Since $\mathbf{H}^p(\psi) : \mathbf{H}^p(\mathbf{B}^\bullet) \xrightarrow{\simeq} \mathbf{H}^p(\mathbf{C}^\bullet)$ is an isomorphism, exactness implies that $\mathbf{H}^p(\theta) = 0 : \mathbf{H}^p(\mathbf{C}^\bullet) \longrightarrow \mathbf{H}^p(\mathbf{M}^\bullet)$. In particular, if $\phi : \mathbf{A}^\bullet \to \mathbf{C}^\bullet$ is any morphism, then $\theta \circ \phi$ induces the zero map $\mathbf{H}^p(\mathbf{A}^\bullet) \to \mathbf{H}^p(\mathbf{M}^\bullet)$. But by the previous Proposition 8.1.8, a morphism $\mathbf{A}^\bullet \to \mathbf{M}^\bullet$ is completely determined on $\mathbf{H}^p$,

$$\mathrm{Hom}_{D^b(X)}(\mathbf{A}^\bullet, \mathbf{M}^\bullet) \cong \mathrm{Hom}_{Sh(X)}(\mathbf{H}^p(\mathbf{A}^\bullet), \mathbf{H}^p(\mathbf{M}^\bullet)).$$

This shows that the map

$$\mathrm{Hom}_{D^b(X)}(\mathbf{A}^\bullet, \mathbf{C}^\bullet) \xrightarrow{\theta\circ-} \mathrm{Hom}_{D^b(X)}(\mathbf{A}^\bullet, \mathbf{M}^\bullet)$$

is zero. As $\mathbf{H}^i(\mathbf{M}^\bullet[-1]) = 0$ for $i \leq p$, we have $\mathrm{Hom}(\mathbf{A}^\bullet, \mathbf{M}^\bullet[-1]) = 0$. The statement follows from the exactness of the sequence

$$0 = \mathrm{Hom}(\mathbf{A}^\bullet, \mathbf{M}^\bullet[-1]) \to \mathrm{Hom}(\mathbf{A}^\bullet, \mathbf{B}^\bullet) \to \mathrm{Hom}(\mathbf{A}^\bullet, \mathbf{C}^\bullet) \xrightarrow{0} \mathrm{Hom}(\mathbf{A}^\bullet, \mathbf{M}^\bullet). \quad \square$$

Let $j_x : \{x\} \hookrightarrow X$ denote the inclusion of a point. If a sheaf $\mathbf{A}^\bullet$ satisfies stalk conditions and a sheaf $\mathbf{B}^\bullet$ satisfies corresponding costalk conditions, then morphisms $\mathbf{A}^\bullet \to \mathbf{B}^\bullet$ in the derived category are determined by their restriction to the top stratum:

Proposition 8.1.10 *Let X^n be a stratified pseudomanifold with singular set Σ, $\mathbf{A}^\bullet, \mathbf{B}^\bullet \in D^b_c(X)$, $\bar{p}$ a perversity* (*in the sense of Definition* 4.1.7). *Assume $\mathbf{A}^\bullet$ satisfies the stalk conditions $\mathbf{H}^i(j_x^*\mathbf{A}^\bullet) = 0$, $i > \bar{p}(k) - n$, $x \in X_{n-k} - X_{n-k-1}$, $k \geq 2$. Then*:

1. *The* weak *costalk conditions $\mathbf{H}^i(j_x^!\mathbf{B}^\bullet) = 0$, $i \leq \bar{p}(k) - k$ imply that restriction to the top stratum induces a monomorphism*

$$\mathrm{Hom}_{D^b(X)}(\mathbf{A}^\bullet, \mathbf{B}^\bullet) \hookrightarrow \mathrm{Hom}_{D^b(X-\Sigma)}(\mathbf{A}^\bullet|_{X-\Sigma}, \mathbf{B}^\bullet|_{X-\Sigma}).$$

2. *The* strong *costalk conditions $\mathbf{H}^i(j_x^!\mathbf{B}^\bullet) = 0$, $i \leq \bar{p}(k) - k + 1$ imply that restriction to the top stratum induces an isomorphism*

$$\mathrm{Hom}_{D^b(X)}(\mathbf{A}^\bullet, \mathbf{B}^\bullet) \xrightarrow{\simeq} \mathrm{Hom}_{D^b(X-\Sigma)}(\mathbf{A}^\bullet|_{X-\Sigma}, \mathbf{B}^\bullet|_{X-\Sigma}).$$

The mentioned stalk and *strong* costalk conditions are precisely the Goresky–MacPherson intersection chain axioms (AX2) and (AX3″), respectively (cf. Sect. 4.1.4).

Proof. (of Proposition 8.1.10) We shall use the notation $U_k = X - X_{n-k}$, $i_k : U_k \hookrightarrow U_{k+1}$, $\mathbf{A}^\bullet_k = \mathbf{A}^\bullet|_{U_k}$, $\mathbf{B}^\bullet_k = \mathbf{B}^\bullet|_{U_k}$. We proceed by induction on the codimension k. The result holds trivially for $\mathbf{A}^\bullet_2$ and $\mathbf{B}^\bullet_2$ over U_2. Assume the result holds over U_k (that is, restriction from U_k to the top stratum induces a mono- or isomorphism). We will use Proposition 8.1.9 applied to the adjunction

$$\psi : \mathbf{B}^\bullet_{k+1} \longrightarrow Ri_{k*}i_k^*\mathbf{B}^\bullet_{k+1} = Ri_{k*}\mathbf{B}^\bullet_k.$$

Let us verify the hypotheses, say under the assumption of strong costalk vanishing. First, ψ is an isomorphism over U_k. Over $U_{k+1} - U_k$, use the distinguished triangle

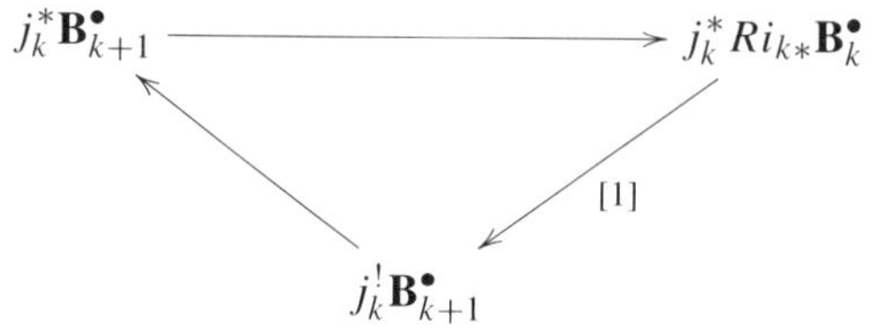

(where $j_k : U_{k+1} - U_k \hookrightarrow U_{k+1}$) to see that $\tau_{\leq \bar{p}(k)-n}\psi$ is an isomorphism. The stalk conditions on $\mathbf{A}^\bullet$ ensure that $\mathbf{H}^i(\mathbf{A}^\bullet_{k+1}) = 0$ for $i > \bar{p}(k) - n$. Thus Proposition 8.1.9 is applicable and asserts that ψ induces an isomorphism

$$\begin{aligned}\operatorname{Hom}_{D^b(U_{k+1})}(\mathbf{A}^\bullet_{k+1}, \mathbf{B}^\bullet_{k+1}) \xrightarrow{\simeq} & \operatorname{Hom}_{D^b(U_{k+1})}(\mathbf{A}^\bullet_{k+1}, Ri_{k*}\mathbf{B}^\bullet_k) \\ & \cong \operatorname{Hom}_{D^b(U_k)}(i_k^*\mathbf{A}^\bullet_{k+1}, \mathbf{B}^\bullet_k) \\ & = \operatorname{Hom}_{D^b(U_k)}(\mathbf{A}^\bullet_k, \mathbf{B}^\bullet_k)\end{aligned}$$

(using adjointness, Proposition 1.1.11). Now apply the induction hypothesis.

Under the weak costalk assumption, the map induced by ψ

$$\mathbf{H}^{\bar{p}(k)-n}(\mathbf{B}^\bullet_{k+1}) \longrightarrow \mathbf{H}^{\bar{p}(k)-n}(Ri_{k*}\mathbf{B}^\bullet_k)$$

is only injective. Thus $\operatorname{Hom}(\mathbf{A}^\bullet_{k+1}, \mathbf{B}^\bullet_{k+1})$ only injects into $\operatorname{Hom}(\mathbf{A}^\bullet_{k+1}, Ri_{k*}\mathbf{B}^\bullet_k)$.
□

The decomposition in Theorem 8.1.7 is achieved up to algebraic bordism of self-dual sheaves. We shall now discuss this concept in detail. To motivate the purely algebraic definition, we consider the geometric situation of an oriented compact manifold W with boundary $W = M \sqcup -N$, i.e. a bordism from M to N. How are the various associated chain complexes related? The inclusion $M \to (W, N)$ induces a map

$$C_*(M) \xrightarrow{v} C_*(W, N).$$

The algebraic mapping cone on v can be identified with $C_*(W, M \cup N)$ and there is a distinguished triangle

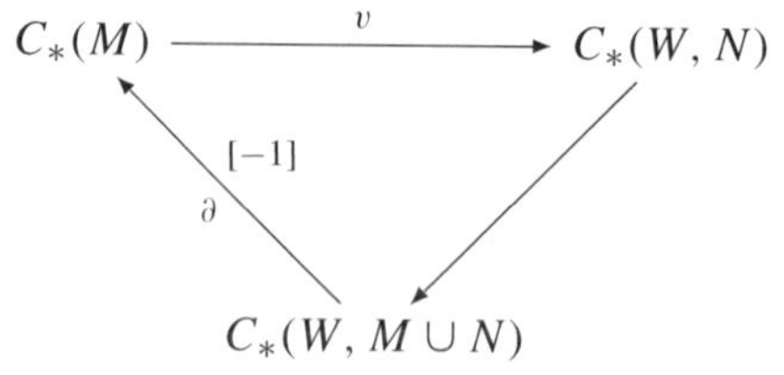

The boundary map

$$C_{*+1}(W, M) \xrightarrow{u} C_*(M)$$

lifts to $C_{*+1}(W, M \cup N)$ with respect to $C_{*+1}(W, M \cup N) \to C_*(M)$; the lift

$$u' : C_{*+1}(W, M) \longrightarrow C_{*+1}(W, M \cup N)$$

is the map induced by the inclusion $(W, M) \to (W, M \cup N)$. The mapping cone on u' however can be identified with $C_*(N)$ and there is a distinguished triangle

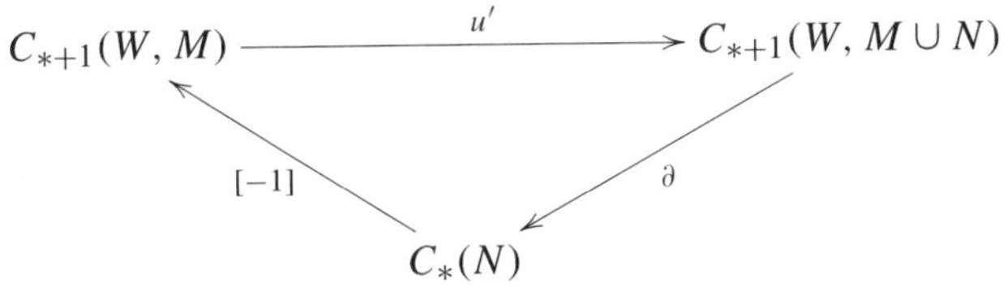

which is just the above triangle with the roles of M and N reversed. Note also that u and v satisfy $vu = 0$ and they are dual to each other under Poincaré duality. Retaining the above algebraic properties will naturally lead us to a definition of algebraic bordism of complexes.

Let $\mathbf{X}^\bullet \xrightarrow{u} \mathbf{Y}^\bullet \xrightarrow{v} \mathbf{Z}^\bullet$ be morphisms in $D_c^b(Y)$ with $vu = 0$. Let $\mathbf{C}_v^\bullet$ be a mapping cone on v:

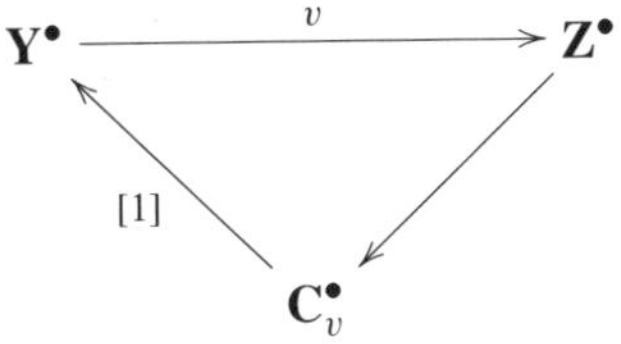

This defines a morphism $\mathbf{C}_v^\bullet[-1] \to \mathbf{Y}^\bullet$, and it follows from the exact sequence

$$\begin{array}{ccccccc} \mathrm{Hom}(\mathbf{X}^\bullet, \mathbf{Z}^\bullet[-1]) & \to & \mathrm{Hom}(\mathbf{X}^\bullet, \mathbf{C}_v^\bullet[-1]) & \to & \mathrm{Hom}(\mathbf{X}^\bullet, \mathbf{Y}^\bullet) & \to & \mathrm{Hom}(\mathbf{X}^\bullet, \mathbf{Z}^\bullet) \\ & & & & u & \mapsto & vu = 0 \end{array}$$

that u can be lifted to $u' : \mathbf{X}^\bullet \to \mathbf{C}_v^\bullet[-1]$:

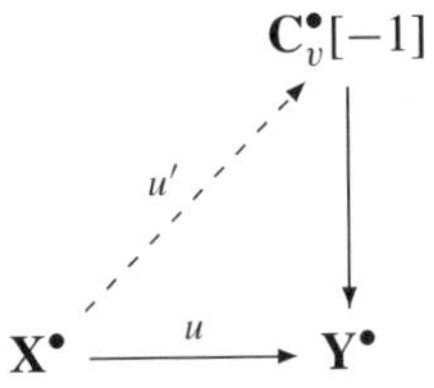

This is well-defined (i.e. u' is unique) if $\mathrm{Hom}(\mathbf{X}^\bullet, \mathbf{Z}^\bullet[-1]) = 0$. Define

$$\mathbf{C}_{u,v}^\bullet = \mathbf{C}_{u'}^\bullet$$

to be a mapping cone of u' so that there is a distinguished triangle

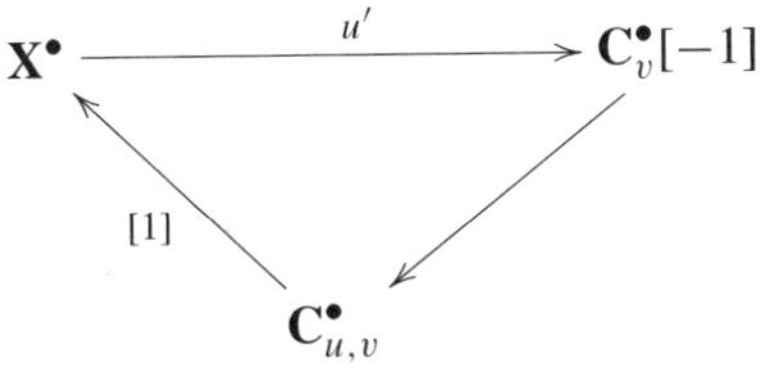

This construction is actually symmetric in u and v, for if we define a morphism $\mathbf{Y}^\bullet \to \mathbf{C}^\bullet_u$ by using the triangle

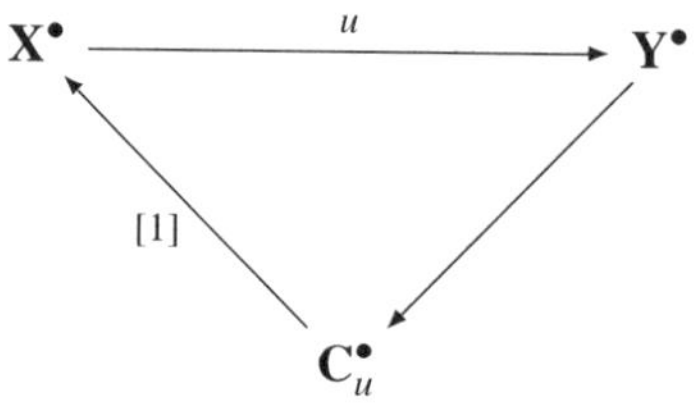

then v can be factored as

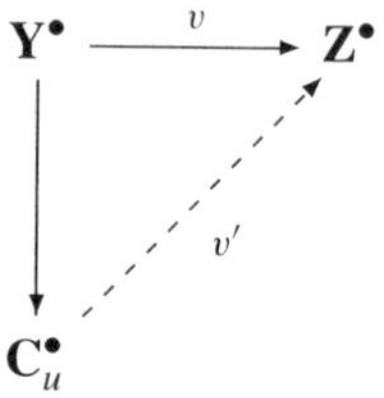

and it is a formal exercise to show that $\mathbf{C}^\bullet_{v'}[-1] \cong \mathbf{C}^\bullet_{u,v}$ (use the octahedral axiom on $\mathbf{Y}^\bullet \to \mathbf{C}^\bullet_u \xrightarrow{v'} \mathbf{Z}^\bullet$). To analyze the dual of $\mathbf{C}^\bullet_{u,v}$ consider the triangle

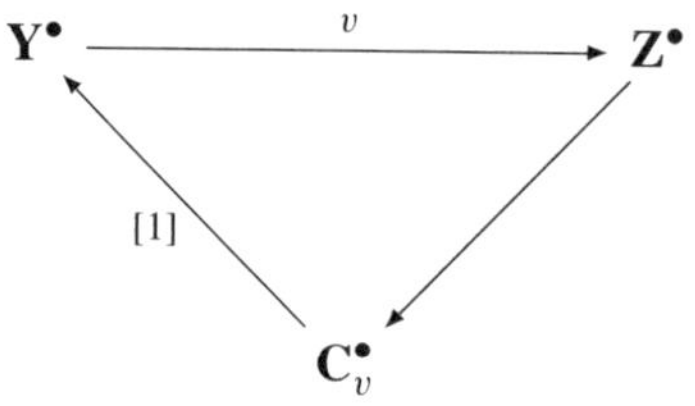

Its dual is

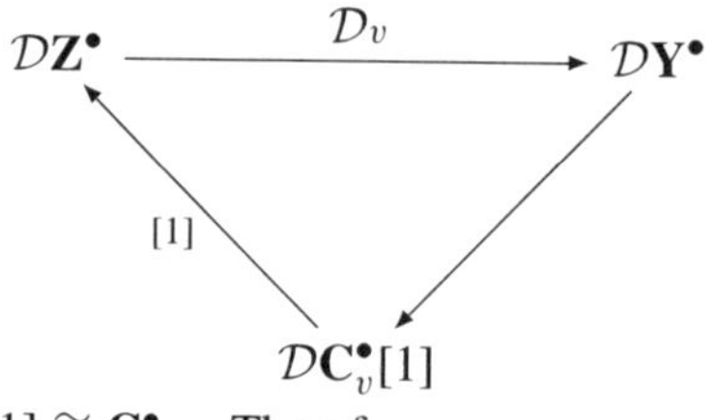

which proves that $\mathcal{D}\mathbf{C}^\bullet_v[1] \cong \mathbf{C}^\bullet_{\mathcal{D}v}$. Therefore,

$$\mathcal{D}\mathbf{C}^\bullet_{u,v} \cong \mathcal{D}\mathbf{C}^\bullet_{u'} \cong \mathbf{C}^\bullet_{\mathcal{D}u'}[-1] \cong \mathbf{C}^\bullet_{\mathcal{D}v,\mathcal{D}u}.$$

Assume now that $\mathbf{Y}^\bullet$ is self-dual on Y^m and that we are given an isomorphism $\mathcal{D}\mathbf{X}^\bullet[m] \cong \mathbf{Z}^\bullet$ such that

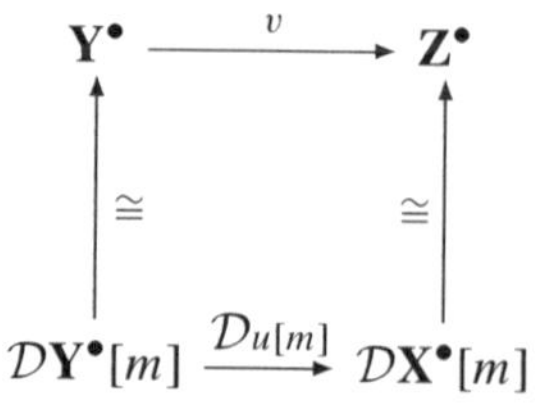

commutes. We will abbreviate such a situation by writing $\mathcal{D}u[m] = v$. Using this notation, we have

$$\mathbf{C}^\bullet_{u,v} \cong \mathcal{D}\mathbf{C}^\bullet_{\mathcal{D}v,\mathcal{D}u} \cong \mathcal{D}(\mathbf{C}^\bullet_{u,v}[-m]) \cong \mathcal{D}\mathbf{C}^\bullet_{u,v}[m],$$

that is, $\mathbf{C}^\bullet_{u,v}$ is again self-dual.

Definition 8.1.11 *We say that* $\mathbf{Y}^\bullet_1 = \mathbf{C}^\bullet_{u,v}$ *is obtained from* $\mathbf{Y}^\bullet$ *by an* elementary bordism. *The sheaf* $\mathbf{Y}^\bullet$ *is* bordant *to* $\widehat{\mathbf{Y}}^\bullet$ *if there exists a sequence* $\mathbf{Y}^\bullet = \mathbf{Y}^\bullet_0, \mathbf{Y}^\bullet_1, \ldots, \mathbf{Y}^\bullet_n = \widehat{\mathbf{Y}}^\bullet$ *such that* $\mathbf{Y}^\bullet_i$ *is obtained from* $\mathbf{Y}^\bullet_{i-1}$ *by an elementary bordism.*

This notion of bordism is an equivalence relation.

We shall describe the version of the decomposition theorem to be proved first. Let Y^m be an oriented pseudomanifold, $\mathbf{S}^\bullet \in D^b_c(Y)$, $d : \mathbf{S}^\bullet \xrightarrow{\cong} \mathcal{D}\mathbf{S}^\bullet[m]$, $V \in \mathcal{V}$, $c = \frac{1}{2}\operatorname{codim} V$. Consider the local coefficient system $\mathcal{Y}^V(\mathbf{S}^\bullet) = \mathbf{H}^{c-m}(j_V^!\mathbf{S}^\bullet)$, where $j_V : V \hookrightarrow Y$, and the composition

$$j_V^!\mathbf{S}^\bullet \longrightarrow j_V^*\mathbf{S}^\bullet \xrightarrow{d} j_V^*\mathcal{D}\mathbf{S}^\bullet[m] \cong \mathcal{D}(j_V^!\mathbf{S}^\bullet)[m]. \tag{8.3}$$

With $j_y : \{y\} \hookrightarrow V$ the inclusion of a point and using $j_y^! = j_y^*[2c-m]$ (as V is nonsingular),

$$\begin{aligned}
\mathbf{H}^{c-m}(\mathcal{D}(j_V^!\mathbf{S}^\bullet)[m])_y &= H^{c-m}(\mathcal{D}_{\{y\}}(j_y^! j_V^!\mathbf{S}^\bullet)[m]) \\
&= \mathcal{H}^c(\{y\}; \mathcal{D}(j_y^! j_V^!\mathbf{S}^\bullet)) \\
&\cong \operatorname{Hom}(\mathcal{H}^{-c}(\{y\}; j_y^! j_V^!\mathbf{S}^\bullet), \mathbb{R}) \\
&= \operatorname{Hom}(H^{-c}(j_y^! j_V^!\mathbf{S}^\bullet), \mathbb{R}) \\
&= \operatorname{Hom}(H^{-c}(j_y^*(j_V^!\mathbf{S}^\bullet)[2c-m]), \mathbb{R}) \\
&= \operatorname{Hom}(\mathbf{H}^{c-m}(j_V^!\mathbf{S}^\bullet)_y, \mathbb{R}).
\end{aligned}$$

Thus on $\mathbf{H}^{c-m}(\cdot)$, (8.3) induces

$$\mathcal{Y}^V(\mathbf{S}^\bullet) \longrightarrow \operatorname{Hom}(\mathcal{Y}^V(\mathbf{S}^\bullet), \mathbb{R}_V) = \mathcal{Y}^V(\mathbf{S}^\bullet)^*.$$

Define

$$\begin{aligned}
\mathcal{X}^V &= \ker(\mathcal{Y}^V(\mathbf{S}^\bullet) \longrightarrow \mathcal{Y}^V(\mathbf{S}^\bullet)^*), \\
\mathcal{Z}^V &= \mathcal{Y}^V/\mathcal{X}^V.
\end{aligned}$$

Then $\mathcal{Y}^V \to \mathcal{Y}^{V*}$ induces a *nonsingular* bilinear pairing

$$\mathcal{P}_V(\mathbf{S}^\bullet) : \mathcal{Z}^V \otimes \mathcal{Z}^V \longrightarrow \mathbb{R}_V.$$

Since V is oriented, $\mathcal{P}_V$ may be viewed as a self-duality isomorphism

$$\mathcal{Z}^V[\dim V] \xrightarrow{\cong} \mathcal{D}(\mathcal{Z}^V[\dim V])[\dim V].$$

By Proposition 8.1.10, this isomorphism has a unique extension over $\overline{V}$,

$$\mathbf{IC}^{\bullet}_{\bar{m}}(\overline{V}; \mathcal{Z}) \xrightarrow{\cong} \mathcal{D}\mathbf{IC}^{\bullet}_{\bar{m}}(\overline{V}; \mathcal{Z})[m-2c]$$

($\dim V = m - 2c$). If $j : \overline{V} \hookrightarrow Y$ denotes the inclusion, then $j_! = j_*$ (since $\overline{V}$ is closed in Y), and thus

$$j_*\mathbf{IC}^{\bullet}_{\bar{m}}(\overline{V}; \mathcal{Z})[c] \cong j_*\mathcal{D}\mathbf{IC}^{\bullet}_{\bar{m}}[m-2c][c] \cong \mathcal{D}(j_!\mathbf{IC}^{\bullet}_{\bar{m}})[m-c] \cong \mathcal{D}(j_*\mathbf{IC}^{\bullet}_{\bar{m}}[c])[m],$$

i.e. $j_*\mathbf{IC}^{\bullet}_{\bar{m}}(\overline{V}; \mathcal{Z})[c]$ is self-dual on Y. Note that at this point it is crucial that V is of even codimension. We will prove the following decomposition theorem:

Theorem 8.1.12 *If Y has only strata of even codimension and $\mathbf{S}^{\bullet} \in D^b_c(Y)$ is self-dual, then there exists an orthogonal*[1] *decomposition*

$$\mathbf{S}^{\bullet} \sim \bigoplus_{V \in \mathcal{V}} j_*\mathbf{IC}^{\bullet}_{\bar{m}}(\overline{V}; \mathcal{Z}^V(\mathbf{S}^{\bullet}))[c(V)],$$

where $\sim$ denotes algebraic bordism of self-dual sheaves.

The strategy is to prove this first for middle perverse sheaves $\mathbf{S}^{\bullet}$, since those are already close to intersection chain sheaves. Then we will obtain the statement for all $\mathbf{S}^{\bullet}$ by constructing an elementary bordism from any given sheaf to a middle perverse one.

Let us briefly review middle perverse sheaves. We have to be somewhat careful with our notation, so as not to confuse the perversity notation of Sect. 4.1.3 (see in particular Examples 4.1.8) with the one customarily used in the theory of perverse t-structures, Sect. 7.3. Thus, let $\bar{\mathrm{m}} : \mathcal{V} \to \mathbb{Z}$ be the function $\bar{\mathrm{m}}(V) = c(V) - m$. We distinguish between $\bar{\mathrm{m}}$ and $\bar{m}$. The latter function was defined in Examples 4.1.8. Its value on an even codimension k is given by $\bar{m}(k) = \frac{k}{2} - 1$. The relation between the two is as follows: If $V \in \mathcal{V}$ has codimension $k = 2c(V)$, then

$$\bar{\mathrm{m}}(V) = c(V) - m = \left(\frac{k}{2} - 1\right) + 1 - m = \bar{m}(k) + 1 - m.$$

This shows that the value of $\bar{\mathrm{m}}$ is precisely one higher than the cut-off value specified in the axiomatics for the middle perversity intersection chain complex. The function $\bar{\mathrm{m}}$ has a corresponding t-structure $({}^{\bar{\mathrm{m}}}D^{\le 0}, {}^{\bar{\mathrm{m}}}D^{\ge 0})$, its heart ${}^{\bar{\mathrm{m}}}P(Y) = {}^{\bar{\mathrm{m}}}D^{\le 0} \cap {}^{\bar{\mathrm{m}}}D^{\ge 0}$ and associated functors ${}^{\bar{\mathrm{m}}}\tau^{\le 0}$, ${}^{\bar{\mathrm{m}}}\tau^{\ge 0}$, ${}^{\bar{\mathrm{m}}}H^0$, ${}^{\bar{\mathrm{m}}}j^!$ and ${}^{\bar{\mathrm{m}}}j^*$. Explicitly, we have

$$\begin{aligned} {}^{\bar{\mathrm{m}}}P(Y) = \{\mathbf{A}^{\bullet} \in D^b_c(Y) \mid\ & \mathbf{H}^n(j_V^*\mathbf{A}^{\bullet}) = 0,\ n > c(V) - m, \\ & \mathbf{H}^n(j_V^!\mathbf{A}^{\bullet}) = 0,\ n < c(V) - m\}, \end{aligned}$$

where $j_V : V \hookrightarrow Y$. In the following, when we say "$\mathbf{A}^{\bullet}$ is perverse on a subspace $A \subset Y$" we mean $\mathbf{A}^{\bullet} \in {}^{\bar{\mathrm{m}}}P(A)$ for the $\bar{\mathrm{m}}$ defined exactly as above, and not that $\mathbf{A}^{\bullet}$ is *intrinsically* perverse on A, which would be equivalent to $\mathbf{A}^{\bullet} \in {}^{\bar{p}}P(A)$, where $\bar{p}(V) = \frac{1}{2}\operatorname{codim}(V \hookrightarrow A) - \dim A$.

[1] Meaning: not only the sheaf complex, but also the self-duality isomorphism decomposes into blocks.

Remark 8.1.13 *If we assume that* $\mathbf{S}^\bullet$ *is self-dual on* Y *and* $\mathbf{S}^\bullet \in {}^{\bar{m}}D^{\leq 0}$, *i.e.* $\mathbf{H}^n(j_V^* \mathbf{S}^\bullet) = 0$, $n > c - m$, *for all* V, *then a duality calculation*[2] *shows that* $\mathbf{H}^n(j_V^! \mathbf{S}^\bullet) = 0$, $n < c - m$. *Hence* $\mathbf{S}^\bullet$ *is already* $\bar{m}$*-perverse,* $\mathbf{S}^\bullet \in {}^{\bar{m}}P(Y)$.

Technically, the splitting itself will emerge through the next simple observation for triangulated categories:

Lemma 8.1.14 *Let* D *be a triangulated category with translation functor* [1] *and* $X, Y, Z \in ObD$. *Let*

$$X \longrightarrow Y \longrightarrow Z \stackrel{\eta}{\longrightarrow} X[1]$$

be a distinguished triangle. If $\eta = 0$, *then*

$$Y \cong X \oplus Z.$$

Proof. Triangles of the form

$$X \stackrel{\text{incl}_X}{\longrightarrow} X \oplus Z \stackrel{\text{proj}_Z}{\longrightarrow} Z \stackrel{\eta'}{\longrightarrow} X[1]$$

are distinguished (see the first bullet in the proof of Theorem 7.1.11). Let $\text{incl}_Z : Z \to X \oplus Z$ be the canonical inclusion so that $\text{proj}_Z \circ \text{incl}_Z = 1$. Then

$$\eta' = \eta' \circ \text{proj}_Z \circ \text{incl}_Z = 0 \circ \text{incl}_Z = 0.$$

By axiom (TR4), the commutative square

$$\begin{array}{ccc}
Z & \xrightarrow{\eta=0} & X[1] \\
\| & & \| \\
Z & \xrightarrow{\eta'=0} & X[1]
\end{array}$$

can be embedded into a morphism of triangles

$$\begin{array}{ccccccc}
X & \longrightarrow & Y & \longrightarrow & Z & \xrightarrow{\eta=0} & X[1] \\
\| & & \downarrow{\scriptstyle \alpha} & & \| & & \| \\
X & \xrightarrow{\text{incl}_X} & X \oplus Z & \xrightarrow{\text{proj}_Z} & Z & \xrightarrow{\eta'=0} & X[1]
\end{array}$$

By Example 2.2.5, α is an isomorphism. □

Definition 8.1.15 *A complex of sheaves* $\mathbf{S}^\bullet \in {}^{\bar{m}}P(Y)$ *is called* locally nonsingular, *if the canonical map*

$$\mathbf{H}^{c(V)-m}(j_V^! \mathbf{S}^\bullet) \longrightarrow \mathbf{H}^{c(V)-m}(j_V^* \mathbf{S}^\bullet)$$

is an isomorphism for all pure strata $V \in \mathcal{V}$.

[2] $\mathbf{H}^n(j_V^! \mathbf{S}^\bullet) = \mathbf{H}^n(j_V^! \mathcal{D}\mathbf{S}^\bullet[m]) = \mathbf{H}^{n+m}(\mathcal{D}(j_V^* \mathbf{S}^\bullet)) =$ etc.

As announced above, we will first construct orthogonal decompositions into twisted intersection chain sheaves for self-dual sheaves which are in addition assumed to be $\bar{\mathrm{m}}$-perverse:

Proposition 8.1.16 *If Y has only strata of even codimension and $\mathbf{S}^\bullet \in {}^{\bar{\mathrm{m}}}P(Y)$ is self-dual, then there exists an orthogonal decomposition*

$$\mathbf{S}^\bullet \sim \bigoplus_{V\in\mathcal{V}} j_*\mathbf{IC}^\bullet_{\bar{m}}(\overline{V}; \mathcal{Z}^V(\mathbf{S}^\bullet))[c(V)].$$

If $\mathbf{S}^\bullet$ is in addition locally nonsingular, then

$$\mathbf{S}^\bullet \cong \bigoplus_{V\in\mathcal{V}} j_*\mathbf{IC}^\bullet_{\bar{m}}(\overline{V}; \mathcal{Z}^V(\mathbf{S}^\bullet))[c(V)].$$

Proof. Let $V \in \mathcal{V}$ be a component of maximal codimension $2c$ such that

$$\mathcal{Y}^V = \mathcal{Y}^V(\mathbf{S}^\bullet) = \mathbf{H}^{c-m}(j_V^!\mathbf{S}^\bullet) \neq 0.$$

(If such V does not exist, jump close to the end of this proof.) Let j denote the closed inclusion $j = j_{\overline{V}} : \overline{V} \hookrightarrow Y$. Using the fact that $\mathbf{S}^\bullet \in {}^{\bar{\mathrm{m}}}P(Y)$ together with the maximality assumption, it follows that $j^!\mathbf{S}^\bullet[-c]$ satisfies the *strong* costalk condition for $\bar{m}$ (see Proposition 8.1.10). Thus by Proposition 8.1.10, there exists a unique morphism

$$\lambda_0 : \mathbf{IC}^\bullet_{\bar{m}}(\overline{V}; \mathcal{Y}^V)[c] \longrightarrow j^!\mathbf{S}^\bullet$$

which induces the identity on stalks over V. By definition, ${}^{\bar{\mathrm{m}}}j^! = {}^{\bar{\mathrm{m}}}H^0 \circ j^! = {}^{\bar{\mathrm{m}}}\tau^{\geq 0}\,{}^{\bar{\mathrm{m}}}\tau^{\leq 0}j^!$. Thus for $\mathbf{S}^\bullet$, ${}^{\bar{\mathrm{m}}}j^!\mathbf{S}^\bullet \cong {}^{\bar{\mathrm{m}}}\tau^{\leq 0}j^!\mathbf{S}^\bullet$ (the other truncation is not needed since clearly $j^!\mathbf{S}^\bullet \in {}^{\bar{\mathrm{m}}}D^{\geq 0}$; use Proposition 7.1.7). The sheaf ${}^{\bar{\mathrm{m}}}j^!\mathbf{S}^\bullet$ is $\bar{\mathrm{m}}$-perverse on $\overline{V}$, but not intrinsically perverse on $\overline{V}$. However, the shifted complex ${}^{\bar{\mathrm{m}}}j^!\mathbf{S}^\bullet[-c]$ *is* intrinsically perverse on $\overline{V}$. Consider the canonical morphism $\omega : {}^{\bar{\mathrm{m}}}j^!\mathbf{S}^\bullet \cong {}^{\bar{\mathrm{m}}}\tau^{\leq 0}(j^!\mathbf{S}^\bullet) \to j^!\mathbf{S}^\bullet$ and embed it into a distinguished triangle

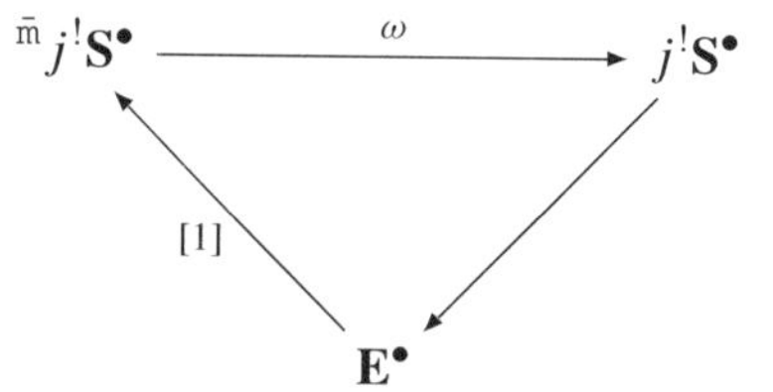

This defines $\mathbf{E}^\bullet$ up to isomorphism in the derived category, and the triangle implies $\mathbf{E}^\bullet \in {}^{\bar{\mathrm{m}}}D^{\geq 1}$ (in fact, we recognize $\mathbf{E}^\bullet$ as $\mathbf{E}^\bullet \cong {}^{\bar{\mathrm{m}}}\tau^{\geq 1}j^!\mathbf{S}^\bullet$). Consequently, if $\mathbf{P}^\bullet \in {}^{\bar{\mathrm{m}}}D^{\leq 0}(Y)$ (in particular for $\mathbf{P}^\bullet \in {}^{\bar{\mathrm{m}}}P(Y)$), then

$$\mathrm{Hom}_{D(Y)}(\mathbf{P}^\bullet, j_*\mathbf{E}^\bullet) = 0 = \mathrm{Hom}_{D(Y)}(\mathbf{P}^\bullet, j_*\mathbf{E}^\bullet[-1])$$

(cf. the axioms for a t-structure, Definition 7.1.1). The triangle induces an exact sequence

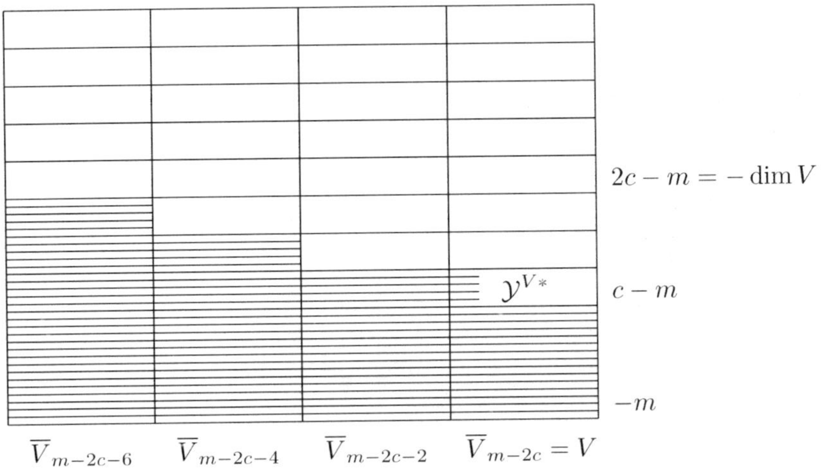

Fig. 8.1. The stalks of $j^*\mathbf{S}^\bullet$ over the strata of $\overline{V}$.

$$0=\mathrm{Hom}(\mathbf{P}^\bullet,\, j_*\mathbf{E}^\bullet[-1]) \longrightarrow \mathrm{Hom}(\mathbf{P}^\bullet,\, j_*{}^{\bar{\mathrm{m}}}j^!\mathbf{S}^\bullet) \xrightarrow[\cong]{\omega\circ-} \mathrm{Hom}(\mathbf{P}^\bullet,\, j_*j^!\mathbf{S}^\bullet)$$

$$\longrightarrow \mathrm{Hom}(\mathbf{P}^\bullet,\, j_*\mathbf{E}^\bullet)=0.$$

Now take $\mathbf{P}^\bullet = j_*\mathbf{IC}^\bullet_{\bar{m}}(\overline{V};\mathcal{Y}^V)[c]$, which is indeed perverse on Y. Then the exact sequence implies the existence of a unique morphism

$$\lambda : \mathbf{IC}^\bullet_{\bar{m}}(\overline{V};\mathcal{Y}^V)[c] \longrightarrow {}^{\bar{\mathrm{m}}}j^!\mathbf{S}^\bullet$$

such that $\lambda_0=\omega\lambda$.

Using the self-duality isomorphism for $\mathbf{S}^\bullet$, we identify

$$\mathbf{H}^{c-m}(j_V^*\mathbf{S}^\bullet)\cong \mathbf{H}^{c-m}(j_V^*\mathcal{D}\mathbf{S}^\bullet[m])\cong \mathbf{H}^c(\mathcal{D}(j_V^!\mathbf{S}^\bullet))\cong(\mathcal{Y}^V)^*.$$

The self-duality of $\mathbf{S}^\bullet$ also implies that V is a component of maximal codimension such that $\mathbf{H}^{c-m}(j_V^*\mathbf{S}^\bullet)\neq 0$. Thus for all lower-dimensional strata of $\overline{V}$ this derived sheaf vanishes and so $j^*\mathbf{S}^\bullet[-c]$ satisfies the $\bar{m}$-stalk condition in Proposition 8.1.10.

The stalks of $j^*\mathbf{S}^\bullet$ and $j^*\mathbf{S}^\bullet[-c]$ over the various strata of $\overline{V}$ are shown in Figs. 8.1 and 8.2, respectively. Figures 8.3 and 8.4 display the stalks of $\mathbf{IC}^\bullet_{\bar{m}}(\overline{V};\mathcal{Y}^{V*})$ and $\mathbf{IC}^\bullet_{\bar{m}}(\overline{V};\mathcal{Y}^{V*})[c]$, respectively. Since of course $\mathbf{IC}^\bullet_{\bar{m}}(\overline{V};\mathcal{Y}^{V*})$ satisfies the strong costalk, the proposition implies that there exists a unique morphism

$$\mu_0 : j^*\mathbf{S}^\bullet \longrightarrow \mathbf{IC}^\bullet_{\bar{m}}(\overline{V};\mathcal{Y}^{V*})[c]$$

which induces the identity on stalks over V. When applied to $\mathbf{S}^\bullet$, the functor ${}^{\bar{\mathrm{m}}}j^* = {}^{\bar{\mathrm{m}}}\tau^{\geq 0}\circ{}^{\bar{\mathrm{m}}}\tau^{\leq 0}\circ j^*$ can be calculated as ${}^{\bar{\mathrm{m}}}j^*\mathbf{S}^\bullet\cong{}^{\bar{\mathrm{m}}}\tau^{\geq 0}(j^*\mathbf{S}^\bullet)$, since $j^*\mathbf{S}^\bullet$ is already in ${}^{\bar{\mathrm{m}}}D^{\leq 0}$. The sheaf ${}^{\bar{\mathrm{m}}}j^*\mathbf{S}^\bullet$ is perverse on $\overline{V}$ (but not intrinsically perverse on $\overline{V}$). For the canonical morphism $\psi : j^*\mathbf{S}^\bullet \to {}^{\bar{\mathrm{m}}}\tau^{\geq 0}(j^*\mathbf{S}^\bullet)\cong{}^{\bar{\mathrm{m}}}j^*\mathbf{S}^\bullet$, an argument similar to the one above shows that there exists a unique morphism

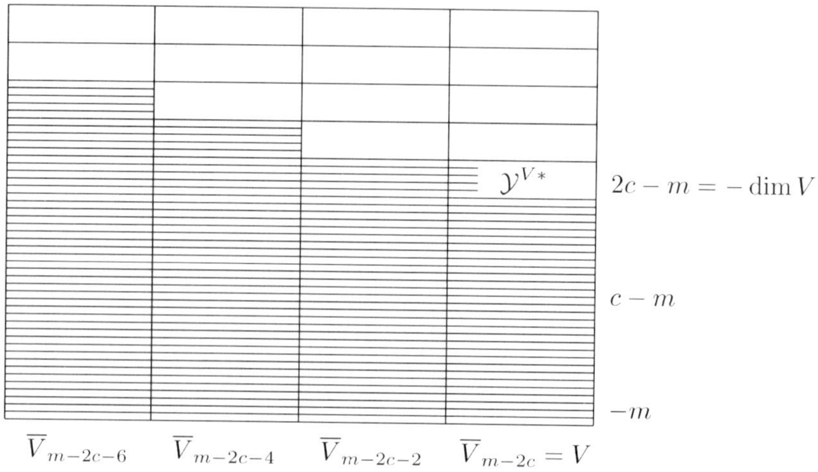

Fig. 8.2. The stalks of $j^*\mathbf{S}^\bullet[-c]$ over the strata of $\overline{V}$.

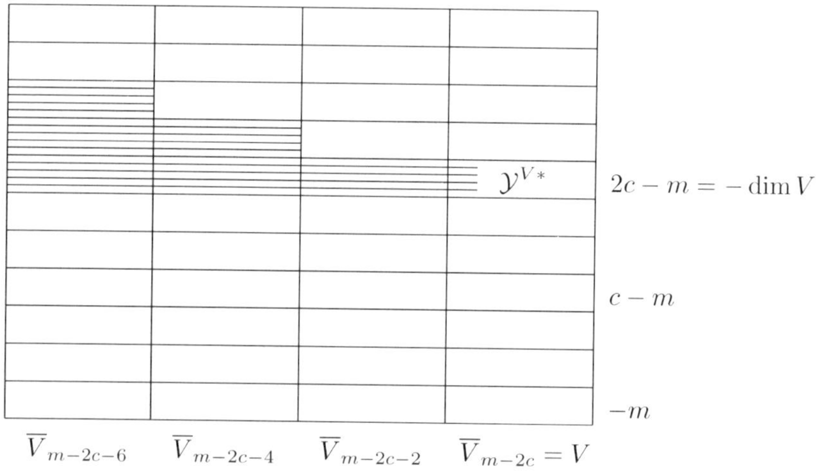

Fig. 8.3. The stalks of $\mathbf{IC}^\bullet_{\bar{m}}(\overline{V}; \mathcal{Y}^{V*})$ over the strata of $\overline{V}$.

$$\mu : {}^{\bar{m}}j^*\mathbf{S}^\bullet \longrightarrow \mathbf{IC}^\bullet_{\bar{m}}(\overline{V}; \mathcal{Y}^{V*})[c]$$

such that $\mu_0 = \mu\psi$.

The square

$$\begin{array}{ccc}
j^*\mathbf{S}^\bullet & \xrightarrow{\quad\mu_0\quad} & \mathbf{IC}^\bullet_{\bar{m}}(\overline{V}; \mathcal{Y}^{V*})[c] \\
\cong \Big\downarrow j^*d & & GM \Big\downarrow \cong \\
\mathcal{D}(j^!\mathbf{S}^\bullet)[m] & \xrightarrow{\mathcal{D}\lambda_0[m]} & \mathcal{D}(\mathbf{IC}^\bullet_{\bar{m}}(\overline{V}; \mathcal{Y}^{V})[c])[m]
\end{array}$$

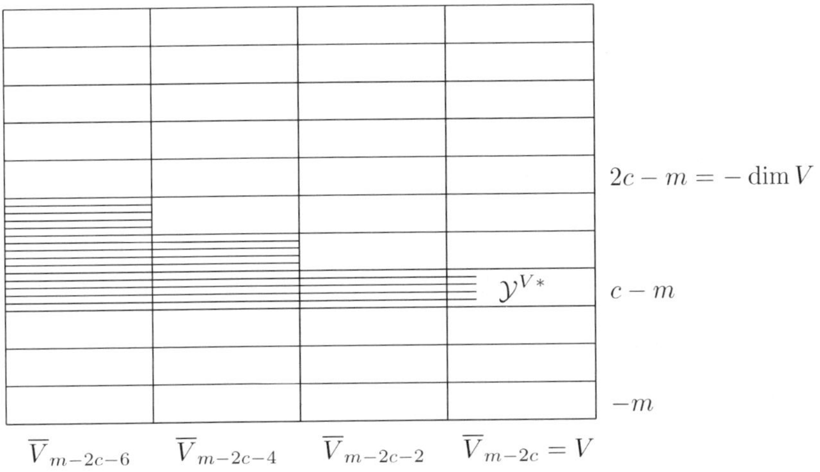

Fig. 8.4. The stalks of $\mathbf{IC}^\bullet_{\bar{m}}(\overline{V}; \mathcal{Y}^{V*})[c]$ over the strata of $\overline{V}$.

commutes, since μ_0 and $\mathcal{D}\lambda_0[m]$ induce the identity on stalks over V, and μ_0 is unique with respect to this property. In the factorization

$$\begin{array}{ccccc}
j^*\mathbf{S}^\bullet & \xrightarrow{\ \psi\ } & {}^{\bar{\mathrm{m}}}j^*\mathbf{S}^\bullet & \xrightarrow{\ \mu\ } & \mathbf{IC}^\bullet_{\bar{m}}(\overline{V}; \mathcal{Y}^{V*})[c] \\
\cong\big\downarrow j^*d & & \cong\big\downarrow {}^{\bar{\mathrm{m}}}j^*d & & GM\big\downarrow\cong \\
\mathcal{D}(j^!\mathbf{S}^\bullet)[m] & \xrightarrow{\mathcal{D}\omega[m]} & \mathcal{D}({}^{\bar{\mathrm{m}}}j^!\mathbf{S}^\bullet)[m] & \xrightarrow{\mathcal{D}\lambda[m]} & \mathcal{D}(\mathbf{IC}^\bullet_{\bar{m}}(\overline{V}; \mathcal{Y}^{V})[c])[m]
\end{array}$$

the left square commutes. Hence

$$\begin{aligned}
GM \circ \mu_0 &= \mathcal{D}\lambda_0[m] \circ j^*d \\
&= \mathcal{D}\lambda[m] \circ \mathcal{D}\omega[m] \circ j^*d \\
&= \mathcal{D}\lambda[m] \circ {}^{\bar{\mathrm{m}}}j^*d \circ \psi,
\end{aligned}$$

and $\mu_0 = GM^{-1} \circ \mathcal{D}\lambda[m] \circ {}^{\bar{\mathrm{m}}}j^*d \circ \psi$. But μ is the unique morphism such that $\mu_0 = \mu\psi$. Thus $\mu = GM^{-1} \circ \mathcal{D}\lambda[m] \circ {}^{\bar{\mathrm{m}}}j^*d$, and the right square commutes as well. Thus, in abbreviated notation,

$$\mu = \mathcal{D}\lambda[m].$$

Applying $\mathbf{H}^{c-m}(\cdot)|_V$ to $j^!\mathbf{S}^\bullet \to j^*\mathbf{S}^\bullet$ defines a map $\mathcal{Y}^V \to \mathcal{Y}^{V*}$ (cf. also the discussion preceding the statement of Theorem 8.1.12). Let $\mathcal{X}^V$ denote the kernel. Then we have an exact sequence

$$0 \longrightarrow \mathcal{X}^V \longrightarrow \mathcal{Y}^V \longrightarrow \mathcal{Y}^{V*} \longrightarrow \mathcal{X}^{V*} \longrightarrow 0.$$

The map $\mathcal{Y}^V \to \mathcal{Y}^{V*}$ induces an isomorphism $\zeta_V : \mathcal{Z}^V \xrightarrow{\cong} \mathcal{Z}^{V*}$, where $\mathcal{Z}^V = \mathcal{Y}^V/\mathcal{X}^V$. If $\mathbf{S}^\bullet$ is locally nonsingular, then $\mathcal{Y}^V \to \mathcal{Y}^{V*}$ is an isomorphism, $\mathcal{X}^V = 0$, and $\mathcal{Z}^V = \mathcal{Y}^V$. The composite

$$\begin{array}{ccc} & & \mathbf{IC}^\bullet_{\bar m}(\overline{V}; \mathcal{Y}^{V*}) \\ & \nearrow & \uparrow {\scriptstyle \mu[-c]} \\ {}^{\bar{\mathrm{m}}}j^! \mathbf{S}^\bullet[-c] & \xrightarrow{\text{canon}} & {}^{\bar{\mathrm{m}}}j^* \mathbf{S}^\bullet[-c] \end{array}$$

induces $\mathcal{Y}^V \to \mathcal{Y}^{V*}$ on $\mathbf{H}|_V$. So the composite

$$ {}^{\bar{\mathrm{m}}}j^! \mathbf{S}^\bullet[-c] \longrightarrow \mathbf{IC}^\bullet_{\bar m}(\overline{V}; \mathcal{Y}^{V*}) \longrightarrow \mathbf{IC}^\bullet_{\bar m}(\overline{V}; \mathcal{X}^{V*}) \tag{8.4}$$

is zero, since it induces

$$\begin{array}{ccc} & & \mathcal{X}^{V*} \\ & \overset{0}{\nearrow} & \uparrow \\ \mathcal{Y}^V & \longrightarrow & \mathcal{Y}^{V*} \end{array}$$

on $\mathbf{H}|_V$ (and using Proposition 8.1.10). Note that Ri_* and $\tau_\leq$ are additive functors. This implies that $\mathbf{IC}^\bullet_{\bar m}(\overline{V}; \cdot)$ is an additive functor by Deligne's formula. Now $\mathcal{Y}^V \to \mathcal{Y}^{V*}$ factors through $\mathcal{Z}^V$,

$$\mathcal{Y}^V \overset{\text{quotient}}{\twoheadrightarrow} \mathcal{Z}^V \dashrightarrow \mathcal{Y}^{V*},$$

and we have the exact sequence

$$0 \longrightarrow \mathcal{Z}^V \longrightarrow \mathcal{Y}^{V*} \longrightarrow \mathcal{X}^{V*} \longrightarrow 0,$$

which splits over the field $\mathbb{R}$. So $\mathcal{Y}^{V*} \cong \mathcal{Z}^V \oplus \mathcal{X}^{V*}$, and consequently

$$\mathbf{IC}^\bullet_{\bar m}(\overline{V}; \mathcal{Y}^{V*}) \cong \mathbf{IC}^\bullet_{\bar m}(\overline{V}; \mathcal{Z}^V) \oplus \mathbf{IC}^\bullet_{\bar m}(\overline{V}; \mathcal{X}^{V*}).$$

Triangles of the form

$$\mathbf{X}^\bullet \overset{\text{incl}}{\longrightarrow} \mathbf{X}^\bullet \oplus \mathbf{Y}^\bullet \overset{\text{proj}}{\longrightarrow} \mathbf{Y}^\bullet \overset{+1}{\longrightarrow}$$

are distinguished (see also the first bullet in the proof of Theorem 7.1.11). This way, we obtain a distinguished triangle

$$\begin{array}{ccc} \mathbf{IC}^\bullet_{\bar m}(\overline{V}; \mathcal{Z}^V) & \longrightarrow & \mathbf{IC}^\bullet_{\bar m}(\overline{V}; \mathcal{Y}^{V*}) \\ {\scriptstyle [1]}\nwarrow & & \swarrow \\ & \mathbf{IC}^\bullet_{\bar m}(\overline{V}; \mathcal{X}^{V*}) & \end{array}$$

(the morphisms are induced uniquely by the maps over V). Consider the induced exact sequence

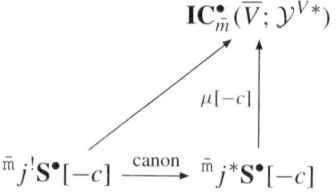

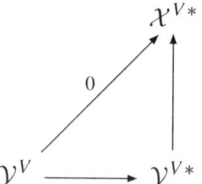

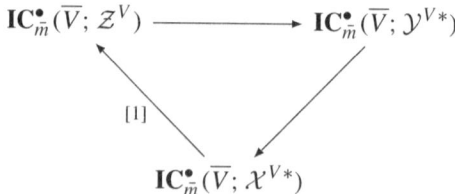

$$
\begin{array}{l}
\operatorname{Hom}({}^{\bar{\mathrm{m}}}j^!\mathbf{S}^\bullet[-c], \mathbf{IC}^\bullet_{\bar{m}}(\mathcal{Z}^V)) \\
\quad \to \operatorname{Hom}({}^{\bar{\mathrm{m}}}j^!\mathbf{S}^\bullet[-c], \mathbf{IC}^\bullet_{\bar{m}}(\mathcal{Y}^{V*})) \to \operatorname{Hom}({}^{\bar{\mathrm{m}}}j^!\mathbf{S}^\bullet[-c], \mathbf{IC}^\bullet_{\bar{m}}(\mathcal{X}^{V*})). \\
\qquad\qquad \mu[-c] \circ \text{canon} \qquad\qquad \mapsto \qquad\qquad 0
\end{array}
$$

The morphism $\mu[-c] \circ$ canon maps to zero, since the composition (8.4) vanishes. Therefore, there exists

$$\lambda_1 : {}^{\bar{\mathrm{m}}}j^!\mathbf{S}^\bullet \longrightarrow \mathbf{IC}^\bullet_{\bar{m}}(\overline{V}; \mathcal{Z}^V)[c],$$

which is the unique morphism inducing the quotient map $\mathcal{Y}^V \twoheadrightarrow \mathcal{Z}^V$ on $\mathbf{H}^{c-m}(\cdot)|_V$. Similarly, there exists a unique morphism

$$\mu_1 : \mathbf{IC}^\bullet_{\bar{m}}(\overline{V}; \mathcal{Z}^{V*})[c] \longrightarrow {}^{\bar{\mathrm{m}}}j^*\mathbf{S}^\bullet$$

inducing the inclusion $\mathcal{Z}^{V*} \hookrightarrow \mathcal{Y}^{V*}$. Using $\mu = \mathcal{D}\lambda[m]$, we have by uniqueness

$$\mu_1 = \mathcal{D}\lambda_1[m].$$

Define $\mathbf{X}^\bullet_1$, $\mathbf{Z}^\bullet_1$, as well as u_1, v_1, by distinguished triangles

$$
\begin{array}{ccc}
\mathbf{X}^\bullet_1 & \xrightarrow{\;u_1\;} & {}^{\bar{\mathrm{m}}}j^!\mathbf{S}^\bullet \\
 & {\scriptstyle [1]}\nwarrow \quad \swarrow{\scriptstyle \lambda_1} & \\
 & \mathbf{IC}^\bullet_{\bar{m}}(\overline{V}; \mathcal{Z}^V)[c] &
\end{array}
\qquad
\begin{array}{ccc}
\mathbf{IC}^\bullet_{\bar{m}}(\overline{V}; \mathcal{Z}^{V*})[c] & \xrightarrow{\;\mu_1\;} & {}^{\bar{\mathrm{m}}}j^*\mathbf{S}^\bullet \\
 & {\scriptstyle [1]}\nwarrow \quad \swarrow{\scriptstyle v_1} & \\
 & \mathbf{Z}^\bullet_1 &
\end{array}
$$

It follows that $\mathbf{Z}^\bullet_1 \cong \mathcal{D}\mathbf{X}^\bullet_1[m]$ and $v_1 = \mathcal{D}u_1[m]$.

We claim that setting $\mathbf{X}^\bullet = j_*\mathbf{X}^\bullet_1$ and $\mathbf{Z}^\bullet = j_*\mathbf{Z}^\bullet_1$ yields perverse sheaves $\mathbf{X}^\bullet, \mathbf{Z}^\bullet \in {}^{\bar{\mathrm{m}}}P(Y)$: The restriction to V of the composition

$$(\lambda_1 \circ \lambda)[-c] : \mathbf{IC}^\bullet_{\bar{m}}(\overline{V}; \mathcal{Y}^V) \longrightarrow {}^{\bar{\mathrm{m}}}j^!\mathbf{S}^\bullet[-c] \longrightarrow \mathbf{IC}^\bullet_{\bar{m}}(\overline{V}; \mathcal{Z}^V)$$

induces $\mathcal{Y}^V \twoheadrightarrow \mathcal{Z}^V$ which splits back. The backwards map $\mathcal{Z}^V \to \mathcal{Y}^V$ has a unique extension to a morphism $\mathbf{IC}^\bullet_{\bar{m}}(\overline{V}; \mathcal{Z}^V) \to \mathbf{IC}^\bullet_{\bar{m}}(\overline{V}; \mathcal{Y}^V)$ (by Proposition 8.1.10), whose composition with $(\lambda_1 \circ \lambda)[-c]$ is the identity. Thus, $(\lambda_1 \circ \lambda)[-c]$ induces surjections on all derived stalks and costalks. Therefore, λ_1 induces surjections on all derived stalks and costalks. Then the left triangle above implies that u_1 induces injections on all derived stalks and costalks. This means that $\mathbf{X}^\bullet_1$ satisfies at least the stalk and costalk axioms that ${}^{\bar{\mathrm{m}}}j^!\mathbf{S}^\bullet$ does, and the latter sheaf is perverse. (In fact, $\mathbf{X}^\bullet_1[-c]$ satisfies the strong $\bar{m}$-costalk condition, i.e. $H^i(j^!_y\mathbf{X}^\bullet_1[-c]) = 0$ for $i \le \bar{m}(k) - k + 1$ and $y \in \overline{V}_{m-2c-k} - \overline{V}_{m-2c-k-1}$.) As j_* is just extension by zero, we have $j_*\mathbf{X}^\bullet_1 \in {}^{\bar{\mathrm{m}}}P(Y)$. The statement for $\mathbf{Z}^\bullet$ follows from duality, $\mathbf{Z}^\bullet_1 \cong \mathcal{D}\mathbf{X}^\bullet_1[m]$. This establishes the claim.

The isomorphism $\zeta_V : \mathcal{Z}^V \to \mathcal{Z}^{V*}$ induces an isomorphism of intersection chain sheaves $\zeta_{\overline{V}} : \mathbf{IC}^\bullet_{\bar{m}}(\overline{V}; \mathcal{Z}^V) \to \mathbf{IC}^\bullet_{\bar{m}}(\overline{V}; \mathcal{Z}^{V*})$. Using Proposition 8.1.10 with the weak costalk assumption, the square

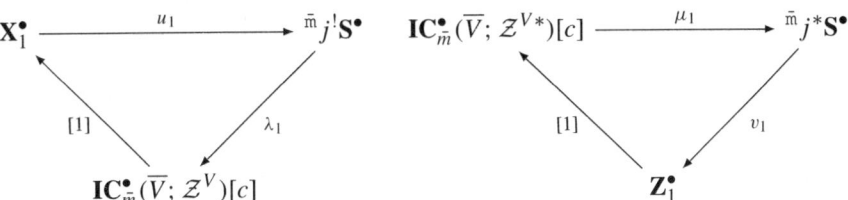

$$\begin{array}{ccc} {}^{\bar{\mathrm{m}}}j^!\mathbf{S}^\bullet & \xrightarrow{\lambda_1} & \mathbf{IC}^\bullet_{\bar{m}}(\overline{V};\mathcal{Z}^V)[c] \\ \Big\downarrow \text{canon} & & \cong \Big\downarrow \zeta_{\overline{V}} \\ {}^{\bar{\mathrm{m}}}j^*\mathbf{S}^\bullet & \xleftarrow{\mu_1} & \mathbf{IC}^\bullet_{\bar{m}}(\overline{V};\mathcal{Z}^{V*})[c] \end{array} \tag{8.5}$$

commutes. Consider the diagram

$$\begin{array}{ccccc} \mathbf{X}^\bullet_1 & \xrightarrow{u_1} & {}^{\bar{\mathrm{m}}}j^!\mathbf{S}^\bullet & \xrightarrow{\lambda_1} & \mathbf{IC}^\bullet_{\bar{m}}(\overline{V};\mathcal{Z}^V)[c] \\ & & \Big\downarrow \text{canon} & & \Big\downarrow \cong \\ \mathbf{Z}^\bullet_1 & \xleftarrow{v_1} & {}^{\bar{\mathrm{m}}}j^*\mathbf{S}^\bullet & \xleftarrow{\mu_1} & \mathbf{IC}^\bullet_{\bar{m}}(\overline{V};\mathcal{Z}^{V*})[c] \end{array}$$

Note that $\lambda_1 u_1 = 0$, since they are adjacent arrows in a distinguished triangle (Proposition 2.2.2). This shows that the composite

$$\mathbf{X}^\bullet_1 \xrightarrow{u_1} {}^{\bar{\mathrm{m}}}j^!\mathbf{S}^\bullet \xrightarrow{\text{canon}} {}^{\bar{\mathrm{m}}}j^*\mathbf{S}^\bullet \xrightarrow{v_1} \mathbf{Z}^\bullet_1 \tag{8.6}$$

is zero.

If $\mathbf{S}^\bullet$ is locally nonsingular, then set $\mathbf{S}^\bullet_0 = \mathbf{S}^\bullet$. Otherwise, we shall now set up an elementary bordism from $\mathbf{S}^\bullet$ to a self-dual sheaf $\mathbf{S}^\bullet_0$ which splits off an intersection chain sheaf. Precisely, it will be shown to have the following properties:

- $\mathbf{S}^\bullet_0|_{Y-\overline{V}} \cong \mathbf{S}^\bullet|_{Y-\overline{V}}$,
- $\mathbf{S}^\bullet_0 \cong j_*\mathbf{IC}^\bullet_{\bar{m}}(\overline{V};\mathcal{Z}^V)[c] \oplus \widehat{\mathbf{S}}^\bullet$,
- $\widehat{\mathbf{S}}^\bullet$ is perverse on Y, and satisfies $\mathbf{H}^{c-m}(j^!_W\widehat{\mathbf{S}}^\bullet) = 0$ for every stratum W of Y such that $c(W) > c(V)$, *as well as for* $W = V$.

The composition $j_*{}^{\bar{\mathrm{m}}}\tau^{\leq 0}j^! \to j_*j^! \to 1$ defines an adjunction $j_*{}^{\bar{\mathrm{m}}}j^!\mathbf{S}^\bullet \to \mathbf{S}^\bullet$. Similarly, we have a morphism $\mathbf{S}^\bullet \to j_*{}^{\bar{\mathrm{m}}}j^*\mathbf{S}^\bullet$. Let u be the composition

$$\mathbf{X}^\bullet = j_*\mathbf{X}^\bullet_1 \xrightarrow{j_*u_1} j_*{}^{\bar{\mathrm{m}}}j^!\mathbf{S}^\bullet \longrightarrow \mathbf{S}^\bullet,$$

and let v be the composition

$$\mathbf{S}^\bullet \longrightarrow j_*{}^{\bar{\mathrm{m}}}j^*\mathbf{S}^\bullet \xrightarrow{j_*v_1} j_*\mathbf{Z}^\bullet_1 = \mathbf{Z}^\bullet.$$

To verify that the diagram

$$\mathbf{X}^\bullet \xrightarrow{u} \mathbf{S}^\bullet \xrightarrow{v} \mathbf{Z}^\bullet$$

defines a bordism, we have to check that $vu = 0$ and that $v = \mathcal{D}u[m]$. Now $vu = 0$ follows from (8.6), since $j_*{}^{\bar{\mathrm{m}}}j^!\mathbf{S}^\bullet \to \mathbf{S}^\bullet \to j_*{}^{\bar{\mathrm{m}}}j^*\mathbf{S}^\bullet$ is j_* of ${}^{\bar{\mathrm{m}}}j^!\mathbf{S}^\bullet \to {}^{\bar{\mathrm{m}}}j^*\mathbf{S}^\bullet$. From $\mathbf{Z}^\bullet_1 \cong \mathcal{D}\mathbf{X}^\bullet_1[m]$, $v_1 = \mathcal{D}u_1[m]$, we conclude $\mathbf{Z}^\bullet \cong \mathcal{D}\mathbf{X}^\bullet[m]$, $v = \mathcal{D}u[m]$ (recall $j_* = j_!$). Hence, we have constructed an elementary bordism from $\mathbf{S}^\bullet$ to $\mathbf{S}^\bullet_0 = \mathbf{C}^\bullet_{u,v}$.

Let us verify $\mathbf{S}_0^\bullet|_{Y-\overline{V}} \cong \mathbf{S}^\bullet|_{Y-\overline{V}}$: Since $\mathbf{Z}^\bullet = j_* \mathbf{Z}_1^\bullet$, its restriction away from $\overline{V}$ is the zero sheaf. Thus the triangle on $v|_{Y-\overline{V}}$ has the form

$$\begin{array}{ccc} \mathbf{S}^\bullet|_{Y-\overline{V}} & \xrightarrow{\;v|=0\;} & \mathbf{Z}^\bullet|_{Y-\overline{V}} = \mathbf{0}^\bullet \\ {\scriptstyle [1]}\nwarrow \cong & & \swarrow {\scriptstyle 0} \\ & \mathbf{C}_v^\bullet|_{Y-\overline{V}} & \end{array}$$

The zero morphism $\mathbf{X}^\bullet|_{Y-\overline{V}} \stackrel{0}{\longrightarrow} \mathbf{C}_v^\bullet[-1]|_{Y-\overline{V}}$ is the lift $u'|$ of $u| = 0 : \mathbf{X}^\bullet|_{Y-\overline{V}} = \mathbf{0}^\bullet \to \mathbf{S}^\bullet|_{Y-\overline{V}}$. Consequently, the triangle on $u'|$ has the form

$$\begin{array}{ccc} \mathbf{0}^\bullet = \mathbf{X}^\bullet|_{Y-\overline{V}} & \xrightarrow{\;u'|=0\;} & \mathbf{C}_v^\bullet[-1]|_{Y-\overline{V}} \\ {\scriptstyle [1]}\nwarrow {\scriptstyle 0} & & \swarrow \cong \\ & \mathbf{S}_0^\bullet|_{Y-\overline{V}} & \end{array}$$

and

$$\mathbf{S}_0^\bullet|_{Y-\overline{V}} \cong \mathbf{C}_v^\bullet[-1]|_{Y-\overline{V}} \cong \mathbf{S}^\bullet|_{Y-\overline{V}}.$$

The sheaf $\mathbf{S}_0^\bullet$ is $\bar{\mathrm{m}}$-perverse on Y: Since $\mathbf{S}_0^\bullet$ is bordant to $\mathbf{S}^\bullet$, it is automatically self-dual. Therefore, it suffices to verify $\mathbf{S}_0^\bullet \in {}^{\bar{\mathrm{m}}}D^{\leq 0}$. Consider the distinguished triangle

$$\mathbf{IC}_{\bar{m}}^\bullet(\overline{V}; \mathcal{Z}^V)[c] \longrightarrow \mathbf{X}_1^\bullet[1] \longrightarrow ({}^{\bar{\mathrm{m}}}j^!\mathbf{S}^\bullet)[1] \stackrel{+1}{\longrightarrow} .$$

We have $\mathbf{IC}_{\bar{m}}^\bullet(\overline{V}; \mathcal{Z}^V)[c] \in {}^{\bar{\mathrm{m}}}D^{\leq 0}(\overline{V})$ and

$$({}^{\bar{\mathrm{m}}}j^!\mathbf{S}^\bullet)[1] \in {}^{\bar{\mathrm{m}}}D^{\leq 0}(\overline{V})[1] = {}^{\bar{\mathrm{m}}}D^{\leq -1}(\overline{V}) \subset {}^{\bar{\mathrm{m}}}D^{\leq 0}(\overline{V}).$$

By Proposition 7.1.8, $\mathbf{X}_1^\bullet[1] \in {}^{\bar{\mathrm{m}}}D^{\leq 0}(\overline{V})$, whence $\mathbf{X}^\bullet[1] = j_*\mathbf{X}_1^\bullet[1] \in {}^{\bar{\mathrm{m}}}D^{\leq 0}$. Using the triangle

$$\mathbf{IC}_{\bar{m}}^\bullet(\overline{V}; \mathcal{Z}^{V*})[c] \longrightarrow {}^{\bar{\mathrm{m}}}j^*\mathbf{S}^\bullet \longrightarrow \mathbf{Z}_1^\bullet \stackrel{+1}{\longrightarrow},$$

one deduces similarly that $\mathbf{Z}^\bullet[-1] \in {}^{\bar{\mathrm{m}}}D^{\leq 0}$. Now $\mathbf{C}_v^\bullet[-1] \in {}^{\bar{\mathrm{m}}}D^{\leq 0}$ follows from the triangle

$$\mathbf{Z}^\bullet[-1] \longrightarrow \mathbf{C}_v^\bullet[-1] \longrightarrow \mathbf{S}^\bullet \stackrel{+1}{\longrightarrow},$$

using $\mathbf{Z}^\bullet[-1], \mathbf{S}^\bullet \in {}^{\bar{\mathrm{m}}}D^{\leq 0}$. Finally, we conclude that $\mathbf{S}_0^\bullet \in {}^{\bar{\mathrm{m}}}D^{\leq 0}$ from the triangle

$$\mathbf{C}_v^\bullet[-1] \longrightarrow \mathbf{S}_0^\bullet \longrightarrow \mathbf{X}^\bullet[1] \stackrel{+1}{\longrightarrow} .$$

Next, we shall produce splittings for the perverse restrictions ${}^{\bar{\mathrm{m}}}j^*\mathbf{S}_0^\bullet$ and ${}^{\bar{\mathrm{m}}}j^!\mathbf{S}_0^\bullet$ of $\mathbf{S}_0^\bullet$. Since $j_*{}^{\bar{\mathrm{m}}}j^!\mathbf{S}_0^\bullet$ is $\bar{\mathrm{m}}$-perverse on Y, the restriction to V satisfies the stalk condition

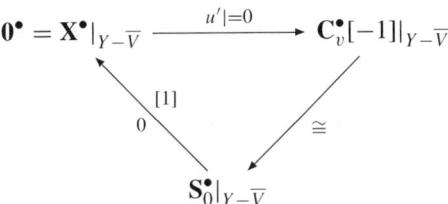

$$\mathbf{H}^i({}^{\bar{\mathrm{m}}}j^!\mathbf{S}_0^\bullet)|_V = \mathbf{H}^i(j_V^* j_*{}^{\bar{\mathrm{m}}}j^!\mathbf{S}_0^\bullet) = 0, \quad i > c - m,$$

and the costalk condition

$$\mathbf{H}^i(j_V^! j_*{}^{\bar{\mathrm{m}}}j^!\mathbf{S}_0^\bullet) = 0, \quad i < c - m.$$

The latter condition is equivalent to $\mathbf{H}^i({}^{\bar{\mathrm{m}}}j^!\mathbf{S}_0^\bullet)|_V = 0$, $i < c-m$, as, with $\bar{\jmath}_V : V \hookrightarrow \overline{V}$ the open inclusion,

$$j_V^! j_* = \bar{\jmath}_V^! j^! j_* = \bar{\jmath}_V^! = \bar{\jmath}_V^*,$$

using property (G5) of Sect. 7.2.1. It follows that $j_V^*{}^{\bar{\mathrm{m}}}j^!\mathbf{S}_0^\bullet$ is concentrated in degree $c - m$. By construction, this local system is $\mathcal{Z}^V$,

$$\mathbf{H}^{c-m}(j_V^*{}^{\bar{\mathrm{m}}}j^!\mathbf{S}_0^\bullet) = \mathcal{Z}^V, \quad j_V^*{}^{\bar{\mathrm{m}}}j^!\mathbf{S}_0^\bullet \cong \mathcal{Z}^V[m-c].$$

Similarly, $j_V^*{}^{\bar{\mathrm{m}}}j^*\mathbf{S}_0^\bullet$ is concentrated in degree $c - m$ and

$$\mathbf{H}^{c-m}(j_V^*{}^{\bar{\mathrm{m}}}j^*\mathbf{S}_0^\bullet) = \mathcal{Z}^{V*}, \quad j_V^*{}^{\bar{\mathrm{m}}}j^*\mathbf{S}_0^\bullet \cong \mathcal{Z}^{V*}[m-c].$$

A morphism

$$\lambda_{10} : {}^{\bar{\mathrm{m}}}j^!\mathbf{S}_0^\bullet \longrightarrow \mathbf{IC}_{\bar{m}}^\bullet(\overline{V}; \mathcal{Z}^V)[c],$$

is obtained in the same way as we obtained λ_1 earlier. It is the unique morphism inducing the identity map $\mathcal{Z}^V \to \mathcal{Z}^V$ on $\mathbf{H}^{c-m}(\cdot)|_V$. Thus $j_V^*\lambda_{10}$ is an isomorphism. Similarly, there exists a unique morphism

$$\mu_{10} : \mathbf{IC}_{\bar{m}}^\bullet(\overline{V}; \mathcal{Z}^{V*})[c] \longrightarrow {}^{\bar{\mathrm{m}}}j^*\mathbf{S}_0^\bullet$$

inducing the identity $\mathcal{Z}^{V*} \to \mathcal{Z}^{V*}$. The restriction $j_V^*\mu_{10}$ is an isomorphism. The commutative diagram

$$\begin{array}{ccc}
{}^{\bar{\mathrm{m}}}j^!\mathbf{S}_0^\bullet & \xrightarrow{\lambda_{10}} & \mathbf{IC}_{\bar{m}}^\bullet(\overline{V}; \mathcal{Z}^V)[c] \\
\Big\downarrow \text{canon} & & \cong \Big\downarrow \zeta_{\overline{V}} \\
{}^{\bar{\mathrm{m}}}j^*\mathbf{S}_0^\bullet & \xleftarrow{\mu_{10}} & \mathbf{IC}_{\bar{m}}^\bullet(\overline{V}; \mathcal{Z}^{V*})[c]
\end{array}$$

(which is the analog of diagram (8.5) above) shows that the canonical map

$$\mathbf{H}^{c-m}({}^{\bar{\mathrm{m}}}j^!\mathbf{S}_0^\bullet)|_V \longrightarrow \mathbf{H}^{c-m}({}^{\bar{\mathrm{m}}}j^*\mathbf{S}_0^\bullet)|_V$$

is the isomorphism $\zeta_V : \mathcal{Z}^V \to \mathcal{Z}^{V*}$. Note that if $\mathbf{S}^\bullet$ is locally nonsingular, then this canonical map is already an isomorphism for $\mathbf{S}_0^\bullet = \mathbf{S}^\bullet$, so that no bordism is required. Define $\mathbf{X}_{10}^\bullet$ and u_{10} by the distinguished triangle

$$\begin{array}{ccc}
\mathbf{X}_{10}^\bullet & \xrightarrow{u_{10}} & {}^{\bar{\mathrm{m}}}j^!\mathbf{S}_0^\bullet \\
 & {}_{[1]}\nwarrow \quad \swarrow_{\lambda_{10}} & \\
 & \mathbf{IC}_{\bar{m}}^\bullet(\overline{V}; \mathcal{Z}^V)[c] &
\end{array} \tag{8.7}$$

Applying j_V^* to this triangle, we obtain

$$\begin{array}{ccc} j_V^*\mathbf{X}_{10}^\bullet & \longrightarrow & \mathcal{Z}^V[m-c] \\ {\scriptstyle [1]}\nwarrow & & \swarrow{\scriptstyle =} \\ & \mathcal{Z}^V[m-c] & \end{array}$$

This implies that $j_V^*\mathbf{X}_{10}^\bullet = \mathbf{0}^\bullet$. As pointed out above, $\mathbf{X}_1^\bullet[-c]$ satisfies the strong $\bar{m}$-costalk condition. So also $\mathbf{X}_{10}^\bullet[-c]$ satisfies the strong costalk condition, and $\mathbf{X}_{10}^\bullet[1-c]$ satisfies the weak costalk condition. By Proposition 8.1.10, restriction to the top stratum V of $\overline{V}$ is an injection

$$\mathrm{Hom}_{D^b(\overline{V})}(\mathbf{IC}^\bullet_{\bar{m}}(\overline{V};\mathcal{Z}^V), \mathbf{X}_{10}^\bullet[1-c]) \hookrightarrow \mathrm{Hom}_{D^b(V)}(\mathbf{IC}^\bullet_{\bar{m}}(\overline{V};\mathcal{Z}^V)|_V, \mathbf{X}_{10}^\bullet[1-c]|_V)$$

$$= \mathrm{Hom}_{D^b(V)}(\mathcal{Z}^V[m-2c], \mathbf{0}^\bullet) = 0.$$

Hence the morphism

$$\mathbf{IC}^\bullet_{\bar{m}}(\overline{V};\mathcal{Z}^V)[c] \longrightarrow \mathbf{X}_{10}^\bullet[1]$$

in triangle (8.7) is zero. By Lemma 8.1.14,

$${}^{\bar{\mathrm{m}}}j^!\mathbf{S}_0^\bullet \cong \mathbf{IC}^\bullet_{\bar{m}}(\overline{V};\mathcal{Z}^V)[c] \oplus \mathbf{X}_{10}^\bullet.$$

Similarly,

$${}^{\bar{\mathrm{m}}}j^*\mathbf{S}_0^\bullet \cong \mathbf{IC}^\bullet_{\bar{m}}(\overline{V};\mathcal{Z}^{V*})[c] \oplus \mathbf{Z}_{10}^\bullet.$$

We shall now use these splittings to obtain the required splitting for $\mathbf{S}_0^\bullet$ itself. Let ψ_0, ω_0 denote the canonical morphisms

$$j^*\mathbf{S}_0^\bullet \xrightarrow{\psi_0} {}^{\bar{\mathrm{m}}}j^*\mathbf{S}_0^\bullet, \qquad {}^{\bar{\mathrm{m}}}j^!\mathbf{S}_0^\bullet \xrightarrow{\omega_0} j^!\mathbf{S}_0^\bullet,$$

and let

$$j_*j^!\mathbf{S}_0^\bullet \xrightarrow{\mathrm{adj}^!} \mathbf{S}_0^\bullet \xrightarrow{\mathrm{adj}^*} j_*j^*\mathbf{S}_0^\bullet$$

be the canonical adjunction morphisms, forming a factorization of j_*c, where $c : j^!\mathbf{S}_0^\bullet \to j^*\mathbf{S}_0^\bullet$ is the canonical morphism. Let

$$\kappa : \mathbf{S}_0^\bullet \longrightarrow j_*\mathbf{IC}^\bullet_{\bar{m}}(\overline{V};\mathcal{Z}^{V*})[c]$$

be the composition

$$\mathbf{S}_0^\bullet \xrightarrow{\mathrm{adj}^*} j_*j^*\mathbf{S}_0^\bullet \xrightarrow{j_*\psi_0} j_*{}^{\bar{\mathrm{m}}}j^*\mathbf{S}_0^\bullet \cong j_*\mathbf{IC}^\bullet_{\bar{m}}(\overline{V};\mathcal{Z}^{V*})[c] \oplus j_*\mathbf{Z}_{10}^\bullet \xrightarrow{\mathrm{proj}} j_*\mathbf{IC}^\bullet_{\bar{m}}(\overline{V};\mathcal{Z}^{V*})[c].$$

We will construct a backwards map $\kappa' : j_*\mathbf{IC}^\bullet_{\bar{m}}(\overline{V};\mathcal{Z}^{V*})[c] \to \mathbf{S}_0^\bullet$, $\kappa\kappa' = 1$. The canonical morphism $c_0 : {}^{\bar{\mathrm{m}}}j^!\mathbf{S}_0^\bullet \to {}^{\bar{\mathrm{m}}}j^*\mathbf{S}_0^\bullet$ splits as

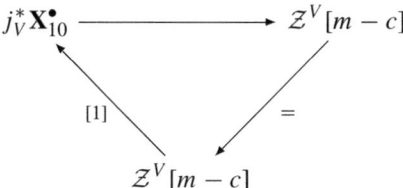

$$\begin{array}{ccc}
{}^{\bar{\mathrm{m}}}j^*\mathbf{S}_0^\bullet & \cong & \mathbf{IC}_{\bar m}^\bullet(\overline{V};\,\mathcal{Z}^{V*})[c] \oplus \mathbf{Z}_{10}^\bullet \\
\Big\uparrow{\scriptstyle c_0} & & \Big\uparrow{\scriptstyle (\zeta_{\overline{V}},0)} \\
{}^{\bar{\mathrm{m}}}j^!\mathbf{S}_0^\bullet & \cong & \mathbf{IC}_{\bar m}^\bullet(\overline{V};\,\mathcal{Z}^{V})[c] \oplus \mathbf{X}_{10}^\bullet
\end{array}$$

(by Proposition 8.1.10). Extension by zero induces the commutative square

$$\begin{array}{ccc}
j_*{}^{\bar{\mathrm{m}}}j^*\mathbf{S}_0^\bullet & \cong & j_*\mathbf{IC}_{\bar m}^\bullet(\overline{V};\,\mathcal{Z}^{V*})[c] \oplus j_*\mathbf{Z}_{10}^\bullet \\
\Big\uparrow{\scriptstyle j_*c_0} & & \Big\uparrow{\scriptstyle (j_*\zeta_{\overline{V}},0)} \\
j_*{}^{\bar{\mathrm{m}}}j^!\mathbf{S}_0^\bullet & \cong & j_*\mathbf{IC}_{\bar m}^\bullet(\overline{V};\,\mathcal{Z}^{V})[c] \oplus j_*\mathbf{X}_{10}^\bullet
\end{array}$$

The diagram

$$\begin{array}{ccc}
j_*\mathbf{IC}_{\bar m}^\bullet(\overline{V};\,\mathcal{Z}^{V*})[c] \oplus j_*\mathbf{Z}_{10}^\bullet & \xrightarrow{\text{proj}} & j_*\mathbf{IC}_{\bar m}^\bullet(\overline{V};\,\mathcal{Z}^{V*})[c] \\
\Big\uparrow{\scriptstyle (j_*\zeta_{\overline{V}},0)} & & \Big\uparrow{\scriptstyle j_*\zeta_{\overline{V}}} \\
j_*\mathbf{IC}_{\bar m}^\bullet(\overline{V};\,\mathcal{Z}^{V})[c] \oplus j_*\mathbf{X}_{10}^\bullet & \xleftarrow{\text{incl}} & j_*\mathbf{IC}_{\bar m}^\bullet(\overline{V};\,\mathcal{Z}^{V})[c]
\end{array}$$

commutes, and

$${}^{\bar{\mathrm{m}}}j^!\mathbf{S}_0^\bullet \xrightarrow{\omega_0} j^!\mathbf{S}_0^\bullet \xrightarrow{c} j^*\mathbf{S}_0^\bullet \xrightarrow{\psi_0} {}^{\bar{\mathrm{m}}}j^*\mathbf{S}_0^\bullet$$

is a factorization of c_0. Thus with

$$\kappa' = \mathrm{adj}^! \circ j_*\omega_0 \circ \mathrm{incl} \circ j_*\zeta_{\overline{V}}^{-1} : j_*\mathbf{IC}_{\bar m}^\bullet(\overline{V};\,\mathcal{Z}^{V*})[c] \longrightarrow \mathbf{S}_0^\bullet$$

we have

$$\begin{aligned}
\kappa\kappa' &= \mathrm{proj} \circ j_*\psi_0 \circ \mathrm{adj}^* \circ \mathrm{adj}^! \circ j_*\omega_0 \circ \mathrm{incl} \circ j_*\zeta_{\overline{V}}^{-1} \\
&= \mathrm{proj} \circ j_*\psi_0 \circ j_*c \circ j_*\omega_0 \circ \mathrm{incl} \circ j_*\zeta_{\overline{V}}^{-1} \\
&= \mathrm{proj} \circ j_*(\psi_0 c \omega_0) \circ \mathrm{incl} \circ j_*\zeta_{\overline{V}}^{-1} \\
&= \mathrm{proj} \circ j_*c_0 \circ \mathrm{incl} \circ j_*\zeta_{\overline{V}}^{-1} \\
&= \mathrm{proj} \circ (j_*\zeta_{\overline{V}}, 0) \circ \mathrm{incl} \circ j_*\zeta_{\overline{V}}^{-1} \\
&= 1.
\end{aligned}$$

Define $\widehat{\mathbf{S}}^\bullet[1]$ as the mapping cone of $j_*\zeta_{\overline{V}}^{-1}\kappa$, so that there is a distinguished triangle

$$\begin{array}{ccc}
\mathbf{S}_0^\bullet & \xrightarrow{\;j_*\zeta_{\overline{V}}^{-1}\kappa\;} & j_*\mathbf{IC}_{\bar m}^\bullet(\overline{V};\,\mathcal{Z}^{V})[c] \\
{\scriptstyle [1]}\nwarrow & & \swarrow{\scriptstyle \eta} \\
& \widehat{\mathbf{S}}^\bullet[1] &
\end{array} \tag{8.8}$$

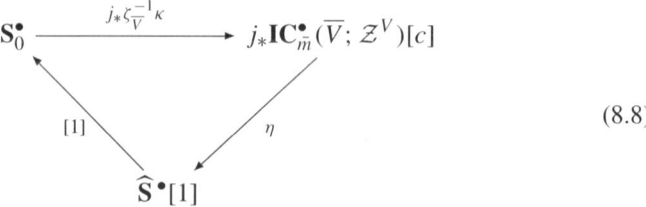

Using the backwards map, we see that η vanishes:

$$\eta = \eta \circ j_*\zeta_{\overline{V}}^{-1}\kappa \circ \kappa' j_*\zeta_{\overline{V}} = 0 \circ \kappa' j_*\zeta_{\overline{V}} = 0.$$

Thus, by Lemma 8.1.14,

$$\mathbf{S}_0^\bullet \cong j_*\mathbf{IC}_{\bar{m}}^\bullet(\overline{V};\, \mathcal{Z}^V)[c] \oplus \widehat{\mathbf{S}}^\bullet.$$

If $\mathbf{S}^\bullet$ is locally nonsingular, then

$$\mathbf{S}^\bullet \cong j_*\mathbf{IC}_{\bar{m}}^\bullet(\overline{V};\, \mathcal{Z}^V)[c] \oplus \widehat{\mathbf{S}}^\bullet.$$

The sheaf $\widehat{\mathbf{S}}^\bullet$ is $\bar{\mathrm{m}}$-perverse on Y, since $\mathbf{S}_0^\bullet$ is $\bar{\mathrm{m}}$-perverse on Y. The long exact sequence of derived sheaves associated to the triangle (8.8) shows that $\mathbf{H}^{c-m}(j_V^!\widehat{\mathbf{S}}^\bullet) = 0$.

Inductively, one obtains thus an orthogonal decomposition

$$\mathbf{S}^\bullet \sim \bigoplus_{V \in \mathcal{V}} j_*\mathbf{IC}_{\bar{m}}^\bullet(\overline{V};\, \mathcal{Z}^V(\mathbf{S}^\bullet))[c(V)] \oplus \mathbf{S}_1^\bullet,$$

such that $\mathbf{S}_1^\bullet$ is $\bar{\mathrm{m}}$-perverse and $\mathbf{H}^{c(V)-m}(j_V^!\mathbf{S}_1^\bullet) = 0$ for *all* $V \in \mathcal{V}$, that is, $\mathbf{S}_1^\bullet$ satisfies the stalk and strong costalk axioms for $\bar{\mathrm{m}}$. Since the top stratum $V = Y - Y_{m-2}$ is open in Y, we have $j_V^!\mathbf{S}_1^\bullet \cong j_V^*\mathbf{S}_1^\bullet$, and therefore

$$\mathbf{H}^i(j_V^*\mathbf{S}_1^\bullet) = \begin{cases} 0, & i > -m, \text{ since } \mathbf{S}_1^\bullet \text{ is perverse,} \\ \mathbf{H}^{c-m}(j_V^*\mathbf{S}_1^\bullet) = \mathbf{H}^{c-m}(j_V^!\mathbf{S}_1^\bullet) = 0, & i = c - m = -m, \\ 0, & i < -m, \text{ since } \mathbf{S}_1^\bullet \text{ is perverse.} \end{cases}$$

Hence $\mathbf{S}_1^\bullet|_{Y-Y_{m-2}} = \mathbf{0}^\bullet$, and, by Proposition 8.1.10, $\mathbf{S}_1^\bullet = \mathbf{0}^\bullet$. □

To complete the proof of Theorem 8.1.12, it remains to reduce the general decomposition problem to the decomposition problem for perverse sheaves. This is carried out in the next proposition.

Proposition 8.1.17 *Any self-dual complex of sheaves* $\mathbf{S}^\bullet \in D_c^b(Y)$ *is bordant to a perverse sheaf* $\mathbf{S}_1^\bullet \in {}^{\bar{\mathrm{m}}}P(Y)$ *with*

$$(\mathcal{Z}^V(\mathbf{S}^\bullet), \mathcal{P}_V(\mathbf{S}^\bullet)) \cong (\mathcal{Z}^V(\mathbf{S}_1^\bullet), \mathcal{P}_V(\mathbf{S}_1^\bullet)).$$

Proof. Set $\mathbf{S}_1^\bullet = {}^{\bar{\mathrm{m}}}H^0(\mathbf{S}^\bullet)$, $d_1 = {}^{\bar{\mathrm{m}}}H^0(d)$. Define $\mathbf{E}^\bullet, \mathbf{F}^\bullet$ as the mapping cones of the canonical morphisms $c_1 : {}^{\bar{\mathrm{m}}}\tau^{\leq 0}\mathbf{S}^\bullet \to \mathbf{S}^\bullet$, $c_2 : \mathbf{S}^\bullet \to {}^{\bar{\mathrm{m}}}\tau^{\geq 0}\mathbf{S}^\bullet$, respectively, so that there are two distinguished triangles

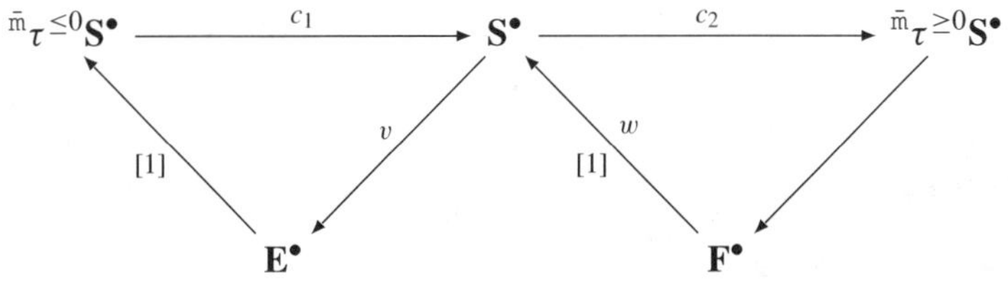

($\mathbf{E}^\bullet \in {}^{\bar{m}}D^{\geq 1}(Y)$). We claim that

$$\mathbf{F}^\bullet[-1] \xrightarrow{u=w[-1]} \mathbf{S}^\bullet \xrightarrow{v} \mathbf{E}^\bullet$$

sets up an elementary bordism on $\mathbf{S}^\bullet$: One has to verify that v is the dual of u and that $vu = 0$. The commutative diagram

$$\begin{array}{ccc} {}^{\bar{m}}\tau^{\leq 0}\mathbf{S}^\bullet & \xrightarrow{c_1} & \mathbf{S}^\bullet \\ \downarrow \cong & & \cong \downarrow d \\ \mathcal{D}({}^{\bar{m}}\tau^{\geq 0}\mathbf{S}^\bullet)[m] & \xrightarrow{\mathcal{D}c_2[m]} & \mathcal{D}\mathbf{S}^\bullet[m] \end{array}$$

implies $\mathbf{E}^\bullet \cong \mathcal{D}(\mathbf{F}^\bullet[-1])[m]$ and $v = \mathcal{D}u[m]$. As $\mathbf{F}^\bullet[-1] \in {}^{\bar{m}}D^{\leq -1} \subset {}^{\bar{m}}D^{\leq 0}$, one has $\operatorname{Hom}(\mathbf{F}^\bullet[-1], \mathbf{E}^\bullet) = 0$, by axiom (2) of the axioms of a t-structure (cf. Definition 7.1.1). Hence $vu = 0$ and $\mathbf{S}^\bullet$ is bordant to $\mathbf{C}^\bullet_{u,v}$. Now

$$\mathbf{C}^\bullet_{u,v} = {}^{\bar{m}}\tau^{\geq 0}\,{}^{\bar{m}}\tau^{\leq 0}\mathbf{S}^\bullet = {}^{\bar{m}}H^0(\mathbf{S}^\bullet) = \mathbf{S}^\bullet_1.$$

The stated isomorphism of local systems and their pairings clearly holds. □

Proposition 8.1.16 together with Proposition 8.1.17 prove Theorem 8.1.12.

Let us now return to the situation of a stratified map $f : X^p \to Y^m$, where X and Y are oriented, Whitney stratified pseudomanifolds, X is compact, Y has only strata of even codimension and $p - m$ is even. Set $t = \frac{1}{2}(p - m)$. Given a self-dual sheaf $\mathbf{S}^\bullet \in D^b_c(X)$, the pushforward $Rf_*\mathbf{S}^\bullet[-t]$ is again self-dual on Y, since f is a proper map and thus $Rf_! \cong Rf_*$ (cf. Sect. 3.1, (3.2)). Applying Theorem 8.1.12 to $Rf_*\mathbf{S}^\bullet[-t]$, we obtain the following result:

Theorem 8.1.18 *Let $f : X^p \to Y^m$ be a stratified map, where X and Y are oriented, Whitney stratified pseudomanifolds, X is compact, Y has only strata of even codimension and $p - m$ is even. Set $t = \frac{1}{2}(p - m)$. If $\mathbf{S}^\bullet \in D^b_c(X)$ is any self-dual complex of sheaves on X, then*

$$Rf_*\mathbf{S}^\bullet[-t] \sim \bigoplus_{V \in \mathcal{V}} j_*\mathbf{IC}^\bullet_{\bar{m}}(\overline{V}; \mathcal{Z}^V(Rf_*\mathbf{S}^\bullet[-t]))[c(V)].$$

It remains to provide a more useful and direct description of the local systems $\mathcal{Z}^V(Rf_*\mathbf{S}^\bullet[-t])$. Recall that for a point $y \in V \in \mathcal{V}$, we have defined the pseudomanifold

$$E_y = f^{-1}N(y) \cup_{f^{-1}L(y)} cf^{-1}L(y)$$

to be the space obtained from the preimage of the normal slice at y by coning off the boundary. (If y is a point in the top stratum, we set $E_y = f^{-1}(y)$.) Furthermore, using the inclusions

$$E_y \xleftarrow{i_y} f^{-1}\overset{\circ}{N}(y) \xrightarrow{\rho_y} X,$$

we have defined a self-dual sheaf

$$\mathbf{S}^\bullet(y) = \tau^{\{cone\}}_{\leq -c-t-1} Ri_{y*}\rho^!_y \mathbf{S}^\bullet$$

on E_y. Then

$$\mathcal{Z}^V(Rf_*\mathbf{S}^\bullet[-t])_y \cong \mathcal{H}^{-c-t}(E_y; \mathbf{S}^\bullet(y)),$$

and indeed $\mathcal{Z}^V(Rf_*\mathbf{S}^\bullet[-t]) \cong \mathfrak{H}_V(\mathbf{S}^\bullet)$. Moreover, the nonsingular pairings

$$\mathcal{P}_V(Rf_*\mathbf{S}^\bullet[-t])_y : \mathcal{Z}^V(Rf_*\mathbf{S}^\bullet[-t])_y \otimes \mathcal{Z}^V(Rf_*\mathbf{S}^\bullet[-t])_y \longrightarrow \mathbb{R}$$

coincide under the above isomorphism with the nonsingular pairings

$$\mathcal{H}^{-c-t}(E_y; \mathbf{S}^\bullet(y)) \otimes \mathcal{H}^{-c-t}(E_y; \mathbf{S}^\bullet(y)) \longrightarrow \mathbb{R}$$

induced on hypercohomology by the self-duality isomorphism $\mathcal{D}\mathbf{S}^\bullet(y)[\dim E_y] \cong \mathbf{S}^\bullet(y)$. This completes the proof of Theorem 8.1.7:

$$Rf_*\mathbf{S}^\bullet[-t] \sim \bigoplus_{V \in \mathcal{V}} j_*\mathbf{IC}^\bullet_{\bar{m}}(\overline{V}; \mathfrak{H}_V(\mathbf{S}^\bullet))[c].$$

8.2 The L-Class of Self-Dual Sheaves

8.2.1 Algebraic Bordism of Self-Dual Sheaves and the Witt Group

Let Y^m, $m = 2n$, be an oriented, stratified pseudomanifold, and let $\mathbf{S}^\bullet \in D^b_c(Y)$ be a self-dual sheaf. We wish to develop the notion "$\mathbf{S}^\bullet$ is an (algebraic) boundary."

Suppose that $\mathbf{S}^\bullet$ is bordant to $\mathbf{0}^\bullet$. This means that we have a diagram

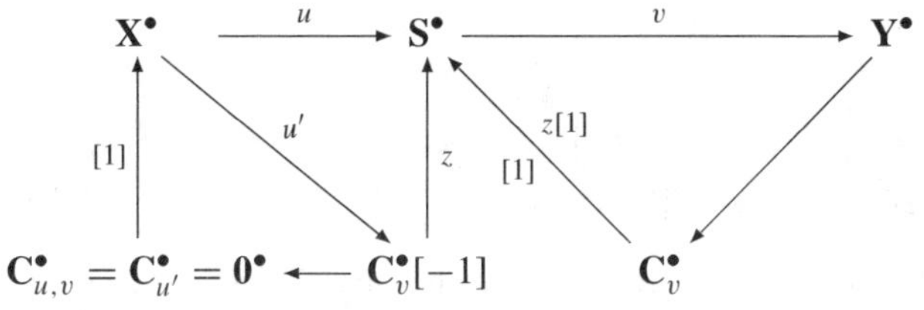

containing two distinguished triangles, one on the lift u' and one on v; $vu = 0$, $\mathcal{D}u[m] = v$, $zu' = u$. Thus u' is an isomorphism,

$$\mathbf{C}^\bullet_v[-1] \cong \mathbf{X}^\bullet \cong \mathcal{D}\mathbf{Y}^\bullet[m],$$

and there is a distinguished triangle

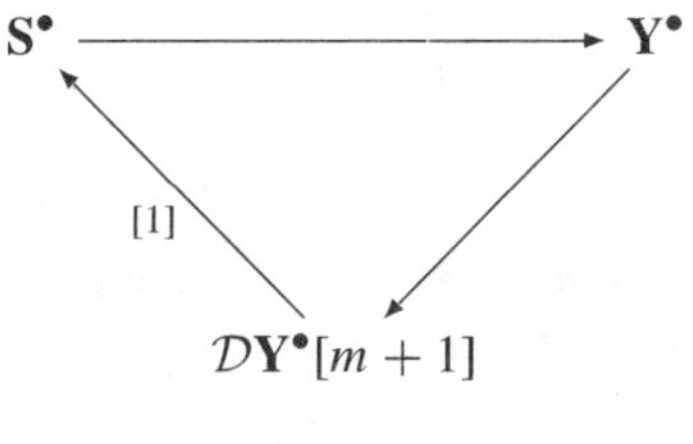

After turning this once and writing $\mathbf{B}^\bullet = \mathbf{Y}^\bullet[-1]$, it becomes

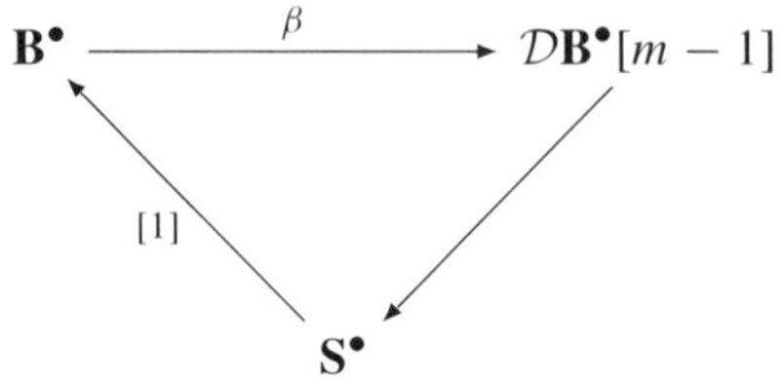

with $\mathcal{D}\beta[m-1] = \beta$.

Conversely, suppose a morphism $\beta : \mathbf{B}^\bullet \to \mathcal{D}\mathbf{B}^\bullet[m-1]$ is given such that $\mathcal{D}\beta[m-1] = \beta$. Embed β into a distinguished triangle

$$\mathbf{B}^\bullet \xrightarrow{\beta} \mathcal{D}\mathbf{B}^\bullet[m-1] \xrightarrow{u} \mathbf{S}^\bullet_\beta \xrightarrow{v} \mathbf{B}^\bullet[1],$$

dualize

$$\mathcal{D}\mathbf{S}^\bullet_\beta \xrightarrow{\mathcal{D}u} \mathcal{D}(\mathcal{D}\mathbf{B}^\bullet[m-1]) \xrightarrow{\mathcal{D}\beta} \mathcal{D}\mathbf{B}^\bullet \xrightarrow{\mathcal{D}v[1]} \mathcal{D}\mathbf{S}^\bullet_\beta[1],$$

turn

$$\mathcal{D}\mathcal{D}\mathbf{B}^\bullet[1-m] \xrightarrow{\mathcal{D}\beta} \mathcal{D}\mathbf{B}^\bullet \xrightarrow{\mathcal{D}v[1]} \mathcal{D}\mathbf{S}^\bullet_\beta[1] \xrightarrow{-\mathcal{D}u[1]} \mathcal{D}\mathcal{D}\mathbf{B}^\bullet[2-m],$$

shift by $[m-1]$

$$\mathcal{D}\mathcal{D}\mathbf{B}^\bullet \xrightarrow{\mathcal{D}\beta[m-1]} \mathcal{D}\mathbf{B}^\bullet[m-1] \xrightarrow{\mathcal{D}v[m]} \mathcal{D}\mathbf{S}^\bullet_\beta[m] \xrightarrow{(-1)^{m-1}(-\mathcal{D}u[1])[m-1]} \mathcal{D}\mathcal{D}\mathbf{B}^\bullet[1].$$

Compare this to the original triangle:

$$\begin{array}{ccccccc}
\mathbf{B}^\bullet & \xrightarrow{\beta} & \mathcal{D}\mathbf{B}^\bullet[m-1] & \xrightarrow{u} & \mathbf{S}^\bullet_\beta & \xrightarrow{v} & \mathbf{B}^\bullet[1] \\
{\scriptstyle \mathrm{can}}\downarrow\cong & & \| & & {\scriptstyle d_\beta}\downarrow\cong & & \cong\downarrow{\scriptstyle \mathrm{can}[1]} \\
\mathcal{D}\mathcal{D}\mathbf{B}^\bullet & \xrightarrow{\mathcal{D}\beta[m-1]} & \mathcal{D}\mathbf{B}^\bullet[m-1] & \xrightarrow{\mathcal{D}v[m]} & \mathcal{D}\mathbf{S}^\bullet_\beta[m] & \xrightarrow{\mathcal{D}u[m]} & \mathcal{D}\mathcal{D}\mathbf{B}^\bullet[1]
\end{array}$$

(Note that the leftmost square commutes because $\mathcal{D}\beta[m-1] = \beta$.) This yields an isomorphism $d_\beta : \mathbf{S}^\bullet_\beta \cong \mathcal{D}\mathbf{S}^\bullet_\beta[m]$. Thus $\mathbf{S}^\bullet_\beta$ is self-dual and obviously nullbordant:

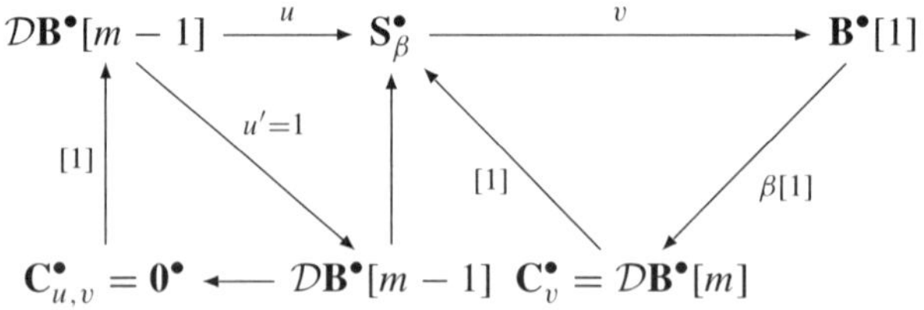

This leads us in a natural way to the following definition:

Definition 8.2.1 *A self-dual sheaf* $(\mathbf{S}^\bullet, d)$ *is the* boundary of $\beta : \mathbf{B}^\bullet \to \mathcal{D}\mathbf{B}^\bullet[m-1]$, $\mathcal{D}\beta[m-1] = \beta$, *if* $(\mathbf{S}^\bullet, d) \cong (\mathbf{S}^\bullet_\beta, d_\beta)$.

This notion will again play an important role in Sect. 9.1, when we define Lagrangian structures on non-Witt spaces.

Lemma 8.2.2 *If* $\mathbf{S}^\bullet \sim \mathbf{S}_1^\bullet$ *and* $\mathbf{S}^\bullet$ *is a boundary, then* $\mathbf{S}_1^\bullet$ *is a boundary.*

We leave the easy proof to the reader.

Now assume that Y is compact. The self-duality isomorphism d of a self-dual sheaf $(\mathbf{S}^\bullet, d)$ on Y induces

$$\mathcal{H}^{-n}(Y; \mathbf{S}^\bullet) \cong \mathcal{H}^n(Y; \mathcal{D}\mathbf{S}^\bullet) \cong \mathrm{Hom}(\mathcal{H}^{-n}(Y; \mathbf{S}^\bullet), \mathbb{R}),$$

i.e. a nonsingular, bilinear pairing

$$Q_{\mathbf{S}^\bullet} : \mathcal{H}^{-n}(Y; \mathbf{S}^\bullet) \otimes \mathcal{H}^{-n}(Y; \mathbf{S}^\bullet) \longrightarrow \mathbb{R}.$$

Proposition 8.2.3 *Let* $W(\mathbb{R})$ *denote the Witt group of symmetric (skew-symmetric) bilinear forms over* $\mathbb{R}$*. If* $\mathbf{S}^\bullet$ *is a boundary, then*

$$Q_{\mathbf{S}^\bullet} = 0 \in W(\mathbb{R}).$$

Proof. Suppose that $\mathbf{S}^\bullet$ is the boundary of β so that there is a distinguished triangle

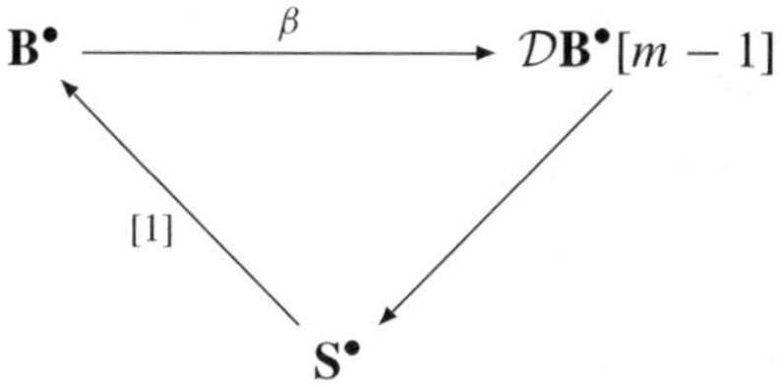

As in the proofs of Theorems 6.1.2 and 6.1.4, we shall establish the statement by constructing a Lagrangian subspace in $\mathcal{H}^{-n}(Y; \mathbf{S}^\bullet)$. The composition

$$\begin{aligned}\mathcal{H}^{-n+1}(Y; \mathbf{B}^\bullet) &\cong \mathcal{H}^{-n+1}(Y; \mathcal{D}\mathcal{D}\mathbf{B}^\bullet) \cong \mathrm{Hom}(\mathcal{H}^{n-1}(Y; \mathcal{D}\mathbf{B}^\bullet), \mathbb{R})\\ &= \mathrm{Hom}(\mathcal{H}^{n-1-(m-1)}(Y; \mathcal{D}\mathbf{B}^\bullet[m-1]), \mathbb{R})\\ &= \mathrm{Hom}(\mathcal{H}^{-n}(Y; \mathcal{D}\mathbf{B}^\bullet[m-1]), \mathbb{R})\end{aligned}$$

defines a nonsingular pairing

$$\mathcal{H}^{-n+1}(Y; \mathbf{B}^\bullet) \otimes \mathcal{H}^{-n}(Y; \mathbf{A}^\bullet) \longrightarrow \mathbb{R},$$

where $\mathbf{A}^\bullet = \mathcal{D}\mathbf{B}^\bullet[m-1]$. Similarly, there is a pairing

$$\mathcal{H}^{-n}(Y; \mathbf{B}^\bullet) \otimes \mathcal{H}^{-n+1}(Y; \mathbf{A}^\bullet) \longrightarrow \mathbb{R},$$

so that there is a commuting diagram of pairings

$$\begin{array}{ccccccccc}
\mathcal{H}^{-n}(Y; \mathbf{B}^\bullet) & \longrightarrow & \mathcal{H}^{-n}(Y; \mathbf{A}^\bullet) & \longrightarrow & \mathcal{H}^{-n}(Y; \mathbf{S}^\bullet) & \longrightarrow & \mathcal{H}^{-n+1}(Y; \mathbf{B}^\bullet) & \longrightarrow & \mathcal{H}^{-n+1}(Y; \mathbf{A}^\bullet)\\
\otimes & & \otimes & & \otimes & & \otimes & & \otimes\\
\mathcal{H}^{-n+1}(Y; \mathbf{A}^\bullet) & \longleftarrow & \mathcal{H}^{-n+1}(Y; \mathbf{B}^\bullet) & \longleftarrow & \mathcal{H}^{-n}(Y; \mathbf{S}^\bullet) & \longleftarrow & \mathcal{H}^{-n}(Y; \mathbf{A}^\bullet) & \longleftarrow & \mathcal{H}^{-n}(Y; \mathbf{B}^\bullet)\\
\downarrow & & \downarrow & & \downarrow Q_{\mathbf{S}^\bullet} & & \downarrow & & \downarrow\\
\mathbb{R} & & \mathbb{R} & & \mathbb{R} & & \mathbb{R} & & \mathbb{R}
\end{array}$$

in which the exact rows are induced by the triangle. Thus $\mathrm{im}(\mathcal{H}^{-n}(Y; \mathbf{A}^\bullet) \to \mathcal{H}^{-n}(Y; \mathbf{S}^\bullet))$ is a Lagrangian subspace for $Q_{\mathbf{S}^\bullet}$. □

Proposition 8.2.4 *If* $\mathbf{S}_1^\bullet$ *and* $\mathbf{S}_2^\bullet$ *are bordant, then* $Q_{\mathbf{S}_1^\bullet} = Q_{\mathbf{S}_2^\bullet} \in W(\mathbb{R})$.

Proof. On the one hand, $(\mathbf{S}_2^\bullet, d_2) \oplus (\mathbf{S}_2^\bullet, -d_2)$ is the boundary of the zero morphism $\mathbf{S}_2^\bullet \xrightarrow{0} \mathcal{D}\mathbf{S}_2^\bullet[m-1]$, on the other hand, $(\mathbf{S}_2^\bullet, d_2) \oplus (\mathbf{S}_2^\bullet, -d_2)$ is bordant to $(\mathbf{S}_1^\bullet, d_1) \oplus (\mathbf{S}_2^\bullet, -d_2)$. By Lemma 8.2.2, $(\mathbf{S}_1^\bullet, d_1) \oplus (\mathbf{S}_2^\bullet, -d_2)$ is a boundary. By Proposition 8.2.3,

$$Q_{\mathbf{S}_1^\bullet} \oplus -Q_{\mathbf{S}_2^\bullet} = Q_{\mathbf{S}_1^\bullet \oplus -\mathbf{S}_2^\bullet} = 0 \in W(\mathbb{R}).$$ □

Corollary 8.2.5 *If* $\mathbf{S}_1^\bullet$ *and* $\mathbf{S}_2^\bullet$ *are bordant, then* $\sigma(\mathbf{S}_1^\bullet) = \sigma(\mathbf{S}_2^\bullet)$.

8.2.2 Geometric Bordism of Spaces Covered by Sheaves

Consider a triple

$$(Y^{n+1}, \mathbf{B}^\bullet, \delta),$$

where

- Y^{n+1} is an $(n+1)$-dimensional, compact, oriented pseudomanifold with boundary (see Definition 6.1.3),
- $\mathbf{B}^\bullet \in D_c^b(\operatorname{int} Y)$,
- $\delta : \mathcal{D}\mathbf{B}^\bullet[n+1] \xrightarrow{\cong} \mathbf{B}^\bullet$.

Let i, j denote the inclusions $i : \operatorname{int} Y \hookrightarrow Y$, $j : \partial Y \hookrightarrow Y$.

Lemma 8.2.6 *If* $\mathbf{B}^\bullet \in D_c^b(\operatorname{int} Y)$, *then there is a natural isomorphism*

$$j^* Ri_* \mathbf{B}^\bullet \xrightarrow{\cong} j^! Ri_! \mathbf{B}^\bullet[1].$$

Proof. Consider the distinguished triangle ($\mathbf{A}^\bullet \in D_c^b(Y)$)

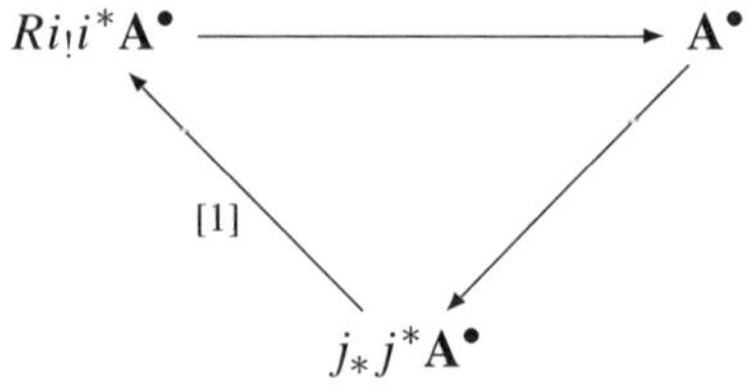

(cf. gluing data axiom (G4) in Sect. 7.2.1. To construct it, take the triangle

$$j_* j^! \mathbf{A}^\bullet \longrightarrow \mathbf{A}^\bullet \longrightarrow Ri_* i^* \mathbf{A}^\bullet \xrightarrow{[1]}$$

constructed in our discussion of the attaching axiom (AX3) in Sect. 4.1.4, and apply the Verdier dualizing functor $\mathcal{D}$ to it.) Apply j^*:

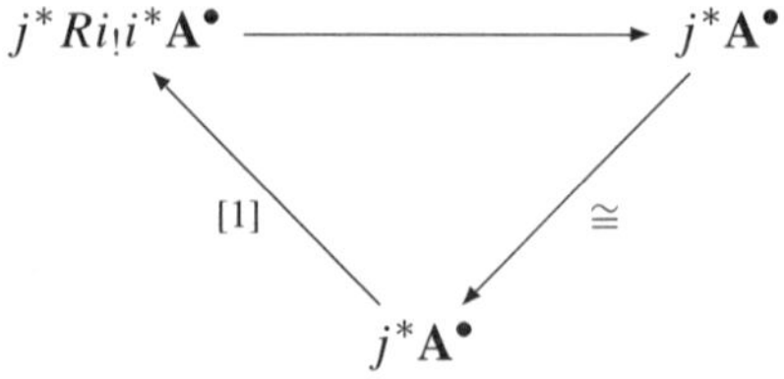

It follows that $j^*Ri_!i^*\mathbf{A}^\bullet \cong \mathbf{0}^\bullet$. Substituting $\mathbf{A}^\bullet = Ri_*\mathbf{B}^\bullet$ and using $i^*Ri_*\mathbf{B}^\bullet \cong \mathbf{B}^\bullet$, we see

$$j^*Ri_!\mathbf{B}^\bullet \cong \mathbf{0}^\bullet.$$

In the triangle

$$\begin{array}{ccc} j^!\mathbf{A}^\bullet & \longrightarrow & j^*\mathbf{A}^\bullet \\ {\scriptstyle [1]}\nwarrow & & \swarrow \\ & j^*Ri_*i^*\mathbf{A}^\bullet & \end{array}$$

set $\mathbf{A}^\bullet = Ri_!\mathbf{B}^\bullet$. We obtain

$$\begin{array}{ccc} j^!Ri_!\mathbf{B}^\bullet & \longrightarrow & j^*Ri_!\mathbf{B}^\bullet \cong \mathbf{0}^\bullet \\ {\scriptstyle [1]}\nwarrow & & \swarrow \\ & j^*Ri_*i^*Ri_!\mathbf{B}^\bullet \cong j^*Ri_*\mathbf{B}^\bullet & \end{array}$$

and hence $j^*Ri_*\mathbf{B}^\bullet \xrightarrow{\cong} j^!Ri_!\mathbf{B}^\bullet[1]$. □

An application of the functor j^*Ri_* to δ produces an induced isomorphism

$$j^*Ri_*(\delta) : j^*Ri_*\mathcal{D}\mathbf{B}^\bullet[n+1] \xrightarrow{\cong} j^*Ri_*\mathbf{B}^\bullet.$$

We have

$$j^*Ri_*\mathcal{D}\mathbf{B}^\bullet[n+1] \cong \mathcal{D}(j^!Ri_!\mathbf{B}^\bullet)[n+1]$$

and according to Lemma 8.2.6,

$$j^*Ri_*\mathbf{B}^\bullet \cong j^!Ri_!\mathbf{B}^\bullet[1].$$

Hence δ induces a self-duality isomorphism d for $j^!Ri_!\mathbf{B}^\bullet$:

$$d : \mathcal{D}(j^!Ri_!\mathbf{B}^\bullet)[n] \xrightarrow{\cong} j^!Ri_!\mathbf{B}^\bullet.$$

Definition 8.2.7 *We call the triple* $(\partial Y, j^!Ri_!\mathbf{B}^\bullet, d)$ *the* boundary of $(Y^{n+1}, \mathbf{B}^\bullet, \delta)$ *and write*

$$\partial(Y^{n+1}, \mathbf{B}^\bullet, \delta) = (\partial Y, j^!Ri_!\mathbf{B}^\bullet, d).$$

Given a triple $(X^{2s}, \mathbf{A}^\bullet, d)$, where

- X^{2s} is a $2s$-dimensional, closed, oriented pseudomanifold,
- $\mathbf{A}^\bullet \in D_c^b(X)$,
- $d : \mathcal{D}\mathbf{A}^\bullet[2s] \xrightarrow{\cong} \mathbf{A}^\bullet$,

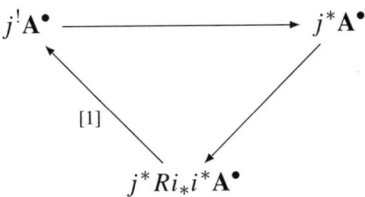

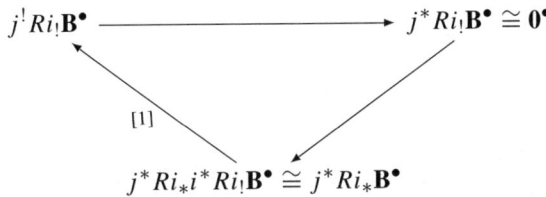

d induces a nonsingular pairing

$$\mathcal{H}^{-s}(X;\mathbf{A}^\bullet)\otimes\mathcal{H}^{-s}(X;\mathbf{A}^\bullet)\longrightarrow\mathbb{R}.$$

Let $\sigma(X,\mathbf{A}^\bullet,d)$ denote the signature of this pairing and set $\sigma(X^n,\mathbf{A}^\bullet,d)=0$ for n odd.

Proposition 8.2.8 *If* $(X^n,\mathbf{A}^\bullet,d)=\partial(Y^{n+1},\mathbf{B}^\bullet,\delta)$, $n=2s$, *then*

$$\sigma(X,\mathbf{A}^\bullet,d)=0.$$

Proof. Let $\mathbf{E}^\bullet=Ri_!\mathbf{B}^\bullet$ and consider the canonical morphism $Ri_*i^*\mathbf{E}^\bullet\to j_*j^!\mathbf{E}^\bullet[1]$. We will show that the image

$$\operatorname{im}(\mathcal{H}^{-s-1}(Y;Ri_*i^*\mathbf{E}^\bullet)\to\mathcal{H}^{-s}(Y;j_*j^!\mathbf{E}^\bullet))$$

is a Lagrangian subspace in $\mathcal{H}^{-s}(Y;j_*j^!\mathbf{E}^\bullet)\cong\mathcal{H}^{-s}(\partial Y;j^!\mathbf{E}^\bullet)$. We have

$$\delta:\mathcal{D}\mathbf{B}^\bullet[n+1]\xrightarrow{\cong}\mathbf{B}^\bullet$$

and

$$\partial\delta=d:\mathcal{D}(j^!\mathbf{E}^\bullet)[n]\xrightarrow{\cong}j^!\mathbf{E}^\bullet.$$

Since $j_*j^!\to 1$ is a natural transformation of functors, we have a commutative diagram

$$\begin{array}{ccc}
j_*j^!(Ri_!\mathbf{B}^\bullet) & \longrightarrow & Ri_!\mathbf{B}^\bullet\\
{\scriptstyle j_*j^!Ri_!\delta}\uparrow & & \uparrow{\scriptstyle Ri_!\delta}\\
j_*j^!(Ri_!\mathcal{D}\mathbf{B}^\bullet[n+1]) & \longrightarrow & Ri_!\mathcal{D}\mathbf{B}^\bullet[n+1]
\end{array}$$

Now

$$Ri_!\mathcal{D}\mathbf{B}^\bullet[n+1]\cong Ri_!\mathcal{D}(i^*\mathbf{E}^\bullet)[n+1]\cong Ri_!i^*\mathcal{D}\mathbf{E}^\bullet[n+1]$$

and using Lemma 8.2.6,

$$j^!Ri_!\mathcal{D}\mathbf{B}^\bullet[n+1]\cong\mathcal{D}(j^*Ri_*\mathbf{B}^\bullet)[n+1]\cong\mathcal{D}(j^!Ri_!\mathbf{B}^\bullet[1])[n+1]\cong\mathcal{D}(j^!Ri_!\mathbf{B}^\bullet)[n].$$

Therefore, we get a commutative diagram

$$\begin{array}{ccc}
j_*j^!\mathbf{E}^\bullet & \longrightarrow & \mathbf{E}^\bullet\\
{\scriptstyle j_*\partial\delta}\uparrow & & \uparrow{\scriptstyle Ri_!\delta}\\
j_*j^*\mathcal{D}\mathbf{E}^\bullet[n] & \longrightarrow & Ri_!i^*\mathcal{D}\mathbf{E}^\bullet[n+1]
\end{array}\tag{8.9}$$

We show that the given data give rise to an isomorphism between the triangle

$$\begin{array}{ccc}
j_*j^!\mathbf{E}^\bullet & \longrightarrow & \mathbf{E}^\bullet\\
& {\scriptstyle [1]}\nwarrow \quad \swarrow & \\
& Ri_*i^*\mathbf{E}^\bullet &
\end{array}$$

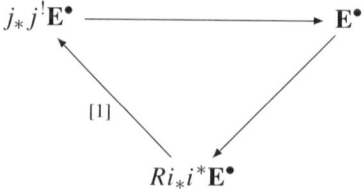

and the triangle

$$\begin{array}{ccc} Ri_!i^*\mathcal{D}\mathbf{E}^\bullet[n+1] & \longrightarrow & \mathcal{D}\mathbf{E}^\bullet[n+1] \\ {\scriptstyle [1]}\nwarrow & & \swarrow \\ & j_*j^*\mathcal{D}\mathbf{E}^\bullet[n+1] & \end{array}$$

Consider the composition

$$\mathcal{D}\mathbf{E}^\bullet[n+1] \cong \mathcal{D}(Ri_!\mathbf{B}^\bullet)[n+1] \cong Ri_*\mathcal{D}\mathbf{B}^\bullet[n+1] \cong Ri_*\mathbf{B}^\bullet \cong Ri_*i^*\mathbf{E}^\bullet$$

and the composition

$$Ri_!i^*\mathcal{D}\mathbf{E}^\bullet[n+1] \cong Ri_!\mathcal{D}(i^*\mathbf{E}^\bullet)[n+1] \cong Ri_!\mathcal{D}\mathbf{B}^\bullet[n+1] \cong Ri_!\mathbf{B}^\bullet \cong \mathbf{E}^\bullet.$$

These fit into a commutative square

$$\begin{array}{ccc} \mathbf{E}^\bullet & \longrightarrow & Ri_*i^*\mathbf{E}^\bullet \\ \cong\uparrow & & \uparrow\cong \\ Ri_!i^*\mathcal{D}\mathbf{E}^\bullet[n+1] & \longrightarrow & \mathcal{D}\mathbf{E}^\bullet[n+1] \end{array}$$

Applying the dualizing functor to diagram (8.9) one obtains

$$\begin{array}{ccc} \mathcal{D}\mathbf{E}^\bullet & \longrightarrow & j_*j^*\mathcal{D}\mathbf{E}^\bullet \\ \cong\downarrow & & \downarrow\cong \\ Ri_*i^*\mathbf{E}^\bullet[-n-1] & \longrightarrow & j_*j^!\mathbf{E}^\bullet[-n] \end{array}$$

Shifting the square by $n+1$ yields

$$\begin{array}{ccc} \mathcal{D}\mathbf{E}^\bullet[n+1] & \longrightarrow & j_*j^*\mathcal{D}\mathbf{E}^\bullet[n+1] \\ \cong\downarrow & & \downarrow\cong \\ Ri_*i^*\mathbf{E}^\bullet & \longrightarrow & j_*j^!\mathbf{E}^\bullet[1] \end{array}$$

Putting our squares together, we obtain the isomorphism between the two triangles:

$$\begin{array}{ccccccc} j_*j^!\mathbf{E}^\bullet & \longrightarrow & \mathbf{E}^\bullet & \longrightarrow & Ri_*i^*\mathbf{E}^\bullet & \longrightarrow & j_*j^!\mathbf{E}^\bullet[1] \\ \uparrow\cong & & \uparrow\cong & & \uparrow\cong & & \uparrow\cong \\ j_*j^*\mathcal{D}\mathbf{E}^\bullet[n] & \longrightarrow & Ri_!i^*\mathcal{D}\mathbf{E}^\bullet[n+1] & \longrightarrow & \mathcal{D}\mathbf{E}^\bullet[n+1] & \longrightarrow & j_*j^*\mathcal{D}\mathbf{E}^\bullet[n+1] \end{array}$$

Taking hypercohomology, we compute using Borel–Moore duality

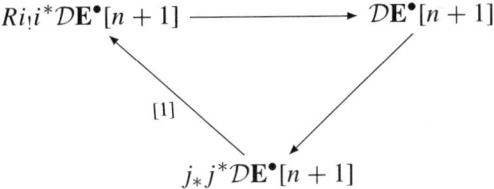

$$\begin{aligned}\mathcal{H}^k(j_*j^!\mathbf{E}^\bullet) &\cong \mathcal{H}^k(j_*j^*\mathcal{D}\mathbf{E}^\bullet[n])\\ &\cong \mathcal{H}^{k+n}(\mathcal{D}(j_*j^!\mathbf{E}^\bullet)) \cong \mathrm{Hom}(\mathcal{H}^{-k-n}(j_*j^!\mathbf{E}^\bullet), \mathbb{R}),\end{aligned}$$

$$\begin{aligned}\mathcal{H}^k(\mathbf{E}^\bullet) &\cong \mathcal{H}^k(Ri_!i^*\mathcal{D}\mathbf{E}^\bullet[n+1]) \cong \mathcal{H}^{k+n+1}(Ri_!i^*\mathcal{D}\mathbf{E}^\bullet)\\ &\cong \mathcal{H}^{k+n+1}(\mathcal{D}(Ri_*i^*\mathbf{E}^\bullet)) \cong \mathrm{Hom}(\mathcal{H}^{-k-n-1}(Ri_*i^*\mathbf{E}^\bullet), \mathbb{R}),\end{aligned}$$

and

$$\begin{aligned}\mathcal{H}^k(Ri_*i^*\mathbf{E}^\bullet) &\cong \mathcal{H}^k(\mathcal{D}\mathbf{E}^\bullet[n+1]) \cong \mathcal{H}^{k+n+1}(\mathcal{D}\mathbf{E}^\bullet)\\ &\cong \mathrm{Hom}(\mathcal{H}^{-k-n-1}(\mathbf{E}^\bullet), \mathbb{R}).\end{aligned}$$

For $k = -s$ we have $-k-n = -s$. Therefore, the long exact hypercohomology sequence associated with the first triangle is dually paired to itself:

$$\begin{array}{cccc} \to \mathcal{H}^{-s-1}(\mathbf{E}^\bullet) & \to \mathcal{H}^{-s-1}(Ri_*i^*\mathbf{E}^\bullet) & \to \mathcal{H}^{-s}(j_*j^!\mathbf{E}^\bullet) \to & \mathcal{H}^{-s}(\mathbf{E}^\bullet) \\ \otimes & \otimes & \otimes & \otimes \\ \leftarrow \mathcal{H}^{-s}(Ri_*i^*\mathbf{E}^\bullet) \leftarrow & \mathcal{H}^{-s}(\mathbf{E}^\bullet) & \leftarrow \mathcal{H}^{-s}(j_*j^*\mathbf{E}^\bullet) & \leftarrow \mathcal{H}^{-s-1}(Ri_*i^*\mathbf{E}^\bullet) \\ \downarrow & \downarrow & \downarrow & \downarrow \\ \mathbb{R} & \mathbb{R} & \mathbb{R} & \mathbb{R} \end{array}$$

Thus,

$$\mathrm{im}(\mathcal{H}^{-s-1}(Ri_*i^*\mathbf{E}^\bullet) \to \mathcal{H}^{-s}(j_*j^!\mathbf{E}^\bullet))$$

is a Lagrangian subspace in $\mathcal{H}^{-s}(\partial Y; j^!\mathbf{E}^\bullet) \cong \mathcal{H}^{-s}(Y; j_*j^!\mathbf{E}^\bullet)$. □

8.2.3 Construction of L-Classes

Let Y^m be an oriented, closed, Whitney stratified pseudomanifold and let $\mathbf{S}^\bullet \in D_c^b(Y)$ be a self-dual sheaf on Y. We shall construct L-classes $L_k(\mathbf{S}^\bullet) \in H_k(Y; \mathbb{Q})$ employing, as in the construction of the Goresky–MacPherson L-class, the method of Thom and Pontrjagin. Let $f : Y^m \to S^k$ be transverse in the sense of Definition 6.3.1. Then $j_f : f^{-1}(N) \hookrightarrow Y$ is normally nonsingular, so $\mathcal{D}(j_f^!\mathbf{S}^\bullet)[m-k] \cong j_f^!\mathbf{S}^\bullet$ by Lemma 8.1.6. (Recall $N \in S^k$ is the north pole.) If $m-k$ is odd (i.e. $f^{-1}(N)$ is odd-dimensional), set $\sigma(j_f^!\mathbf{S}^\bullet) = 0$. If $m-k = 2l$, we have a middle-dimensional nonsingular pairing

$$\mathcal{H}^{-l}(f^{-1}(N); j_f^!\mathbf{S}^\bullet) \otimes \mathcal{H}^{-l}(f^{-1}(N); j_f^!\mathbf{S}^\bullet) \longrightarrow \mathbb{R},$$

and $\sigma(j_f^!\mathbf{S}^\bullet)$ denotes the signature of this pairing. The assignment

$$\begin{aligned}\sigma : \pi^k(Y) &\longrightarrow \mathbb{Z}\\ [f] &\mapsto \sigma(j_f^!\mathbf{S}^\bullet)\end{aligned}$$

is a well-defined homomorphism ($2k-1 > m$) by Proposition 8.2.8. By Serre's theorem, the Hurewicz map $\pi^k(Y) \otimes \mathbb{Q} \to H^k(Y; \mathbb{Q})$ is an isomorphism for $2k-1 > m$.

Thus, the above homomorphism is a linear functional $\sigma \otimes \mathbb{Q} \in \mathrm{Hom}(H^k(Y), \mathbb{Q}) \cong H_k(Y)$. We define

$$L_k(\mathbf{S}^\bullet) = \sigma \otimes \mathbb{Q} \in H_k(Y; \mathbb{Q}).$$

In order to remove the dimensional restriction, we will proceed as in Sect. 5.7, however, we will need the following fact about pullbacks of self-dual sheaves under projections with nonsingular fiber:

Lemma 8.2.9 *Let M^m be an orientable manifold, X^n an oriented pseudomanifold, $\mathbf{S}^\bullet \in D_c^b(X)$ a self-dual sheaf, and $\pi_1 : X \times M \to X$ the first-factor projection. Then $\pi_1^! \mathbf{S}^\bullet$ is self-dual on $X \times M$,*

$$\mathcal{D}(\pi_1^! \mathbf{S}^\bullet)[n+m] \cong \pi_1^! \mathbf{S}^\bullet.$$

Proof. Let $\pi_2 : X \times M \to M$ denote the second-factor projection. Choose an orientation for M, $\mathbb{D}_M^\bullet[-m] \cong \mathcal{O}_M \cong \mathbb{R}_M$. (Recall $\mathcal{O}_M$ denotes the orientation sheaf.) Then the dualizing complex on the product is the pullback of the dualizing complex on X:

$$\mathbb{D}_{X\times M}^\bullet \cong \pi_1^* \mathbb{D}_X^\bullet \otimes \pi_2^* \mathbb{D}_M^\bullet \cong \pi_1^* \mathbb{D}_X^\bullet \otimes \pi_2^* \mathbb{R}_M[m] \cong \pi_1^* \mathbb{D}_X^\bullet[m].$$

So for any $\mathbf{A}^\bullet \in D_c^b(X)$,

$$\begin{aligned} \pi_1^*(\mathcal{D}\mathbf{A}^\bullet)[m] &\cong \pi_1^* R\mathbf{Hom}^\bullet(\mathbf{A}^\bullet, \mathbb{D}_X^\bullet)[m] \\ &\cong R\mathbf{Hom}^\bullet(\pi_1^*\mathbf{A}^\bullet, \pi_1^*\mathbb{D}_X^\bullet[m]) \\ &\cong R\mathbf{Hom}^\bullet(\pi_1^*\mathbf{A}^\bullet, \mathbb{D}_{X\times M}^\bullet) \\ &= \mathcal{D}(\pi_1^*\mathbf{A}^\bullet) \\ &\cong \pi_1^!(\mathcal{D}\mathbf{A}^\bullet). \end{aligned}$$

There is thus an isomorphism

$$\pi_1^! \cong \pi_1^*[m].$$

It follows that

$$\mathcal{D}(\pi_1^! \mathbf{S}^\bullet)[n+m] \cong \pi_1^* \mathcal{D}\mathbf{S}^\bullet[n][m] \cong \pi_1^* \mathbf{S}^\bullet[m] \cong \pi_1^! \mathbf{S}^\bullet. \qquad \square$$

Form the product $Y \times S^N$, where S^N is a sphere of sufficiently large dimension N. By Lemma 8.2.9, $\pi_1^! \mathbf{S}^\bullet \in D_c^b(Y \times S^N)$ is self-dual, $\pi_1 : Y \times S^N \to Y$. The class

$$L_{k+N}(\pi_1^! \mathbf{S}^\bullet) \in H_{k+N}(Y \times S^N)$$

is defined, since $m + N < 2(k+N) - 1$ for sufficiently large N. Using

$$H_{k+N}(Y \times S^N) \cong H_{k+N}(Y) \oplus H_k(Y) \xrightarrow{p_2} H_k(Y),$$

we define

$$L_k(\mathbf{S}^\bullet) = p_2(L_{k+N}(\pi_1^! \mathbf{S}^\bullet)).$$

This completes the construction of $L_k(\mathbf{S}^\bullet)$ for all $k \geq 0$. Note that $L_k(\mathbf{S}^\bullet) = 0$ for $m - k$ odd.

Let X^n be a closed, oriented, Whitney stratified pseudomanifold and let

$$j : Y^m \hookrightarrow X^n$$

be a normally nonsingular inclusion of an oriented, stratified pseudomanifold Y^m. Consider an open neighborhood $E \subset X$ of Y, the total space of an $\mathbb{R}^{n-m}$-vector bundle over Y, and put $E_0 = E - Y$, the total space with the zero section removed. Let $u \in H^{n-m}(E, E_0)$ denote the Thom class. If $\pi : E \to Y$ denotes the projection, then the composition

$$H_k(X) \xrightarrow{i_*} H_k(X, X - Y) \underset{\cong}{\xleftarrow{e_*}} H_k(E, E_0) \underset{\cong}{\xrightarrow{u\cap -}} H_{k-n+m}(E) \underset{\cong}{\xrightarrow{\pi_*}} H_{k-n+m}(Y)$$

defines a map

$$j^! : H_k(X) \longrightarrow H_{k-n+m}(Y).$$

Let $\iota \in H^{n-m}(S^{n-m})$ denote the generator such that $\langle \iota, [S^{n-m}] \rangle = 1$.

Lemma 8.2.10 *Let X^n be a closed, oriented, Whitney stratified pseudomanifold, and let $f : X \to S^{n-m}$ be transverse to $N \in S^{n-m}$. Then*

$$f^*(\iota) \cap x = j_{f*} j_f^!(x)$$

for any $x \in H_*(X)$, where $j_f : f^{-1}(N) \hookrightarrow X$.

Proof. Set $Y = f^{-1}(N)$, let $x \in H_k(X)$, and put $u' = (e^*)^{-1}(u)$, where $e^* : H^{n-m}(X, X-Y) \xrightarrow{\cong} H^{n-m}(E, E_0)$ is the excision isomorphism. The commutative square

$$\begin{array}{ccc} H^{n-m}(X, X - Y) \otimes H_k(X, X - Y) & \xrightarrow{e^* \otimes e_*^{-1}} & H^{n-m}(E, E_0) \otimes H_k(E, E_0) \\ \downarrow \cap & & \downarrow \cap \\ H_{k-n+m}(X) & \xleftarrow{j_{f*}} & H_{k-n+m}(Y) \end{array}$$

yields the relation

$$u' \cap i_*(x) = j_{f*}(u \cap e_*^{-1} i_*(x)) = j_{f*} j_f^!(x).$$

Viewing ι as a class in $H^{n-m}(S^{n-m}, S^{n-m} - N)$, we can interpret it as the Thom class of $\{N\} \hookrightarrow S^{n-m}$. Thus, by naturality of the Thom class,

$$i^*(u') = f^*(\iota).$$

The commutativity of

$$\begin{array}{ccc} H^{n-m}(X, X-Y) \otimes H_k(X) & \xrightarrow{1 \otimes i_*} & H^{n-m}(X, X-Y) \otimes H_k(X, X-Y) \\ \Big\downarrow{\scriptstyle i^* \otimes 1} & & \Big\downarrow{\scriptstyle \cap} \\ H^{n-m}(X) \otimes H_k(X) & \xrightarrow{\quad \cap \quad} & H_{k-n+m}(X) \end{array}$$

implies

$$u' \cap i_*(x) = i^*(u') \cap x = f^*(\iota) \cap x. \qquad \square$$

The construction of $L_*(\mathbf{S}^\bullet)$ implies the following existence and uniqueness result for L-classes of self-dual complexes of sheaves:

Proposition 8.2.11 *Let $\mathbf{S}^\bullet \in D_c^b(X)$ be a self-dual complex of sheaves. There exist unique classes $L_k(\mathbf{S}^\bullet) \in H_k(X; \mathbb{Q})$, $k \geq 0$, with the following properties:*

(i) $\epsilon_* L_0(\mathbf{S}^\bullet) = \sigma(\mathbf{S}^\bullet)$,
(ii) *If $j : Y^m \hookrightarrow X^n$ is a normally nonsingular inclusion with trivial normal bundle, then*
$$L_{k-n+m}(j^! \mathbf{S}^\bullet) = j^! L_k(\mathbf{S}^\bullet).$$

Proof. For the existence part, we shall verify that the classes constructed above indeed enjoy properties (i) and (ii). Note that (i) and (ii) are equivalent to the statement

$$\text{(iii)} \quad \epsilon_* j^! L_{n-m}(\mathbf{S}^\bullet) = \sigma(j^! \mathbf{S}^\bullet)$$

for every normally nonsingular inclusion $j : Y^m \hookrightarrow X^n$ with trivial normal bundle. Thus it suffices to prove (iii). First, let us construct a map $f : X \to S^{n-m}$, transverse to N, such that $Y = f^{-1}(N)$, $j = j_f$. A neighborhood of Y in X is a product $Y \times D^{n-m}$. Let $T(Y \times D^{n-m})$ denote the Thom space of the trivial normal bundle. Collapsing the complement of the tube to a point defines a map

$$X \longrightarrow T(Y \times D^{n-m}).$$

More precisely, this map is the identity on $Y \times \operatorname{int}(D^{n-m})$ and maps the complement of $Y \times \operatorname{int}(D^{n-m})$ to the point in $T(Y \times D^{n-m})$ which is the image of $Y \times \partial D^{n-m}$. Composing this with the map

$$\begin{aligned} T(Y \times D^{n-m}) &\cong Y \times D^{n-m} \cup_{Y \times S^{n-m-1}} c(Y \times S^{n-m-1}) \\ &\xrightarrow{\text{proj}_2} D^{n-m} \cup_{S^{n-m-1}} cS^{n-m-1} \cong S^{n-m} \end{aligned}$$

induced by the second-factor projection yields the desired map $f : X \to S^{n-m}$. Under the homeomorphism $D^{n-m} \cup_{S^{n-m-1}} cS^{n-m-1} \cong S^{n-m}$, the north pole $N \in S^{n-m}$ is to correspond to the center of the disk D^{n-m}. Then $Y = f^{-1}(N)$ and $j = j_f$.

By construction,

$$\langle f^*(\iota), L_{n-m}(\mathbf{S}^\bullet) \rangle = \sigma(j_f^! \mathbf{S}^\bullet),$$

cf. also the proof of Proposition 5.7.2. Thus, with the aid of Lemma 8.2.10,

$$\begin{aligned}\sigma(j_f^! \mathbf{S}^\bullet) &= \epsilon_*(f^*(\iota) \cap L_{n-m}(\mathbf{S}^\bullet)) \\ &= \epsilon_* j_{f*} j_f^! L_{n-m}(\mathbf{S}^\bullet) \\ &= \epsilon_* j_f^! L_{n-m}(\mathbf{S}^\bullet),\end{aligned}$$

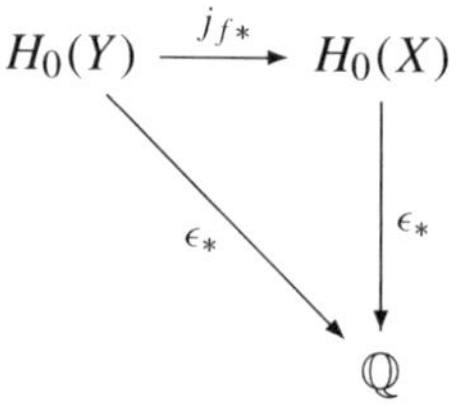

As for uniqueness, assume that $L_*(\mathbf{S}^\bullet)$ and $L'_*(\mathbf{S}^\bullet)$ are homology classes such that

$$\epsilon_* j^! L_{n-m}(\mathbf{S}^\bullet) = \sigma(j^! \mathbf{S}^\bullet) = \epsilon_* j^! L'_{n-m}(\mathbf{S}^\bullet) \tag{8.10}$$

for every normally nonsingular inclusion j with trivial normal bundle. Let $f : X \to S^k$ be an arbitrary transverse map; $j_f : f^{-1}(N) \hookrightarrow X$ is a normally nonsingular inclusion with trivial normal bundle. According to Lemma 8.2.10,

$$\langle f^*(\iota), L_k(\mathbf{S}^\bullet)\rangle = \epsilon_* j_f^! L_k(\mathbf{S}^\bullet),$$

$$\langle f^*(\iota), L'_k(\mathbf{S}^\bullet)\rangle = \epsilon_* j_f^! L'_k(\mathbf{S}^\bullet).$$

By (8.10),

$$\langle f^*(\iota), L_k(\mathbf{S}^\bullet)\rangle = \langle f^*(\iota), L'_k(\mathbf{S}^\bullet)\rangle,$$

and this holds for every transverse map $f : X \to S^k$. This implies $L_k(\mathbf{S}^\bullet) = L'_k(\mathbf{S}^\bullet)$. □

What can we say about the top L-class, $L_n(\mathbf{S}^\bullet)$? Let us assume that X^n is connected, $n \geq 2$. Take $\xi \in H^n(X^n)$ to be the cohomology class with $\langle \xi, [X]\rangle = 1$. For $k = n$, the inequality $2k-1 > n$ holds, whence the Hurewicz map is an isomorphism and ξ can be represented by a map $f : X^n \to S^n$ such that $\xi = f^*(\iota)$ ($\iota \in H^n(S^n)$, $\langle \iota, [S^n]\rangle = 1$). Since $H_n(S^n)$ has rank one, $f_*[X]$ is some scalar multiple of $[S^n]$: $f_*[X] = d \cdot [S^n]$, $d = \deg f$. The calculation

$$\begin{aligned} d = d\langle \iota, [S^n]\rangle &= \langle \iota, d[S^n]\rangle \\ &= \langle \iota, f_*[X]\rangle = \langle f^*(\iota), [X]\rangle \\ &= \langle \xi, [X]\rangle = 1\end{aligned}$$

shows that f has degree one. The map f is the restriction of a smooth map $F : M \to S^n$, where $M \supset X$ is the ambient manifold. Since f has degree one, $f^{-1}(N) = \{p\}$. As F is transverse to $N \in S^n$, we have for every pure stratum $V \subset X$ containing p,

$$dF(p)(T_p V) = T_N S^n.$$

This can only be satisfied if p is located in the top stratum of X. Therefore, $j_f^! \mathbf{S}^\bullet = j_f^* \mathbf{S}^\bullet = \mathbf{S}_p^\bullet$ and with $L_n(\mathbf{S}^\bullet) = \lambda[X] \in H_n(X^n)$ for some $\lambda \in \mathbb{Q}$, we have

$$\begin{aligned}\lambda &= \lambda\langle \iota, [S^n]\rangle \\ &= \lambda\langle \iota, f_*[X]\rangle \\ &= \langle f^*(\iota), \lambda[X]\rangle \\ &= \langle f^*(\iota), L_n(\mathbf{S}^\bullet)\rangle \\ &= \sigma(j_f^! \mathbf{S}^\bullet) \\ &= \sigma(\mathbf{S}_p^\bullet).\end{aligned}$$

We obtain the following formula for the top-dimensional L-class of a self-dual complex of sheaves:

$$L_n(\mathbf{S}^\bullet) = \sigma(\mathbf{S}_p^\bullet) \cdot [X] \in H_n(X^n),$$

where $p \in X$ is a point in the top stratum.

Proposition 8.2.12 *If $\mathbf{S}_1^\bullet, \mathbf{S}_2^\bullet \in D_c^b(X)$ are bordant, self-dual sheaves on X, then $L_k(\mathbf{S}_1^\bullet) = L_k(\mathbf{S}_2^\bullet)$ for all $k \geq 0$.*

Proof. Let $j : Y^m \hookrightarrow X^n$ be a normally nonsingular inclusion with trivial normal bundle. Since $j^!$ is a functor of triangulated categories, i.e. takes distinguished triangles over X to distinguished triangles over Y, we may apply it to a bordism between $\mathbf{S}_1^\bullet$ and $\mathbf{S}_2^\bullet$ and obtain a bordism between $j^!\mathbf{S}_1^\bullet$ and $j^!\mathbf{S}_2^\bullet$. By Corollary 8.2.5, $\sigma(j^!\mathbf{S}_1^\bullet) = \sigma(j^!\mathbf{S}_2^\bullet)$. Hence, by uniqueness of classes satisfying (iii) in (the proof of) Proposition 8.2.11, the two sets of classes must be identical. □

The proofs of the next three propositions are obtained similarly by appealing to the uniqueness statement of Proposition 8.2.11. We leave the details to the reader.

Proposition 8.2.13 *Suppose X has only strata of even codimension. Let V be a connected component of the stratum of codimension $2c$ of X, and let $\mathbf{S}^\bullet \in D_c^b(\overline{V})$ be a self-dual sheaf on the closure $\overline{V}$. If $j : \overline{V} \hookrightarrow X$ is the inclusion, then $j_*\mathbf{S}^\bullet[c]$ is self-dual on X and*

$$L_k(j_*\mathbf{S}^\bullet[c]) = j_* L_k(\mathbf{S}^\bullet), \quad k \geq 0.$$

Proposition 8.2.14 *For self-dual sheaves $\mathbf{S}_1^\bullet, \mathbf{S}_2^\bullet \in D_c^b(X)$,*

$$L_k(\mathbf{S}_1^\bullet \oplus \mathbf{S}_2^\bullet) = L_k(\mathbf{S}_1^\bullet) + L_k(\mathbf{S}_2^\bullet).$$

Proposition 8.2.15 *Let $f : X^p \to Y^m$ be a stratified map of even relative dimension between closed, oriented, Whitney stratified pseudomanifolds, and let $\mathbf{S}^\bullet \in D_c^b(X)$ be self-dual. Then*

$$L_k(Rf_*\mathbf{S}^\bullet[-t]) = f_* L_k(\mathbf{S}^\bullet), \quad k \geq 0,$$

$t = \frac{1}{2}(p - m) \in \mathbb{Z}$.

8.2.4 Poincaré Local Systems

This section discusses in some more detail intersection chain sheaves with coefficients in a local system (= locally constant sheaf) given on the top stratum ("*twisted* intersection sheaves"). In order to get Verdier self-duality, and consequently generalized Poincaré duality, for twisted intersection chain sheaves, the local system has to be equipped with a self-duality isomorphism on the top stratum, that is, a nonsingular, symmetric or antisymmetric, bilinear pairing. This leads us to the notion of a *Poincaré local system.* (For untwisted intersection sheaves with $\mathbb{R}$-coefficients, this pairing is $1 : \mathbb{R}_{X-\Sigma} \otimes \mathbb{R}_{X-\Sigma} \to \mathbb{R}_{X-\Sigma}$ given by $v \otimes w \mapsto vw$.) We will see that on a space with only even-codimensional strata, the duality of the Poincaré local system extends to a duality for the corresponding twisted intersection chain sheaf. Thus one obtains in particular a twisted signature. We shall define morphisms of Poincaré local systems and prove that isomorphic systems have equal twisted signatures. Next, we will show that if the system is untwisted (but the pairing is still arbitrary), then the "twisted" signature is the product of the signature of the pairing and the signature of the space. This implies a corresponding multiplicative formula for L-classes.

Definition 8.2.16 *A* Poincaré local system *on a topological space X is a pair $(\mathcal{S}, \phi)$, where $\mathcal{S}$ is a locally constant sheaf $\mathcal{S} \in Sh(X)$ whose stalks are real vector spaces, and ϕ is a nondegenerate, symmetric or antisymmetric, bilinear pairing $\phi : \mathcal{S} \otimes \mathcal{S} \to \mathbb{R}_X$.*

We have seen in Examples 1.1.6 how a linear monodromy representation of $\pi_1(X)$ gives rise to a locally constant sheaf on a path connected space X. Let $(\mathcal{S}, \phi)$ be a Poincaré local system of stalk dimension m on the space X and let $\Pi_1(X)$ denote the fundamental groupoid of X. By $\mathfrak{Vect}_m$ denote the category whose objects are pairs (V, ψ), with V an m-dimensional real vector space and $\psi : V \otimes V \to \mathbb{R}$ a nondegenerate, symmetric or antisymmetric, bilinear pairing, and whose morphisms are linear maps preserving the pairings:

$$\begin{aligned}&\mathrm{Hom}_{\mathfrak{Vect}_m}((V_1, \psi_1), (V_2, \psi_2))\\ &\quad = \{A : V_1 \to V_2 \text{ linear} \mid \psi_2(Av, Aw) = \psi_1(v, w),\ v, w \in V_1\}.\end{aligned}$$

The system $(\mathcal{S}, \phi)$ induces a covariant *monodromy functor*

$$\mu(\mathcal{S}) : \Pi_1(X) \longrightarrow \mathfrak{Vect}_m$$

as follows: For $x \in X$, let

$$\mu(\mathcal{S})(x) = (\mathcal{S}_x, \phi_x)$$

and for a path class $[\omega] \in \pi_1(X, x_1, x_2) = \mathrm{Hom}_{\Pi_1(X)}(x_2, x_1)$, $\omega : I \to X$, $\omega(0) = x_1$, $\omega(1) = x_2$, define the linear operator

$$\mu(\mathcal{S})[\omega] : \mu(\mathcal{S})(x_2) \longrightarrow \mu(\mathcal{S})(x_1)$$

to be the composition

$$\mu(\mathcal{S})(x_2) = \mathcal{S}_{\omega(1)} \cong (\omega^*\mathcal{S})_1 \underset{\text{restr}}{\overset{\cong}{\leftarrow}} \Gamma(I;\omega^*\mathcal{S}) \underset{\text{restr}}{\overset{\cong}{\rightarrow}} (\omega^*\mathcal{S})_0 \cong \mathcal{S}_{\omega(0)} = \mu(\mathcal{S})(x_1).$$

If we choose a base point $x \in X$, then restricting $\mu(\mathcal{S})$ to the fundamental group $\pi_1(X,x) = \mathrm{Hom}_{\Pi_1(X)}(x,x)$ gives an assignment of a linear automorphism on the stalk $\mathcal{S}_x$,

$$\mu(\mathcal{S})_x(g) : \mathcal{S}_x \longrightarrow \mathcal{S}_x,$$

preserving the pairing $\phi_x : \mathcal{S}_x \otimes \mathcal{S}_x \to \mathbb{R}$, to each $g \in \pi_1(X,x)$. Thus one obtains the *monodromy representation*

$$\mu(\mathcal{S})_x : \pi_1(X,x) \longrightarrow O(p,q;\mathbb{R})$$

in the symmetric case ($p+q=m$ is the rank of $\mathcal{S}$, $p-q$ the signature of ϕ_x), and

$$\mu(\mathcal{S})_x : \pi_1(X,x) \longrightarrow Sp(2r;\mathbb{R})$$

in the antisymmetric case ($m = 2r$ is the rank of $\mathcal{S}$).

Conversely, as already indicated in Examples 1.1.6, a given functor $\mu : \Pi_1(X) \to \mathfrak{Vect}_m$ determines a Poincaré local system: Let X_0 be a path component of X, and $x_0 \in X_0$. Then $\pi(X_0,x_0)$ acts on $\mu(x_0) = (V,\phi)$ by the restriction μ_{x_0} and we have the associated local system

$$\mathcal{S}|_{X_0} = \widetilde{X}_0 \times_{\pi_1(X_0,x_0)} V$$

over X_0 with an induced pairing ϕ, where $\widetilde{X}_0$ denotes the universal cover of X_0.

In our notation, we will frequently omit the pairing and simply write $\mathcal{S}$ for a Poincaré local system. As we have seen in Sect. 8.1, geometric mapping situations involving stratified maps generally lead to Poincaré local systems that are only defined on the top stratum of a stratified space X. Let X^n be an orientable, topological stratified pseudomanifold (without boundary) with singular set Σ, and let $(\mathcal{S},\phi)$ be a Poincaré local system on the top stratum $X-\Sigma$. The pairing ϕ, being nondegenerate, induces an isomorphism

$$d_\phi : \mathbf{Hom}(\mathcal{S},\mathbb{R}_{X-\Sigma}) \overset{\cong}{\longrightarrow} \mathcal{S}.$$

As $X-\Sigma$ is a manifold,

$$\mathcal{D}\mathcal{S}[-n] = \mathbf{Hom}(\mathcal{S},\mathbb{R}_{X-\Sigma}) \otimes \mathcal{O}_{X-\Sigma},$$

where $\mathcal{O}_{X-\Sigma}$ is the orientation sheaf on $X-\Sigma$ and $\mathcal{D}$ the Borel–Moore–Verdier dualizing functor. Thus ϕ induces an isomorphism

$$\mathcal{D}\mathcal{S}[-n] \cong \mathcal{S} \otimes \mathcal{O}_{X-\Sigma}.$$

Orient X by choosing an isomorphism $\mathcal{O}_{X-\Sigma} \cong \mathbb{R}_{X-\Sigma}$. Using the orientation, ϕ induces a self-duality isomorphism

$$d_\phi : \mathcal{D}\mathcal{S}[-n] \cong \mathcal{S}. \tag{8.11}$$

The twisted intersection chain complex $\mathbf{IC}^\bullet_{\bar{m}}(X;\mathcal{S})$ can be defined by the Goresky–MacPherson–Deligne extension process, i.e.

$$\mathbf{IC}^\bullet_{\bar{m}}(X;\mathcal{S}) = \tau_{\leq \bar{m}(n)-n} Ri_{n*} \dots \tau_{\leq \bar{m}(2)-n} Ri_{2*}\mathcal{S}[n]$$

($i_k : X - X_{n-k} \hookrightarrow X - X_{n-k-1}$ is the inclusion). (Note that if we construct $\mathbf{IC}^\bullet_{\bar{m}}$ this way, then $\mathbf{IC}^\bullet_{\bar{m}}(X;-)$ becomes a functor from locally constant sheaves on $X - \Sigma$ to $D^b_c(X)$.) For certain classes of spaces, the self-duality isomorphism (8.11) will extend to a self-duality isomorphism for $\mathbf{IC}^\bullet_{\bar{m}}(X;\mathcal{S})$. Consider the case of a space X with only even-codimensional strata. Then d_ϕ extends to

$$\bar{d}_\phi : \mathcal{D}\mathbf{IC}^\bullet_{\bar{m}}(X;\mathcal{S})[n] \xrightarrow{\cong} \mathbf{IC}^\bullet_{\bar{m}}(X;\mathcal{S})$$

without any further obstruction, as follows from Proposition 8.1.10. Alternatively, $\bar{d}_\phi$ can be constructed inductively through the Goresky–MacPherson–Deligne extension process:

$$\begin{aligned} &\mathcal{D}(\tau_{\leq \bar{m}(2)-n} Ri_{2*}\mathcal{S}[n])[n] \\ &\qquad \cong \tau_{\leq \bar{m}(2)-n} Ri_{2*}(\mathcal{D}\mathcal{S}[-n])[n] \xrightarrow{\tau_{\leq \bar{m}(2)-n} Ri_{2*} d_\phi[n]} \tau_{\leq \bar{m}(2)-n} Ri_{2*}\mathcal{S}[n], \end{aligned}$$

etc. If X possesses strata of odd codimension as well, then such an extension need not exist.

Definition 8.2.17 *Let $(\mathcal{S},\phi)$ and $(\mathcal{T},\psi)$ be Poincaré local systems on a space X. A* morphism $(\mathcal{S},\phi) \to (\mathcal{T},\psi)$ of Poincaré local systems *is a morphism $\gamma : \mathcal{S} \to \mathcal{T}$ of sheaves such that*

$$\begin{array}{ccc} \mathcal{S} \otimes \mathcal{S} & \xrightarrow{\phi} & \mathbb{R}_X \\ {\scriptstyle \gamma\otimes\gamma}\downarrow & \nearrow{\scriptstyle \psi} & \\ \mathcal{T} \otimes \mathcal{T} & & \end{array}$$

commutes.

In order to explain how an isomorphism of Poincaré local systems on the top stratum induces a commutative duality diagram between the associated twisted intersection chain sheaves, let us first recall some basic linear algebra. Let V, W be vector spaces over some field k and let $V^* = \mathrm{Hom}(V,k)$, W^* denote the linear duals. Let $\phi : V \otimes V \to k$, $\psi : W \otimes W \to k$ be nondegenerate forms, and let $\gamma : V \xrightarrow{\cong} W$ be a linear isomorphism such that

$$\begin{array}{ccc} V \otimes V & \xrightarrow{\phi} & k \\ {\scriptstyle \gamma\otimes\gamma}\downarrow & \nearrow{\scriptstyle \psi} & \\ W \otimes W & & \end{array}$$

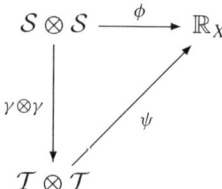

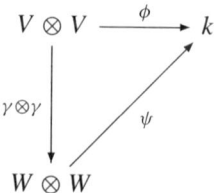

commutes. What does the corresponding diagram, reformulated in terms of duality isomorphisms, look like? The form ϕ induces a duality isomorphism

$$d_\phi : V^* \xrightarrow{\cong} V$$

given by

$$d_\phi(l) = v_l,$$

where v_l is the unique vector such that $\phi(v_l, v) = l(v)$ for all $v \in V$. Similarly, ψ induces

$$d_\psi : W^* \xrightarrow{\cong} W$$

such that $\psi(d_\psi(l), w) = l(w)$ for all $w \in W$. The dual of γ,

$$\gamma^* : W^* \longrightarrow V^*,$$

is given by

$$\gamma^*(l)(v) = l(\gamma(v)).$$

Let $l \in W^*$. Then

$$\phi(d_\phi(\gamma^* l), v) = (\gamma^* l)(v) = l(\gamma(v))$$

for all $v \in V$, and

$$\begin{aligned} \psi(\gamma d_\phi(\gamma^* l), w) &= \psi(\gamma d_\phi(\gamma^* l), \gamma(\gamma^{-1} w)) \\ &= \phi(d_\phi(\gamma^* l), \gamma^{-1} w) \\ &= l(\gamma(\gamma^{-1} w)) \\ &= l(w) \\ &= \psi(d_\psi(l), w) \end{aligned}$$

for all $w \in W$. Since ψ is nondegenerate, this implies

$$(\gamma d_\phi \gamma^*)(l) = d_\psi(l).$$

In other words, the square

$$\begin{array}{ccc} V^* & \xrightarrow{d_\phi} & V \\ \uparrow{\scriptstyle \gamma^*} & & \downarrow{\scriptstyle \gamma} \\ W^* & \xrightarrow[d_\psi]{} & W \end{array}$$

commutes.

Let $(\mathcal{S}, \phi)$, $(\mathcal{T}, \psi)$ be Poincaré local systems on the top stratum $X - \Sigma$ of a pseudomanifold X^n, and let $\gamma : (\mathcal{S}, \phi) \xrightarrow{\cong} (\mathcal{T}, \psi)$ be an isomorphism. Then the square

$$\begin{array}{ccc}
\mathcal{S}^* & \xrightarrow{d_\phi} & \mathcal{S} \\
\uparrow{\scriptstyle \gamma^*} & & \downarrow{\scriptstyle \gamma} \\
\mathcal{T}^* & \xrightarrow[d_\psi]{} & \mathcal{T}
\end{array}$$

commutes in $Sh(X - \Sigma)$, and

$$\begin{array}{ccc}
\mathcal{DS}[-n] & \xrightarrow{d_\phi} & \mathcal{S} \\
\uparrow{\scriptstyle \mathcal{D}\gamma[-n]} & & \downarrow{\scriptstyle \gamma} \\
\mathcal{DT}[-n] & \xrightarrow[d_\psi]{} & \mathcal{T}
\end{array}$$

commutes in $D_c^b(X - \Sigma)$. The induced diagram of extensions

$$\begin{array}{ccc}
\mathcal{D}\mathbf{IC}^\bullet_{\bar{m}}(X; \mathcal{S})[n] & \xrightarrow{\bar{d}_\phi} & \mathbf{IC}^\bullet_{\bar{m}}(X; \mathcal{S}) \\
\uparrow{\scriptstyle \mathcal{D}\bar{\gamma}[n]} & & \downarrow{\scriptstyle \bar{\gamma}=\mathbf{IC}^\bullet_{\bar{m}}(X;\gamma)} \\
\mathcal{D}\mathbf{IC}^\bullet_{\bar{m}}(X; \mathcal{T})[n] & \xrightarrow[\bar{d}_\psi]{} & \mathbf{IC}^\bullet_{\bar{m}}(X; \mathcal{T})
\end{array}$$

commutes in $D_c^b(X)$ by Proposition 8.1.10.

If X is compact, then an application of middle-dimensional hypercohomology to this diagram, together with Borel–Moore duality (3.5), Sect. 3.4, shows that the self-dual sheaves $(\mathbf{IC}^\bullet_{\bar{m}}(X; \mathcal{S}), \bar{d}_\phi)$ and $(\mathbf{IC}^\bullet_{\bar{m}}(X; \mathcal{T}), \bar{d}_\psi)$ have the same signature. Summarizing, we have shown:

Proposition 8.2.18 *Let X^n be a closed, oriented, topological stratified pseudomanifold with singular set Σ having only even-codimensional strata. Let $(\mathcal{S}, \phi)$ and $(\mathcal{T}, \psi)$ be isomorphic Poincaré local systems on the top stratum $X - \Sigma$. Then ϕ, ψ induce self-duality isomorphisms*

$$\bar{d}_\phi : \mathcal{D}\mathbf{IC}^\bullet_{\bar{m}}(X; \mathcal{S})[n] \cong \mathbf{IC}^\bullet_{\bar{m}}(X; \mathcal{S}), \qquad \bar{d}_\psi : \mathcal{D}\mathbf{IC}^\bullet_{\bar{m}}(X; \mathcal{T})[n] \cong \mathbf{IC}^\bullet_{\bar{m}}(X; \mathcal{T}),$$

and

$$\sigma(\mathbf{IC}^\bullet_{\bar{m}}(X; \mathcal{S}), \bar{d}_\phi) = \sigma(\mathbf{IC}^\bullet_{\bar{m}}(X; \mathcal{T}), \bar{d}_\psi).$$

For untwisted Poincaré local systems, the signature is multiplicative:

Proposition 8.2.19 *Let X^n be a closed, oriented pseudomanifold with only even-codimensional strata and top stratum $X - \Sigma$. Let $(\mathbb{R}^r_{X-\Sigma}, \phi)$ be an untwisted, symmetric Poincaré local system. Then*

$$\sigma(X; (\mathbb{R}^r_{X-\Sigma}, \phi)) = \sigma(\phi_{pt})\sigma(X).$$

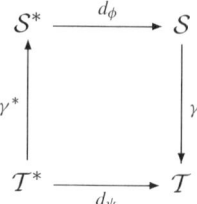

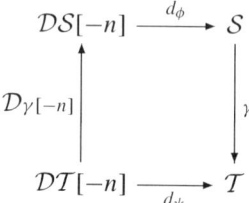

Proof. The symmetric, bilinear, nondegenerate form ϕ can be diagonalized, i.e. there exists a linear isomorphism $\gamma : \mathbb{R}^r_{X-\Sigma} \to \mathbb{R}^r_{X-\Sigma}$ such that the composition ψ,

$$\begin{array}{ccc} \mathbb{R}^r_{X-\Sigma} \otimes \mathbb{R}^r_{X-\Sigma} & \xrightarrow{\psi} & \mathbb{R}_{X-\Sigma} \\ {\scriptstyle \gamma\otimes\gamma}\downarrow\cong & \nearrow{\scriptstyle \phi} & \\ \mathbb{R}^r_{X-\Sigma} \otimes \mathbb{R}^r_{X-\Sigma} & & \end{array}$$

is given at any point by the matrix

$$\begin{pmatrix} +1 & & & & & \\ & \ddots & & & & \\ & & +1 & & & \\ & & & -1 & & \\ & & & & \ddots & \\ & & & & & -1 \end{pmatrix}$$

in the standard basis of $\mathbb{R}^r$. In other words,

$$(\mathbb{R}^r_{X-\Sigma}, \psi) = \bigoplus^P (\mathbb{R}_{X-\Sigma}, 1) \oplus \bigoplus^N (\mathbb{R}_{X-\Sigma}, -1),$$

where P is the number of positive entries, N the number of negative entries, and $1 : \mathbb{R}_{X-\Sigma} \otimes \mathbb{R}_{X-\Sigma} \to \mathbb{R}_{X-\Sigma}$ denotes the form $v \otimes w \mapsto vw$. The pairings ϕ, ψ, 1, -1 induce isomorphisms

$$d_\phi, d_\psi : \mathcal{D}\mathbb{R}^r_{X-\Sigma}[-n] \cong \mathbb{R}^r_{X-\Sigma},$$

$$d_1, d_{-1} : \mathcal{D}\mathbb{R}_{X-\Sigma}[-n] \cong \mathbb{R}_{X-\Sigma}$$

on the top stratum. Since X has only strata of even codimension, these isomorphisms have extensions

$$\bar{d}_\phi, \bar{d}_\psi : \mathcal{D}\mathbf{IC}^\bullet_{\bar{m}}(X; \mathbb{R}^r_{X-\Sigma})[n] \cong \mathbf{IC}^\bullet_{\bar{m}}(X; \mathbb{R}^r_{X-\Sigma}),$$

$$\bar{d}_1, \bar{d}_{-1} : \mathcal{D}\mathbf{IC}^\bullet_{\bar{m}}(X; \mathbb{R}_{X-\Sigma})[n] \cong \mathbf{IC}^\bullet_{\bar{m}}(X; \mathbb{R}_{X-\Sigma}),$$

where $\bar{d}_\psi$ has an orthogonal splitting

$$(\mathbf{IC}^\bullet_{\bar{m}}(X; \mathbb{R}^r_{X-\Sigma}), \bar{d}_\psi) = \bigoplus^P (\mathbf{IC}^\bullet_{\bar{m}}(X; \mathbb{R}_{X-\Sigma}), \bar{d}_1) \oplus \bigoplus^N (\mathbf{IC}^\bullet_{\bar{m}}(X; \mathbb{R}_{X-\Sigma}), \bar{d}_{-1}).$$

By Proposition 8.2.18,

$$
\begin{aligned}
\sigma(X; (\mathbb{R}^r_{X-\Sigma}, \phi)) &= \sigma(\mathbf{IC}^\bullet_{\bar{m}}(X; \mathbb{R}^r_{X-\Sigma}), \bar{d}_\phi) \\
&= \sigma(\mathbf{IC}^\bullet_{\bar{m}}(X; \mathbb{R}^r_{X-\Sigma}), \bar{d}_\psi) \\
&= \sigma\left(\bigoplus^P (\mathbf{IC}^\bullet_{\bar{m}}(X; \mathbb{R}_{X-\Sigma}), \bar{d}_1) \oplus \bigoplus^N (\mathbf{IC}^\bullet_{\bar{m}}(X; \mathbb{R}_{X-\Sigma}), \bar{d}_{-1})\right) \\
&= \sum^P \sigma(X) - \sum^N \sigma(X) \\
&= (P - N) \cdot \sigma(X) \\
&= \sigma(\phi_{pt})\sigma(X).
\end{aligned}
$$

□

This result for the signature implies a multiplicative formula for the L-class with coefficients in an untwisted Poincaré local system:

Proposition 8.2.20 *Let X^n be a closed, oriented, Whitney stratified pseudomanifold with only even-codimensional strata and top stratum $X - \Sigma$. Let $(\mathbb{R}^r_{X-\Sigma}, \phi)$ be an untwisted, symmetric Poincaré local system. Then*

$$L_k(X; (\mathbb{R}^r_{X-\Sigma}, \phi)) = \sigma(\phi_{pt})L_k(X) \in H_k(X; \mathbb{Q}).$$

Proof. Set $\mathbf{S}^\bullet = \mathbf{IC}^\bullet_{\bar{m}}(X; \mathbb{R}^r_{X-\Sigma})$. This is a self-dual sheaf complex with duality isomorphism $\bar{d}_\phi : \mathcal{D}\mathbf{S}^\bullet[n] \cong \mathbf{S}^\bullet$, since X has only even-codimensional strata. For any normally nonsingular inclusion $j : Y^m \hookrightarrow X^n$ with trivial normal bundle, define

$$L'_k(j^!\mathbf{S}^\bullet) = \sigma(\phi_{pt})L_k(Y).$$

(In particular, $L'_k(\mathbf{S}^\bullet) = \sigma(\phi_{pt})L_k(X)$ for j the identity map.) Then

$$
\begin{aligned}
\epsilon_* j^! L'_{n-m}(\mathbf{S}^\bullet) &= \sigma(\phi_{pt}) \cdot \epsilon_* j^! L_{n-m}(X) \\
&= \sigma(\phi_{pt}) \cdot \epsilon_* j^! L_{n-m}(\mathbf{IC}^\bullet_{\bar{m}}(X)) \\
&= \sigma(\phi_{pt}) \cdot \sigma(j^!\mathbf{IC}^\bullet_{\bar{m}}(X)) \\
&= \sigma(\phi_{pt}) \cdot \sigma(\mathbf{IC}^\bullet_{\bar{m}}(Y)) && \text{(as } j \text{ norm. nonsing.)} \\
&= \sigma(\phi_{pt}) \cdot \sigma(Y) \\
&= \sigma(Y; (\mathbb{R}^r_{Y-\Sigma}, \phi)) && \text{(by Prop. 8.2.19)} \\
&= \sigma(j^!\mathbf{S}^\bullet).
\end{aligned}
$$

(During this calculation, the base point is of course chosen in the top stratum of Y.) By the uniqueness statement with respect to property (iii) of Proposition 8.2.11,

$$L'_k(\mathbf{S}^\bullet) = L_k(\mathbf{S}^\bullet) = L_k\big(X; (\mathbb{R}^r_{X-\Sigma}, \phi)\big).$$

□

If a Poincaré local system on the top stratum is not constant, then it may or may not extend *as a local system* (not via the Goresky–MacPherson–Deligne extension as a *complex* of sheaves) to the entire stratified space. Let us thus seek a characterization of those systems that do extend.

Definition 8.2.21 *Let X be a stratified pseudomanifold with singular set Σ and let $\mathcal{V}$ denote the set of components of open strata of X of codimension at least 2. Each $V \in$*

$\mathcal{V}$ has a link $Lk(V)$. Call a Poincaré local system $\mathcal{S}$ on $X - \Sigma$ strongly transverse to *Σ if the composite functor*

$$\Pi_1(Lk(V) - \Sigma) \xrightarrow{incl_*} \Pi_1(X - \Sigma) \xrightarrow{\mu(\mathcal{S})} \mathfrak{Vect}_m$$

is isomorphic to the trivial functor for all $V \in \mathcal{V}$.

On normal spaces (in the sense of Remark 6.6.4), strong transversality of local systems characterizes those systems that extend as local systems over the whole space:

Proposition 8.2.22 *Let X^n be normal. A Poincaré local system $\mathcal{S}$ on $X - \Sigma$ is strongly transverse to Σ if and only if it extends as a Poincaré local system over all of X. Such an extension is unique.*

See [BCS03] for a proof. The normality assumption is not necessary for the "if"-direction. The following examples show that the normality assumption can not be omitted in the "only if"-direction and in the uniqueness statement.

Example 8.2.23 (*We suppress mentioning pairings ϕ.*) *Let X^3 be the pseudomanifold $X = S^1 \times S^1 \times \mathbb{R}/\sim$, where $(x_1, y_1, z_1) \sim (x_2, y_2, z_2)$ iff $x_1 = x_2$ and $z_1 = z_2 = 0$. We denote the image of (x, y, z) under the collapse $S^1 \times S^1 \times \mathbb{R} \to X$ by $(x, [y], z)$. X is stratified as $X \supset \Sigma \supset \varnothing$, with singular set $\Sigma = S^1 \times [S^1] \times \{0\}$. Note that X is not normal, the link of Σ at the point $(x, [y], 0)$ is the disjoint union $\{x\} \times S^1 \times \{-1\} \sqcup \{x\} \times S^1 \times \{+1\}$. We define a local system $\mathcal{S}$ on $X - \Sigma$ as follows: Over $S^1 \times S^1 \times (-\infty, 0)$, let $\mathcal{S}$ be the constant sheaf with stalk $\mathbb{R}$. Let $\mathcal{M}$ denote the Möbius sheaf with stalk $\mathbb{R}$ over a circle. Over $S^1 \times S^1 \times (0, +\infty)$, let $\mathcal{S}$ be $\pi_1^*\mathcal{M}$, $\pi_1 : S^1 \times S^1 \times (0, +\infty) \to S^1$ projection to the first coordinate. Now as $\mathcal{S}_{\{x\}\times S^1\times\{-1\}\sqcup\{x\}\times S^1\times\{+1\}}$ is constant for all $x \in S^1$, $\mathcal{S}$ is strongly transverse to Σ. However, an extension $\bar{\mathcal{S}}$ to all of X does clearly not exist, as the restriction of $\bar{\mathcal{S}}$ to say $S^1 \times \{[y]\} \times \mathbb{R}$ would have to be of the form $(\bar{\mathcal{S}}|_{S^1\times\{[y]\}\times\{0\}}) \times \mathbb{R}$, which is impossible.*

Example 8.2.24 *Consider the projection to the first coordinate $\pi_1 : S^1 \times S^1 \to S^1$ and fix $p \in S^1$. Let X^2 be the singular space obtained from the torus by collapsing $\pi_1^{-1}(p) = \{p\} \times S^1$ to a point. The projection induces a map $f : X \to S^1$ so that*

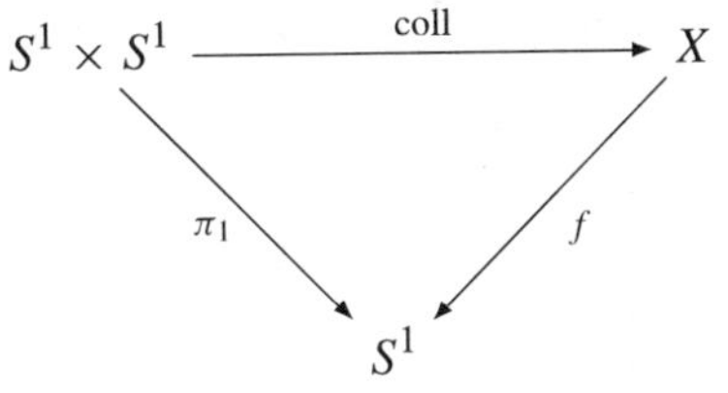

commutes. The local system $\bar{\mathcal{S}} = f^*\mathcal{M}$, with $\mathcal{M}$ as in the previous example, is nontrivial on X as it has nontrivial monodromy around the generator of $\pi_1(X)$. On the other hand, the restriction away from the singularity, $\bar{\mathcal{S}}|_{X-\Sigma}$, is trivial, and so can also be extended trivially into Σ. This shows that if X is not normal, then uniqueness of extensions fails as well.

Proposition 8.2.22 allows us to state yet another useful characterization of strongly transverse local systems:

Corollary 8.2.25 *Let X^n be normal. A Poincaré local system $\mathcal{S}$ on $X - \Sigma$ is strongly transverse to Σ if and only if its monodromy functor $\mu(\mathcal{S}) : \Pi_1(X - \Sigma) \to \mathfrak{Vect}_m$ factors (up to isomorphism of functors) through $\Pi_1(X)$:*

$$\begin{array}{ccc} \Pi_1(X - \Sigma) & \xrightarrow{incl_*} & \Pi_1(X) \\ & \underset{\mu(\mathcal{S})}{\searrow} & \downarrow \\ & & \mathfrak{Vect}_m \end{array}$$

Remark 8.2.26 *The normality assumption is not necessary for the "if"-direction.*

Corollary 8.2.27 *Let X^n be normal. A Poincaré local system $(\mathcal{S}, \phi)$ on $X^n - \Sigma$ strongly transverse to Σ has a K-theory signature*

$$[\mathcal{S}]_K \in \begin{cases} KO(X), & \text{if } n \equiv 0(4), \\ KU(X), & \text{if } n \equiv 2(4). \end{cases}$$

Proof. By Proposition 8.2.22, $(\mathcal{S}, \phi)$ has a unique extension to a Poincaré local system $(\bar{\mathcal{S}}, \bar{\phi})$ on X. We now proceed as in [Mey72]. Let $\mathcal{S}^c$ denote the flat vector bundle associated to the locally constant sheaf $\bar{\mathcal{S}}$, that is

$$\mathcal{S}^c|_{X_0} = \widetilde{X}_0 \times_\pi \mathbb{R}^m$$

over a path component X_0 of X, where $\mathbb{R}^m$ is given the usual topology, $\pi = \pi_1(X_0)$, and π acts on $\mathbb{R}^m$ by means of the monodromy $\mu(\bar{\mathcal{S}})$ of $\bar{\mathcal{S}}$. A suitable choice of Euclidean metric on $\mathcal{S}^c$ induces (using $\bar{\phi}$) a vector bundle automorphism

$$A : \mathcal{S}^c \longrightarrow \mathcal{S}^c$$

such that $A^2 = 1$ (if $\bar{\phi}$ is symmetric) or $A^2 = -1$ (if $\bar{\phi}$ is anti-symmetric). Thus in the case $n \equiv 0(4)$, $\mathcal{S}^c$ decomposes as a direct sum of vector bundles

$$\mathcal{S}^c = \mathcal{S}_+ \oplus \mathcal{S}_-$$

corresponding to the ± 1-eigenspaces of A. Put

$$[\mathcal{S}]_K = [\mathcal{S}_+] - [\mathcal{S}_-] \in KO(X).$$

In the case $n \equiv 2(4)$, A defines a complex structure on $\mathcal{S}^c$ and we obtain the complex vector bundle $\mathcal{S}_\mathbb{C}$ and its conjugate bundle $\mathcal{S}^*_\mathbb{C}$; we put

$$[\mathcal{S}]_K = [\mathcal{S}^*_\mathbb{C}] - [\mathcal{S}_\mathbb{C}] \in KU(X).$$

□

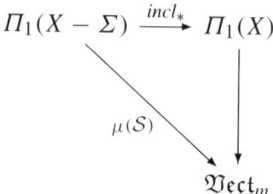

8.2.5 Computation of L-Classes

Combining the sheaf-theoretic decomposition of Theorem 8.1.7 with the properties of L-classes of self-dual sheaves as developed in the previous section, we obtain a formula that computes the pushforward of L-classes under a stratified map. Let $f : X^p \to Y^n$ be a stratified map, X compact, Y has only even-codimensional strata and $t = \frac{1}{2}(p-n) \in \mathbb{Z}$. We have seen that the pushforward under f of a self-dual sheaf $\mathbf{S}^\bullet \in D_c^b(X)$ decomposes, up to bordism, as an orthogonal sum of intersection chain complexes,

$$Rf_*\mathbf{S}^\bullet[-t] \sim \bigoplus_{V \in \mathcal{V}} j_*\mathbf{IC}^\bullet_{\bar{m}}(\overline{V}; \mathfrak{H}_V(\mathbf{S}^\bullet))[c].$$

Let us introduce the notation $L_k(\overline{V}; \mathfrak{H}_V(\mathbf{S}^\bullet))$ for the L-classes of the self-dual sheaf $\mathbf{IC}^\bullet_{\bar{m}}(\overline{V}; \mathfrak{H}_V(\mathbf{S}^\bullet))$,

$$L_k(\overline{V}; \mathfrak{H}_V(\mathbf{S}^\bullet)) = L_k(\mathbf{IC}^\bullet_{\bar{m}}(\overline{V}; \mathfrak{H}_V(\mathbf{S}^\bullet))).$$

The image $f_*L_k(\mathbf{S}^\bullet)$ can then be calculated as follows:

$$\begin{aligned} f_*L_k(\mathbf{S}^\bullet) &= L_k(Rf_*\mathbf{S}^\bullet[-t]), && \text{by Prop. 8.2.15,} \\ &= L_k(\bigoplus j_*\mathbf{IC}^\bullet_{\bar{m}}(\overline{V}; \mathfrak{H}_V(\mathbf{S}^\bullet))[c]), && \text{by Prop. 8.2.12,} \\ &= \sum L_k(j_*\mathbf{IC}^\bullet_{\bar{m}}(\overline{V}; \mathfrak{H}_V(\mathbf{S}^\bullet))[c]), && \text{by Prop. 8.2.14,} \\ &= \sum j_*L_k(\overline{V}; \mathfrak{H}_V(\mathbf{S}^\bullet)), && \text{by Prop. 8.2.13.} \end{aligned}$$

We have proved:

Theorem 8.2.28 (*Cappell–Shaneson*)

$$f_*L_k(\mathbf{S}^\bullet) = \sum_{V \in \mathcal{V}} j_*L_k(\overline{V}; \mathfrak{H}_V(\mathbf{S}^\bullet)).$$

This statement is already very useful for f the identity map. For $k = 0$, we obtain the signature formula

$$\sigma(\mathbf{S}^\bullet) = \sum_{V \in \mathcal{V}} \sigma(\overline{V}; \mathfrak{H}_V(\mathbf{S}^\bullet)).$$

Another important instance of this result is the case where X is a Witt space. We have seen in Sect. 6.4 that on such a space $\mathbf{S}^\bullet = \mathbf{IC}^\bullet_{\bar{m}}(X)$ is a self-dual sheaf. The associated L-classes $L_*(\mathbf{IC}^\bullet_{\bar{m}}(X))$ are the Goresky–MacPherson–Siegel L-classes $L_*(X)$ of X. For a point $y \in V \in \mathcal{V}$, the self-dual sheaf

$$\mathbf{S}^\bullet(y) = \tau^{\{cone\}}_{\leq -c-t-1} Ri_{y*}\rho_y^!\mathbf{S}^\bullet$$

$$(E_y = f^{-1}N(y) \cup_{f^{-1}L(y)} cf^{-1}L(y),$$

$$E_y \overset{i_y}{\hookleftarrow} f^{-1}\mathring{N}(y) \overset{\rho_y}{\hookrightarrow} X)$$

on E_y is simply $\mathbf{IC}^\bullet_{\bar{m}}(E_y)$ (check this by verifying the axioms [AX]). The group

$$\mathcal{H}^{-c-t}(E_y; \mathbf{S}^\bullet(y)) = IH^{\bar{m}}_{c+t}(E_y)$$

is the stalk at y of the local system $\mathfrak{H}_V = \mathfrak{H}_V(\mathbf{S}^\bullet)$. The nonsingular pairing

$$\mathcal{P}_V : \mathfrak{H}_V \otimes \mathfrak{H}_V \longrightarrow \mathbb{R}_V$$

is given at y by the intersection pairing

$$IH^{\bar{m}}_{c+t}(E_y) \otimes IH^{\bar{m}}_{c+t}(E_y) \longrightarrow \mathbb{R}$$

with signature $\sigma(E_y)$. Now suppose that V happens to be simply connected. Then $(\mathfrak{H}_V, \mathcal{P}_V)$ is a constant Poincaré local system, and the term $\sigma(\overline{V}; \mathfrak{H}_V)$ can be computed as

$$\sigma(\overline{V}; \mathfrak{H}_V) = \sigma(E_y)\sigma(\overline{V}),$$

according to Proposition 8.2.19. Thus

$$\sigma(X) = \sum_{V \in \mathcal{V}} \sigma(E_{y_V})\sigma(\overline{V})$$

($y_V \in V$ for all $V \in \mathcal{V}$), provided all components V of strata of Y are simply connected. This proves Corollary 8.1.5. More generally, of course, this formula holds if all the Poincaré local systems $\mathfrak{H}_V$ are constant coefficient systems. As for the L-classes, Proposition 8.2.20 shows that

$$L_k(\overline{V}; \mathfrak{H}_V) = \sigma(E_y)L_k(\overline{V}).$$

Therefore, provided all coefficient systems are untwisted,

$$f_* L_k(X) = \sum_{V \in \mathcal{V}} \sigma(E_{y_V}) j_* L_k(\overline{V}),$$

and we have established Theorem 8.1.4.

At this point, two natural questions present themselves:

- How does one compute the twisted L-classes arising e.g. in the formula of Theorem 8.2.28 if the local system is not constant?
- What if the space has strata of odd codimension?

We will attack the former question in the next section for the case of a Witt space. For a general pseudomanifold, the question is settled in Sect. 9.4. The latter question is answered in Chap. 9.

8.3 The Presence of Monodromy

Let E be an oriented closed smooth manifold, the total space of a fiber bundle $F \to E \to B$. It is a classical result of Chern, Hirzebruch and Serre [CHS57] that if B is simply connected (or the fundamental group of B acts trivially on the cohomology of F), then the signatures satisfy the multiplicative formula

$$\sigma(E) = \sigma(B)\sigma(F).$$

Kodaira [Kod67], Atiyah [Ati69] and Hirzebruch [Hir69] constructed various examples of fiber bundles in which $\pi_1(B)$ acts nontrivially on $H^*(F; \mathbb{R})$ and the signature is not multiplicative. (See also Example 8.3.1 below.) In the case where both B and F are even-dimensional, the Hirzebruch signature theorem and the Atiyah–Singer index theorem were used by Atiyah [Ati69] to obtain a characteristic class formula for $\sigma(E)$ involving a contribution from the $\pi_1(B)$-action on $H^*(F)$: Let $k = \dim(F)/2$. The flat (symmetric if k even, anti-symmetric if k odd) bundle $\mathcal{S}$ over B with fibers $H^k(F_x; \mathbb{R})$ $(x \in B)$, has a real (resp. complex) K-theory signature $[\mathcal{S}]_K \in KO(B)$ for k even (resp. $KU(B)$ for k odd) and the twisted signature theorem is

$$\sigma(E) = \langle \mathrm{ch}([\mathcal{S}]_K) \cup \widetilde{\ell}(B), [B] \rangle$$

with $\widetilde{\ell}$ the modification of the Hirzebruch ℓ-genus defined by

$$\widetilde{\ell}(B) = \prod_{i=1}^{r} \frac{y_i}{\tanh y_i/2} \in H^{4*}(B; \mathbb{Q})$$

and $y_1, \ldots, y_r$ notional elements of degree 2 such that the i-th Pontrjagin class of the tangent bundle of B is the i-th elementary symmetric function in $y_1^2, \ldots, y_r^2$.

8.3.1 Meyer's Generalization

For a complex vector bundle ξ over a base space B, let

$$\widetilde{\mathrm{ch}}(\xi) \in H^{2*}(B; \mathbb{Q})$$

denote the modified Chern character $\widetilde{\mathrm{ch}} = \mathrm{ch} \circ \psi^2$ obtained by composition with the second Adams operation ψ^2, that is, if ξ has total Chern class

$$c(\xi) = (1 + y_1) \cdots (1 + y_r)$$

$(r = \mathrm{rank}\xi)$, then

$$\mathrm{ch}(\xi) = \sum_{i=1}^{r} e^{y_i} \in H^{2*}(B; \mathbb{Q})$$

and $\psi^2(\xi)$ is a bundle with

$$\mathrm{ch}(\psi^2(\xi)) = \sum_{i=1}^{r} e^{2y_i}.$$

Hence,

$$\widetilde{\mathrm{ch}}(\xi)_{2k} = 2^k \,\mathrm{ch}(\xi)_{2k} \in H^{2k}(B;\mathbb{Q})$$

(see [HBJ92]).

Now W. Meyer [Mey72] considers a locally constant sheaf $\mathcal{S}$, not necessarily arising from a fiber bundle projection, over a closed oriented smooth manifold B of even dimension such that the stalks $\mathcal{S}_x$ $(x \in B)$ are nondegenerate (anti-) symmetric bilinear forms over $\mathbb{R}$. The twisted signature $\sigma(B;\mathcal{S}) \in \mathbb{Z}$ is defined to be the signature of the nondegenerate form on the sheaf cohomology group $H^{\dim(B)/2}(B;\mathcal{S})$. The twisted signature formula of [Mey72] is

$$\sigma(B;\mathcal{S}) = \langle \widetilde{\mathrm{ch}}([\mathcal{S}]_K) \cup \ell(B), [B]\rangle,$$

where ℓ is the original Hirzebruch ℓ-genus.

Example 8.3.1 *The following example of a Poincaré local system $\mathcal{S}$ over a surface Σ_2 of genus 2 such that $\sigma(\Sigma_2;\mathcal{S}) \neq \sigma(\mathcal{S}_x)\sigma(\Sigma_2) = 0$ is due to Meyer, [Mey72]. Define $\mathcal{S}$ on Σ_2 by the representation*

$$\mu : \pi_1\Sigma_2 \longrightarrow Sp(2;\mathbb{R}),$$

$$\mu(a_1) = \begin{pmatrix} -5 & 1 \\ -\frac{27}{2} & \frac{5}{2} \end{pmatrix},\ \mu(a_2) = \begin{pmatrix} -4 & -1 \\ 33 & 8 \end{pmatrix},\ \mu(b_1) = \begin{pmatrix} -4 & 1 \\ -33 & 8 \end{pmatrix},\ \mu(b_2) = \begin{pmatrix} -5 & -1 \\ \frac{27}{2} & \frac{5}{2} \end{pmatrix},$$

where $\pi_1\Sigma_2 = \langle a_1, a_2, b_1, b_2 | [a_1, b_1][a_2, b_2] = 1\rangle$. If $H^2(\Sigma_2) = \mathbb{Q}\langle\gamma\rangle$, $\langle\gamma, [\Sigma_2]\rangle = 1$, then $\mathcal{S}$ has the property that the first Chern class

$$c_1(\mathcal{S}_{\mathbb{C}}) = -\gamma,$$

where $[\mathcal{S}]_K = [\mathcal{S}^*_{\mathbb{C}}] - [\mathcal{S}_{\mathbb{C}}] \in KU(\Sigma_2)$, [Mey72]. If ξ is any complex line bundle on Σ_2, then

$$\widetilde{\mathrm{ch}}(\xi) = e^{2c_1(\xi)} = 1 + 2c_1(\xi).$$

Hence

$$\begin{aligned}\widetilde{\mathrm{ch}}[\mathcal{S}]_K &= \widetilde{\mathrm{ch}}([\mathcal{S}^*_{\mathbb{C}}] - [\mathcal{S}_{\mathbb{C}}]) \\ &= 1 + 2c_1(\mathcal{S}^*_{\mathbb{C}}) - 1 - 2c_1(\mathcal{S}_{\mathbb{C}}) \\ &= -2c_1(\mathcal{S}_{\mathbb{C}}) - 2c_1(\mathcal{S}_{\mathbb{C}}) \\ &= -4c_1(\mathcal{S}_{\mathbb{C}}) \\ &= 4\gamma\end{aligned}$$

and by Meyer's formula

$$\sigma(\Sigma_2;\mathcal{S}) = \langle \widetilde{\mathrm{ch}}([\mathcal{S}]_K) \cup \ell(\Sigma_2), [\Sigma_2]\rangle = \langle 4\gamma, [\Sigma_2]\rangle = 4.$$

8.3.2 L-Classes for Singular Spaces with Boundary

The present section reviews the construction of homology L-classes of stratified spaces with boundary as well as their relation to the L-classes of the boundary. Recall that every self-dual complex of sheaves $\mathbf{S}^\bullet \in D_c^b(X)$ on a closed oriented pseudomanifold X of dimension n has a set of L-classes

$$L_i(\mathbf{S}^\bullet) \in H_i(X; \mathbb{Q}),$$

uniquely determined by the requirements $L_0(\mathbf{S}^\bullet) = \sigma(\mathbf{S}^\bullet) \in H_0(X) \cong \mathbb{Z}$ (X connected, σ the signature of $\mathbf{S}^\bullet$) and $L_{i-n+m}(j^! \mathbf{S}^\bullet) = j^! L_i(\mathbf{S}^\bullet)$ for $j : Y^m \hookrightarrow X^n$ a normally nonsingular inclusion with trivial normal bundle, $j^! : H_i(X; \mathbb{Q}) \to H_{i-n+m}(Y; \mathbb{Q})$ the map given by intersection of cycles with Y (see Sect. 8.2.3, note that $j^! \mathbf{S}^\bullet$ is a self-dual sheaf on Y). The intersection chain sheaf $\mathbf{S}^\bullet = \mathbf{IC}^\bullet_{\bar{m}}(X)$ is self-dual for X a Witt space and we obtain L-classes

$$L_i(X) = L_i(\mathbf{IC}^\bullet_{\bar{m}}(X)) \in H_i(X; \mathbb{Q})$$

for Witt spaces X. If $n - i$ is odd, then $L_i = 0$.

Let $(M^n, \partial M)$ be a compact oriented smooth manifold with boundary. The Hirzebruch L-classes ℓ_i of M are the L-classes of the tangent bundle TM of M in cohomology

$$\ell_i(M) = \ell_i(TM) \in H^{4i}(M; \mathbb{Q}).$$

By Poincaré duality $H^{4i}(M) \cong H_{n-4i}(M, \partial M)$, and we have dual homology L-classes

$$L_i(M) \in H_i(M, \partial M; \mathbb{Q}),$$

so that $L_{n-4i} = \ell_i \cap [M]$, where $[M] \in H_n(M, \partial M)$ is the fundamental class. This illustrates that the correct address for L-classes of singular spaces X with boundary is relative homology $H_*(X, \partial X)$.

Now let $(X^n, \partial X)$ be a compact oriented Whitney stratified Witt space with boundary and S^k be a k-sphere with base point $p \in S^k$. The cohomotopy set $\pi^k(X, \partial X) = [(X, \partial X), (S^k, p)]$ is a group for $2k > n + 1$ and in that range the Hurewicz map is rationally an isomorphism

$$\pi^k(X, \partial X) \otimes \mathbb{Q} \cong H^k(X, \partial X; \mathbb{Q}). \tag{8.12}$$

Fix a point $q \in S^k$, $q \neq p$. A given continuous map $f : (X, \partial X) \to (S^k, p)$ is homotopic rel ∂X to a map $\tilde{f}$, the restriction of a smooth map on an open neighborhood of X in the ambient manifold implicit in the Whitney stratification, such that $\tilde{f}$ is transverse regular to q and $\tilde{f}^{-1}(q) \subset \operatorname{int} X$ is transverse to each stratum of X (in fact the modification of f takes place only in $f^{-1}(U_q) \subset \operatorname{int} X$, where $U_q \subset S^k$ is a small open neighborhood of q). Transversality implies that $\tilde{f}^{-1}(q)$ is a Witt space. Thus the intersection chain sheaf $\mathbf{IC}^\bullet_{\bar{m}}(\tilde{f}^{-1}(q))$ is self-dual and $\tilde{f}^{-1}(q)$

has a well-defined signature $\sigma(\tilde{f}^{-1}(q))$. Alternatively, observe that as transversality implies the normal nonsingularity of the inclusion $i : \tilde{f}^{-1}(q) \hookrightarrow X$, the restriction $i^!\mathbf{IC}^\bullet_{\bar{m}}(X)$ is a self-dual complex of sheaves on $\tilde{f}^{-1}(q)$ and hence has a signature. If $f, g : (X, \partial X) \to (S^k, p)$ are homotopic transverse maps, then the pre-image $H^{-1}(q) \subset \operatorname{int} X$ under a transverse homotopy rel ∂X, $H : X \times [0, 1] \to S^k$ is a Witt bordism between $f^{-1}(q)$ and $g^{-1}(q)$, so that the map

$$\begin{aligned} \lambda_k(X) : \pi^k(X, \partial X) &\to \quad \mathbb{Z} \\ [f] &\mapsto \sigma(\tilde{f}^{-1}(q)) \end{aligned}$$

is a well-defined homomorphism. Under the identification (8.12), $\lambda_k(X)$ induces a map

$$\lambda_k(X) \otimes \mathbb{Q} : H^k(X, \partial X; \mathbb{Q}) \longrightarrow \mathbb{Q}$$

defining an element $L_k(X) \in H_k(X, \partial X; \mathbb{Q}) \cong \operatorname{Hom}(H^k(X, \partial X; \mathbb{Q}), \mathbb{Q})$, the *$L$-class of the Witt space* $(X, \partial X)$. The restriction $2k > n+1$ is removed by considering products of $(X, \partial X)$ with spheres.

The main purpose of this section is to establish a relation between the L-class of X and the L-class of the boundary ∂X. What we will prove is that $L_{k+1}(X)$ hits $L_k(\partial X)$ under the boundary homomorphism on homology. Consequently, the push-forward of the L-class of the boundary into $H_*(X)$ vanishes.

Proposition 8.3.2 *Let $(X, \partial X)$ be a compact oriented Whitney stratified Witt space with boundary, $L_{k+1}(X)$ its $(k+1)$-th L-class and $L_k(\partial X)$ the k-th L-class of the boundary. With $\partial_* : H_{k+1}(X, \partial X; \mathbb{Q}) \to H_k(\partial X; \mathbb{Q})$ the homology boundary operator, we have*

$$\partial_* L_{k+1}(X) = L_k(\partial X).$$

Proof. Given $f : \partial X \to S^k$ transverse to $p \in S^k$. We shall describe how the cohomotopy coboundary operator

$$\delta^* : \pi^k(\partial X) \longrightarrow \pi^{k+1}(X, \partial X)$$

acts on $[f]$.

Write $c\partial X$ for the cone on ∂X and view $D^{k+1} \cong cS^k$. Then f extends over the cones as

$$cf : c\partial X \longrightarrow cS^k \cong D^{k+1}.$$

Let $q = (p, \frac{1}{2}) \in cS^k$ and N be an open collar neighborhood of ∂X in X. Consider the collapse maps

$$X \longrightarrow X/(X - N) \cong c\partial X$$

and

$$D^{k+1} \longrightarrow D^{k+1}/S^k \cong S^{k+1}.$$

Denote the images of p and q under the latter collapse again by $p, q \in S^{k+1}$. Then $\delta^*[f]$ is represented by the composition

$$g : (X, \partial X) \to (c\partial X, \partial X) \xrightarrow{cf} (D^{k+1}, S^k) \to (S^{k+1}, p).$$

Observe that g is transverse to q, since in fact $g^{-1}(q) = f^{-1}(p)\times\{\frac{1}{2}\}$ when regarded as a subvariety of the collar $N \cong \partial X \times [0, 1)$. If

$$\lambda_k(\partial X) : \pi^k(\partial X) \longrightarrow \mathbb{Z}$$

is the L-class of ∂X and

$$\lambda_{k+1}(X) : \pi^{k+1}(X, \partial X) \longrightarrow \mathbb{Z}$$

is the L-class of X, then

$$\lambda_{k+1}(X)(\delta^*[f]) = \lambda_{k+1}(X)[g] = \sigma(g^{-1}(q)) = \sigma(f^{-1}(p)) = \lambda_k(\partial X)[f]$$

and so

$$\lambda_{k+1}(X) \circ \delta^* = \lambda_k(\partial X).$$

The commutative diagram

$$\begin{array}{ccc} \pi^k(\partial X)\otimes\mathbb{Q} & \xrightarrow{\delta^*\otimes\mathbb{Q}} & \pi^{k+1}(X,\partial X)\otimes\mathbb{Q} \\ \cong\downarrow & & \downarrow\cong \\ H^k(\partial X;\mathbb{Q}) & \xrightarrow[\delta^*_H]{} & H^{k+1}(X,\partial X;\mathbb{Q}) \end{array}$$

shows

$$(\lambda_{k+1}(X)\otimes\mathbb{Q}) \circ \delta^*_H = \lambda_k(\partial X)\otimes\mathbb{Q}.$$

In other words, if $L_{k+1}(X) \in \mathrm{Hom}(H^{k+1}(X, \partial X), \mathbb{Q})$ is the element defined by $\lambda_{k+1}(X)\otimes\mathbb{Q}$ and $L_k(\partial X) \in \mathrm{Hom}(H^k(\partial X), \mathbb{Q})$ is the element defined by $\lambda_k(\partial X)\otimes \mathbb{Q}$, then

$$\mathrm{Hom}(\delta^*_H, \mathbb{Q})(L_{k+1}(X)) = L_k(\partial X)$$

and the commutative square

$$\begin{array}{ccc} \mathrm{Hom}(H^{k+1}(X,\partial X),\mathbb{Q}) & \xrightarrow{\mathrm{Hom}(\delta^*_H,\mathbb{Q})} & \mathrm{Hom}(H^k(\partial X),\mathbb{Q}) \\ \cong\uparrow & & \uparrow\cong \\ H_{k+1}(X,\partial X;\mathbb{Q}) & \xrightarrow[\partial_*]{} & H_k(\partial X;\mathbb{Q}) \end{array}$$

implies

$$\partial_* L_{k+1}(X) = L_k(\partial X).$$

□

8.3.3 Representability of Witt Spaces

We show that results of Sullivan and Siegel imply that at odd primes, Witt bordism is representable by smooth oriented bordism. This fact will be used subsequently to pull back calculations on singular spaces to calculations on smooth spaces.

Let $\Omega^{SO}_*(X, A)$ denote bordism of smooth oriented manifolds and let $\Omega^{Witt}_*(X, A)$ denote Witt space bordism.

Proposition 8.3.3 *For compact PL-pairs (X, A), the natural map*

$$\Omega_*^{SO}(X, A) \otimes \mathbb{Z}[\tfrac{1}{2}] \longrightarrow \Omega_*^{Witt}(X, A) \otimes \mathbb{Z}[\tfrac{1}{2}]$$

is surjective.

Proof. Considering the signature as a map $\sigma : \Omega_*^{SO}(pt) \to \mathbb{Z}[\frac{1}{2}]$ makes $\mathbb{Z}[\frac{1}{2}]$ into an $\Omega_*^{SO}(pt)$-module and we can form the homology theory

$$\Omega_*^{SO}(X, A) \otimes_{\Omega_*^{SO}(pt)} \mathbb{Z}[\tfrac{1}{2}].$$

Let $ko_*(X, A)$ denote connected KO homology. Sullivan [Sul70b] constructs a natural isomorphism of homology theories

$$\Omega_*^{SO}(X, A) \otimes_{\Omega_*^{SO}(pt)} \mathbb{Z}[\tfrac{1}{2}] \xrightarrow{\simeq} ko_*(X, A) \otimes \mathbb{Z}[\tfrac{1}{2}]$$

(for compact PL-pairs). Siegel [Sie83] shows that Witt spaces provide a geometric description of connected KO homology at odd primes: He constructs a natural isomorphism of homology theories

$$\Omega_*^{Witt}(X, A) \otimes \mathbb{Z}[\tfrac{1}{2}] \xrightarrow{\simeq} ko_*(X, A) \otimes \mathbb{Z}[\tfrac{1}{2}].$$

Now $\Omega_*^{SO}(X, A) \otimes_{\Omega_*^{SO}(pt)} \mathbb{Z}[\frac{1}{2}]$ being a quotient of $\Omega_*^{SO}(X, A) \otimes \mathbb{Z}[\frac{1}{2}]$ yields a natural surjection

$$\Omega_*^{SO}(X, A) \otimes \mathbb{Z}[\tfrac{1}{2}] \longrightarrow \Omega_*^{SO}(X, A) \otimes_{\Omega_*^{SO}(pt)} \mathbb{Z}[\tfrac{1}{2}].$$

The statement follows from the commutative diagram

$$\begin{array}{ccc} \Omega_*^{SO}(X, A) \otimes_{\Omega_*^{SO}(pt)} \mathbb{Z}[\frac{1}{2}] & \xrightarrow{\simeq} & ko_*(X, A) \otimes \mathbb{Z}[\frac{1}{2}] \\ \uparrow & & \uparrow \cong \\ \Omega_*^{SO}(X, A) \otimes \mathbb{Z}[\frac{1}{2}] & \longrightarrow & \Omega_*^{Witt}(X, A) \otimes \mathbb{Z}[\frac{1}{2}] \end{array}$$

□

Theorem 8.3.4 *Let X^n be a closed oriented Whitney stratified normal Witt space of even dimension with singular set Σ, and let $(\mathcal{S}, \phi)$ be a Poincaré local system on $X - \Sigma$, strongly transverse to Σ. Then*

$$\sigma(X; \mathcal{S}) = \epsilon_*(\widetilde{\mathrm{ch}}([\mathcal{S}]_K) \cap L(X)) \in \mathbb{Z}$$

where $L(X) \in H_{2*}(X; \mathbb{Q})$ is the total L-class of X.

Proof. First note that $\sigma(X; \mathcal{S})$ is indeed defined: By Proposition 8.2.22, $\mathcal{S}$ extends as a Poincaré local system over all of X. Thus if Lk is the link of an odd-codimensional stratum of X, then the restriction $\mathcal{S}|_{Lk-\Sigma}$ is a constant sheaf, and

$$IH_k^{\bar{m}}(Lk; \mathcal{S}) = 0,$$

since X is a Witt space ($\dim Lk = 2k$). Therefore (as pointed out in Sect. 8.2.4)

$$\phi : \mathcal{D}\mathcal{S}[-n] \xrightarrow{\simeq} \mathcal{S}$$

extends to a self-duality isomorphism

$$\bar{\phi} : \mathcal{D}\mathbf{IC}^{\bullet}_{\bar{m}}(X; \mathcal{S})[n] \xrightarrow{\simeq} \mathbf{IC}^{\bullet}_{\bar{m}}(X; \mathcal{S})$$

and the twisted signature $\sigma(X; \mathcal{S})$ is defined.

Next, we show that the expression

$$\epsilon_*(\widetilde{\mathrm{ch}}([\mathcal{S}]_K) \cap L(X)) \tag{8.13}$$

is a bordism invariant of $(X^n, \mathcal{S}, \phi)$ for globally defined Poincaré local systems $(\mathcal{S}, \phi)$. Let $(Y^{n+1}, \partial Y)$ be a compact oriented Witt space with boundary ∂Y, singular set Σ_Y, and $(\mathcal{T}, \psi)$ a Poincaré local system on all of Y. By naturality,

$$\widetilde{\mathrm{ch}}([\mathcal{T}|_{\partial Y}]_K) = j^*\widetilde{\mathrm{ch}}([\mathcal{T}]_K) \in H^{2*}(\partial Y; \mathbb{Q})$$

with $j : \partial Y \hookrightarrow Y$ the inclusion. Thus,

$$\begin{aligned} \epsilon_*(\widetilde{\mathrm{ch}}([\mathcal{T}|_{\partial Y}]_K) \cap L(\partial Y)) &= \langle \widetilde{\mathrm{ch}}([\mathcal{T}|_{\partial Y}]_K), L(\partial Y)\rangle \\ &= \langle j^*\widetilde{\mathrm{ch}}([\mathcal{T}]_K), L(\partial Y)\rangle \\ &= \langle \widetilde{\mathrm{ch}}([\mathcal{T}]_K), j_*L(\partial Y)\rangle. \end{aligned}$$

Now $j_*L(\partial Y) = 0$ by Proposition 8.3.2 and we have

$$\epsilon_*(\widetilde{\mathrm{ch}}([\mathcal{T}|_{\partial Y}]_K) \cap L(\partial Y)) = 0,$$

proving (8.13) to be bordism invariant.

The twisted signature $\sigma(X; \mathcal{S})$ is a bordism invariant as well: We prove this using geometric bordism of spaces covered by self-dual sheaves as developed in Sect. 8.2.2. Let $(Y^{n+1}, \partial Y)$, Σ_Y be as above and $(\mathcal{T}, \psi)$ be a Poincaré local system on $Y - \Sigma_Y$, strongly transverse to Σ_Y. Consider the diagram of inclusions

$$\begin{array}{ccccc} \partial Y - \Sigma_Y & \xrightarrow{j|} & Y - \Sigma_Y & \xleftarrow{i|} & (\mathrm{int}\, Y) - \Sigma_Y \\ \downarrow & & \downarrow & & \downarrow \\ \partial Y & \xrightarrow{j} & Y & \xleftarrow{i} & \mathrm{int}\, Y \end{array}$$

The restriction

$$j|^*\psi : j|^*\mathcal{T} \times j|^*\mathcal{T} \longrightarrow \mathbb{R}_{\partial Y - \Sigma_Y}$$

induces a self-duality isomorphism in the derived category $D^b_c(\partial Y)$

$$\bar{\psi}_\partial : \mathcal{D}\mathbf{IC}^{\bullet}_{\bar{m}}(\partial Y; j|^*\mathcal{T})[n] \xrightarrow{\simeq} \mathbf{IC}^{\bullet}_{\bar{m}}(\partial Y; j|^*\mathcal{T})$$

as $\mathcal{T}$ being strongly transverse to Σ_Y implies that $\mathcal{T}$ extends as a Poincaré local system over Y (Proposition 8.2.22), so is constant on links of odd-codimensional strata, and

$$\sigma(\partial Y; \mathcal{T}|_{\partial Y - \Sigma_Y}) = \sigma(\mathbf{IC}^\bullet_{\bar{m}}(\partial Y; j|^*\mathcal{T}), \bar{\psi}_\partial).$$

The triple $(\partial Y, \mathbf{IC}^\bullet_{\bar{m}}(\partial Y; j|^*\mathcal{T}), \bar{\psi}_\partial)$ is of the type considered in Sect. 8.2.2. The restriction to the interior

$$i|^*\psi : i|^*\mathcal{T} \times i|^*\mathcal{T} \longrightarrow \mathbb{R}_{\mathrm{int}\, Y - \Sigma_Y}$$

induces a self-duality isomorphism in $D^b_c(\mathrm{int}\, Y)$

$$\bar{\psi}_0 : \mathcal{D}\mathbf{IC}^\bullet_{\bar{m}}(\mathrm{int}\, Y; i|^*\mathcal{T})[n+1] \xrightarrow{\simeq} \mathbf{IC}^\bullet_{\bar{m}}(\mathrm{int}\, Y; i|^*\mathcal{T}).$$

Thus, the triple $(Y^{n+1}, \mathbf{IC}^\bullet_{\bar{m}}(\mathrm{int}\, Y; i|^*\mathcal{T}), \bar{\psi}_0)$ is an admissible bordism in the sense of Sect. 8.2.2. The boundary of such a bordism was defined to be (see Definition 8.2.7)

$$\partial(Y^{n+1}, \mathbf{IC}^\bullet_{\bar{m}}(\mathrm{int}\, Y; i|^*\mathcal{T}), \bar{\psi}_0) = (\partial Y, j^! Ri_! \mathbf{IC}^\bullet_{\bar{m}}(\mathrm{int}\, Y; i|^*\mathcal{T}), \partial\bar{\psi}_0),$$

where $\partial\bar{\psi}_0$ is induced by $j^* Ri_*(\bar{\psi}_0)$ under the canonical identification

$$j^* Ri_* \mathbf{A}^\bullet \xrightarrow{\simeq} j^! Ri_! \mathbf{A}^\bullet[1],$$

for any $\mathbf{A}^\bullet \in D^b_c(\mathrm{int}\, Y)$, Lemma 8.2.6. Now

$$\partial(Y^{n+1}, \mathbf{IC}^\bullet_{\bar{m}}(\mathrm{int}\, Y; i|^*\mathcal{T}), \bar{\psi}_0) = (\partial Y, \mathbf{IC}^\bullet_{\bar{m}}(\partial Y; j|^*\mathcal{T}), \bar{\psi}_\partial)$$

(cf. also [Ban02, Lemma 4.4]). Thus Proposition 8.2.8 implies

$$\sigma(\partial Y; \mathcal{T}|_{\partial Y - \Sigma_Y}) = \sigma(\mathbf{IC}^\bullet_{\bar{m}}(\partial Y; j|^*\mathcal{T}), \bar{\psi}_\partial) = 0.$$

We return to the given X^n, $(\mathcal{S}, \phi)$ on $X - \Sigma$, strongly transverse to Σ. Consider the identity map

$$[X \xrightarrow{1} X] \otimes 1 \in \Omega_n^{Witt}(X) \otimes \mathbb{Z}[\tfrac{1}{2}].$$

By Proposition 8.3.3,

$$\Omega_*^{SO}(X) \otimes \mathbb{Z}[\tfrac{1}{2}] \longrightarrow \Omega_*^{Witt}(X) \otimes \mathbb{Z}[\tfrac{1}{2}]$$

is onto. Hence there exists a smooth oriented manifold M^n, a continuous map $f : M \to X$ and $r, s \in \mathbb{Z}$ such that

$$[M \xrightarrow{f} X] \otimes \frac{r}{2^s} = [X \xrightarrow{1} X] \otimes 1 \in \Omega_n^{Witt}(X) \otimes \mathbb{Z}[\tfrac{1}{2}]$$

and we have

$$r[M \xrightarrow{f} X] = 2^s[X \xrightarrow{1} X] \in \Omega_n^{Witt}(X).$$

Let $(\bar{\mathcal{S}}, \bar{\phi})$ denote the extension of $(\mathcal{S}, \phi)$ to X, Proposition 8.2.22. On M, we consider the Poincaré local system $(f^*\bar{\mathcal{S}}, f^*\bar{\phi})$. Then

$$\begin{aligned}
2^s\sigma(X; \mathcal{S}) &= r\sigma(M; f^*\bar{\mathcal{S}}) && \text{(bordism invariance)}\\
&= r\epsilon_*(\widetilde{\mathrm{ch}}([f^*\bar{\mathcal{S}}]_K) \cap L(M)) && \text{(by Atiyah and Meyer)}\\
&= 2^s\epsilon_*(\widetilde{\mathrm{ch}}([\mathcal{S}]_K) \cap L(X)) && \text{(bordism invariance).}
\end{aligned}$$

□

9

Invariants of Non-Witt Spaces

9.1 Duality on Non-Witt Spaces: Lagrangian Structures

In Sect. 6.4, we have defined a Witt space to be a pseudomanifold such that the middle-perversity, middle-dimensional intersection homology groups of all links of odd-codimensional strata vanish, that is, if $IH_l^{\bar{m}}(Link(x)) = 0$ for all $x \in X_{n-2l-1} - X_{n-2l-2}$ and all $l \geq 1$. As we have seen, this condition characterizes the self-duality of $\mathbf{IC}^\bullet_{\bar{m}}(X)$. If X has only strata of even codimension, for example if X is complex algebraic, then it is automatically a Witt space. The suspension $X^3 = \Sigma T^2$ has two singular points which form a stratum of odd codimension 3. The space X^3 is not Witt, since the middle Betti number of the link T^2 is 2. Consequently, the sheaf complex $\mathbf{IC}^\bullet_{\bar{m}}(X^3)$ is not self-dual. It was the self-duality of $\mathbf{IC}^\bullet_{\bar{m}}$ that allowed us to construct invariants such as the signature $\sigma(X)$ and the L-class $L(X)$ of a Witt space X. Thus the following fundamental task arises:

> Develop an obstruction theory for the existence of self-dual intersection homology theories for non-Witt spaces, and, provided the obstructions vanish, construct these theories.

Let $\mathcal{LK}$ be a collection of closed, oriented, topological pseudomanifolds. We can then form the bordism group $\Omega_i^{\mathcal{LK}}$ whose elements are represented by closed, oriented i-dimensional pseudomanifolds whose links are all homeomorphic to (finite disjoint unions of) elements of $\mathcal{LK}$. Two spaces X and X' represent the same bordism class, $[X] = [X']$, if there exists an $(i+1)$-dimensional, oriented, compact pseudomanifold-with-boundary Y^{i+1} such that all links of $Y - \partial Y$ are in $\mathcal{LK}$ and $\partial Y \cong X \sqcup -X'$ under an orientation-preserving homeomorphism. (The boundary is, as always, collared in a stratum-preserving way.) If, for instance,

$$\mathcal{LK} = \{S^1, S^2, S^3, \dots\},$$

then $\Omega_*^{\mathcal{LK}}$ is bordism of manifolds. If

$$\mathcal{LK} = Odd \cup \{L^{2l} \mid IH_l^{\bar{m}}(L) = 0\},$$

where *Odd* is the collection of all homeomorphism types of odd-dimensional, oriented, closed pseudomanifolds, then $\Omega_*^{\mathcal{LK}} = \Omega_*^{Witt}$. The question is: Which other spaces can one throw into this $\mathcal{LK}$, yielding an enlarged collection $\mathcal{LK}' \supset \mathcal{LK}$, such that one can still define a bordism invariant signature

$$\sigma : \Omega_*^{\mathcal{LK}'} \longrightarrow \mathbb{Z}$$

so that the diagram

$$\begin{array}{ccc} \Omega_*^{Witt} & \xrightarrow{\sigma} & \mathbb{Z} \\ \downarrow & \nearrow_{\sigma} & \\ \Omega_*^{\mathcal{LK}'} & & \end{array}$$

commutes, where

$$\Omega_*^{Witt} \longrightarrow \Omega_*^{\mathcal{LK}'}$$

is the canonical map induced by the inclusion $\mathcal{LK} \subset \mathcal{LK}'$? Note that $\sigma(L) = 0$ for every $L \in \mathcal{LK}$. Suppose we took an $\mathcal{LK}'$ that contains a manifold P with $\sigma(P) \neq 0$, e.g. $P = \mathbb{C}P^{2n}$. Then

$$[P] = 0 \in \Omega_*^{\mathcal{LK}'}$$

since P is the boundary of the cone on P, and the cone on P is an admissible bordism in $\Omega_*^{\mathcal{LK}'}$, as the link of the cone-point is P and $P \in \mathcal{LK}'$. Thus, in the above diagram,

$$\begin{array}{ccc} [P] & \stackrel{\sigma}{\longmapsto} & \sigma(P) \neq 0 \\ \downarrow & \nearrow_{\sigma} & \\ 0 & & \end{array}$$

This argument shows that the desired diagonal arrow cannot exist for any collection $\mathcal{LK}'$ that contains any manifolds with nonzero signature. Thus we are naturally lead to consider only links with zero signature, that is, links whose intersection form on middle-dimensional homology possesses a Lagrangian subspace. In the following, we shall formalize the notion of a *Lagrangian structure* along strata of odd codimension. We will show that every Verdier self-dual sheaf close to $\mathbf{IC}^\bullet_{\bar{m}}$ and $\mathbf{IC}^\bullet_{\bar{n}}$ possesses Lagrangian structures, and conversely, that Lagrangian structures serve as the building blocks for such self-dual sheaves. This can be carried out functorially, so that ultimately one obtains an equivalence of categories between self-dual sheaves close to $\mathbf{IC}^\bullet_{\bar{m}}$ and $\mathbf{IC}^\bullet_{\bar{n}}$ and a fibered product of categories of Lagrangian structures. The equivalence thus encodes both the obstruction theory and the constructive technology that we set out to develop in the task stated above. This theory was introduced in [Ban02], and the present section is a summary thereof.

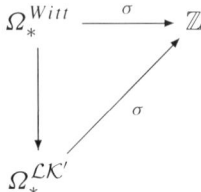

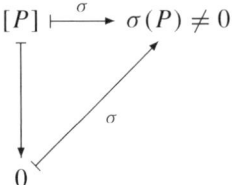

Let X^n be an n-dimensional pseudomanifold with a fixed stratification $X = X_n \supset X_{n-2} \supset X_{n-3} \supset \cdots \supset X_0 \supset X_{-1} = \varnothing$ such that X_j is closed in X and $X_j - X_{j-1}$ is an open manifold of dimension j. Set $U_k = X - X_{n-k}$ and let $i_k : U_k \hookrightarrow U_{k+1}$ denote the inclusion. Assume k is odd and $\mathbf{A}^\bullet \in D_c^b(U_k)$. Note that $\bar{n}(k) = \bar{m}(k) + 1$. We will work here with real coefficients.

9.1.1 The Lifting Obstruction

Definition 9.1.1 *The* lifting obstruction $\mathcal{O}(\mathbf{A}^\bullet)$ *associated with* $\mathbf{A}^\bullet$ *is defined to be*

$$\mathcal{O}(\mathbf{A}^\bullet) = \mathbf{H}^{\bar{n}(k)-n}(Ri_{k*}\mathbf{A}^\bullet)[n - \bar{n}(k)] \in D_c^b(U_{k+1})$$

Thus, $\mathcal{O}(\mathbf{A}^\bullet)$ is a complex concentrated in dimension $\bar{n}(k) - n$. If $\mathbf{A}^\bullet$ satisfies a stalk condition $\mathbf{H}^i(\mathbf{A}^\bullet) = 0$ for $i > \bar{n}(k-1) - n$, then the support of the lifting obstruction $\mathcal{O}(\mathbf{A}^\bullet)$ is contained in $U_{k+1} - U_k = X_{n-k} - X_{n-k-1}$. We shall henceforth use the shorthand notation

$${}^{\bar{m}}\mathbf{A}^\bullet = \tau_{\leq \bar{m}(k)-n} Ri_{k*}\mathbf{A}^\bullet, \qquad {}^{\bar{n}}\mathbf{A}^\bullet = \tau_{\leq \bar{n}(k)-n} Ri_{k*}\mathbf{A}^\bullet, \qquad s = \bar{n}(k) - n.$$

Consider the mapping cone $\mathbf{C}^\bullet$ of the canonical morphism ${}^{\bar{m}}\mathbf{A}^\bullet \to {}^{\bar{n}}\mathbf{A}^\bullet$. The distinguished triangle

$$\begin{array}{ccc} {}^{\bar{m}}\mathbf{A}^\bullet & \longrightarrow & {}^{\bar{n}}\mathbf{A}^\bullet \\ {\scriptstyle [1]}\nwarrow & & \swarrow \\ & \mathbf{C}^\bullet & \end{array}$$

induces a long exact sequence of cohomology sheaves, which shows that $\mathbf{C}^\bullet \cong \mathcal{O}(\mathbf{A}^\bullet)$. Thus the lifting obstruction fits into a distinguished triangle

$$\begin{array}{ccc} {}^{\bar{m}}\mathbf{A}^\bullet & \longrightarrow & {}^{\bar{n}}\mathbf{A}^\bullet \\ {\scriptstyle [1]}\nwarrow & & \swarrow \\ & \mathcal{O}(\mathbf{A}^\bullet) & \end{array} \tag{9.1}$$

This triangle induces a long exact sequence on hypercohomology groups so that $\mathcal{H}^*(X; \mathcal{O}(\mathbf{A}^\bullet))$ can be used to decide whether a class in $\mathcal{H}^*(X; {}^{\bar{n}}\mathbf{A}^\bullet)$ can be lifted to a class in $\mathcal{H}^*(X; {}^{\bar{m}}\mathbf{A}^\bullet)$. The lifting obstruction is obviously a covariant functor

$$\mathcal{O} : D_c^b(U_k) \longrightarrow D_c^b(U_{k+1}).$$

9.1.2 The Category $SD(X)$ of Self-Dual Sheaves Compatible with IH

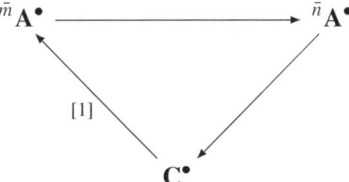

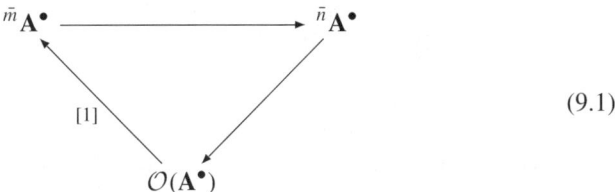

Assume X is oriented, i.e. assume we can and do fix an isomorphism $\mathbb{D}^\bullet_{U_2} \xrightarrow{\cong} \mathbb{R}_{U_2}[n]$. We assume furthermore that the strata and links are also given orientations that are compatible with the orientation of X under the stratum-preserving homeomorphism $U \times c^\circ L_x \xrightarrow{\cong} N_x$, where $x \in X_{n-k} - X_{n-k-1}$, U is an open neighborhood of x in $X_{n-k} - X_{n-k-1}$, $c^\circ L_x$ is the open cone on the link L_x at x and N_x is a neighborhood of x in X. We define the category of complexes of sheaves suitable for studying intersection homology type invariants on non-Witt spaces. The objects of this category should satisfy two properties: On the one hand, they should be self-dual, on the other hand, they should be as close to the middle perversity intersection chain sheaves as possible, that is, interpolate between $\mathbf{IC}^\bullet_{\bar{m}}(X)$ and $\mathbf{IC}^\bullet_{\bar{n}}(X)$. Given these specifications, we adopt the following definition:

Definition 9.1.2 *Let $SD(X)$ be the full subcategory of $D^b_c(X)$ whose objects $\mathbf{L}^\bullet$ satisfy the following axioms:*

(SD1): *(Normalization) $\mathbf{L}^\bullet$ has an associated isomorphism $\nu : \mathbb{R}_{U_2}[n] \xrightarrow{\cong} \mathbf{L}^\bullet|_{U_2}$.*
(SD2): *(Lower bound) $\mathbf{H}^i(\mathbf{L}^\bullet) = 0$, for $i < -n$.*
(SD3): *(Vanishing condition for the upper middle perversity $\bar{n}$) $\mathbf{H}^i(\mathbf{L}^\bullet|_{U_{k+1}}) = 0$, for $i > \bar{n}(k) - n$, $k \geq 2$.*
(SD4): *(Self-Duality) $\mathbf{L}^\bullet$ has an associated isomorphism $d : \mathcal{D}\mathbf{L}^\bullet[n] \xrightarrow{\cong} \mathbf{L}^\bullet$ such that $\mathcal{D}d[n] = (-1)^n d$ and $d|_{U_2}$ is compatible to the orientation under normalization so that*

$$\begin{array}{ccc}
\mathbb{R}_{U_2}[n] & \xrightarrow[\cong]{\nu} & \mathbf{L}^\bullet|_{U_2} \\
\text{orient}\uparrow\cong & & \cong\uparrow d|_{U_2} \\
\mathbb{D}^\bullet_{U_2} & \xrightarrow[\cong]{\mathcal{D}\nu^{-1}[n]} & \mathcal{D}\mathbf{L}^\bullet|_{U_2}[n]
\end{array}$$

commutes.

The reader may want to compare this set of axioms with [AX] of Definition 4.1.27: The axioms for $SD(X)$ equal the axioms for $\mathbf{IC}^\bullet_{\bar{n}}$, except that the costalk vanishing axiom (or attaching condition) has been replaced by the self-duality requirement. Note that if X is a Witt space and $\mathbf{L}^\bullet \in SD(X)$, then $\mathbf{IC}^\bullet_{\bar{m}}(X) \cong \mathbf{L}^\bullet \cong \mathbf{IC}^\bullet_{\bar{n}}(X)$. In general, $SD(X)$ may or may not be empty, and we have to analyze the structure of $SD(X)$ more closely to be able to decide for which spaces X the category $SD(X)$ is not empty. Axioms (SD3) and (SD4) imply that any $\mathbf{L}^\bullet \in SD(X)$ satisfies the $\bar{m}$-costalk condition, i.e.

$$\mathbf{H}^i(j_x^! \mathbf{L}^\bullet) = 0 \quad \text{for } i \leq \bar{m}(k) - k + 1, x \in X_{n-k} - X_{n-k-1},$$

where $j_x : \{x\} \hookrightarrow X$ is the inclusion of a point. This fact, together with (SD3) and Proposition 8.1.10, shows that for $\mathbf{E}^\bullet, \mathbf{F}^\bullet \in SD(U_{k+1})$, restriction induces an injection

$$\operatorname{Hom}_{D^b_c(U_{k+1})}(\mathbf{E}^\bullet, \mathbf{F}^\bullet) \hookrightarrow \operatorname{Hom}_{D^b_c(U_k)}(\mathbf{E}^\bullet|_{U_k}, \mathbf{F}^\bullet|_{U_k}).$$

This establishes the important principle that morphisms between objects in $SD(X)$ are uniquely determined by their restrictions to the top stratum. Let $\mathbf{A}^\bullet \in SD(U_k)$. Then d induces isomorphisms

$$\mathcal{D}^{\bar{m}}\mathbf{A}^\bullet[n] \xrightarrow{\cong} {}^{\bar{n}}\mathbf{A}^\bullet, \qquad \mathcal{D}^{\bar{n}}\mathbf{A}^\bullet[n] \xrightarrow{\cong} {}^{\bar{m}}\mathbf{A}^\bullet$$

which are dual to each other. Thus, applying the dualizing functor $\mathcal{D}$ to the lifting obstruction triangle (9.1), d induces a self-duality isomorphism δ for the lifting obstruction:

$$\delta : \mathcal{D}\mathcal{O}(\mathbf{A}^\bullet)[n+1] \xrightarrow{\cong} \mathcal{O}(\mathbf{A}^\bullet).$$

Let us describe the relation of objects in $SD(X)$ to $\mathbf{IC}^\bullet_{\bar{m}}(X)$ and $\mathbf{IC}^\bullet_{\bar{n}}(X)$. Given $\mathbf{L}^\bullet \in SD(U_{k+1})$, we shall construct certain natural morphisms

$${}^{\bar{m}}(\mathbf{L}^\bullet|_{U_k}) \xrightarrow{a} \mathbf{L}^\bullet \xrightarrow{b} {}^{\bar{n}}(\mathbf{L}^\bullet|_{U_k}).$$

Consider the adjunction $\mathbf{L}^\bullet \to Ri_{k*}i_k^*\mathbf{L}^\bullet$. Over U_k, this is an isomorphism. For the inclusion $j_k : U_{k+1} - U_k \hookrightarrow U_{k+1}$, we have that

$$\mathbf{H}^i(j_k^*\mathbf{L}^\bullet) \to \mathbf{H}^i(j_k^*Ri_{k*}i_k^*\mathbf{L}^\bullet)$$

is an isomorphism when $i \leq \bar{m}(k) - n$, as $\mathbf{L}^\bullet$ satisfies the $\bar{m}$-costalk condition. It follows that

$$\tau_{\leq \bar{m}(k)-n}\mathbf{L}^\bullet \xrightarrow{\cong} \tau_{\leq \bar{m}(k)-n}Ri_{k*}i_k^*\mathbf{L}^\bullet$$

is an isomorphism.

We define a to be the composition of the inverse morphism ${}^{\bar{m}}\mathbf{L}^\bullet|_{U_k} \xrightarrow{\cong} \tau_{\leq \bar{m}(k)-n}\mathbf{L}^\bullet$ with the canonical map $\tau_{\leq \bar{m}(k)-n}\mathbf{L}^\bullet \to \mathbf{L}^\bullet$:

$$\begin{array}{ccc} {}^{\bar{m}}\mathbf{L}^\bullet|_{U_k} & \xrightarrow{\cong} & \tau_{\leq \bar{m}(k)-n}\mathbf{L}^\bullet \\ & \searrow^{a} & \downarrow \\ & & \mathbf{L}^\bullet \end{array}$$

Note that on U_k, a is the identity: $a|_{U_k} = 1_{\mathbf{L}^\bullet|U_k}$.

To construct b, consider the canonical map $\tau_{\leq \bar{n}(k)-n}\mathbf{L}^\bullet \to \mathbf{L}^\bullet$. This is a quasi-isomorphism by (SD3). Adjunction induces $\tau_{\leq \bar{n}(k)-n}\mathbf{L}^\bullet \to \tau_{\leq \bar{n}(k)-n}Ri_{k*}i_k^*\mathbf{L}^\bullet$ and we define b to be the composition

$$\begin{array}{ccc} \mathbf{L}^\bullet & \xrightarrow{\cong} & \tau_{\leq \bar{n}(k)-n}\mathbf{L}^\bullet \\ & \searrow^{b} & \downarrow \\ & & {}^{\bar{n}}\mathbf{L}^\bullet|_{U_k} \end{array}$$

Note that on U_k, b is the identity: $b|_{U_k} = 1_{\mathbf{L}^\bullet|U_k}$.

Now given $\mathbf{L}^\bullet \in SD(X)$, we can apply this method inductively, starting with

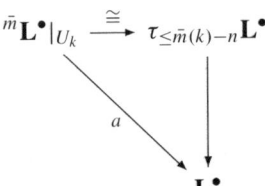

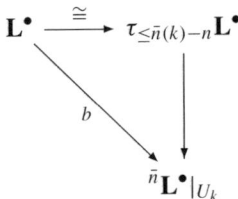

$$^{\bar{m}}(\mathbf{L}^\bullet|_{X-X_0}) \xrightarrow{a} \mathbf{L}^\bullet \xrightarrow{b} {}^{\bar{n}}(\mathbf{L}^\bullet|_{X-X_0}),$$

to obtain morphisms

$$\mathbf{IC}^\bullet_{\bar{m}}(X) \xrightarrow{\alpha} \mathbf{L}^\bullet \xrightarrow{\beta} \mathbf{IC}^\bullet_{\bar{n}}(X),$$

factoring the canonical morphism $\mathbf{IC}^\bullet_{\bar{m}}(X) \to \mathbf{IC}^\bullet_{\bar{n}}(X)$, such that

$$\begin{array}{ccc} \mathbf{IC}^\bullet_{\bar{m}}(X) & \xrightarrow{\alpha} & \mathbf{L}^\bullet \\ \cong\uparrow & & \cong\uparrow d \\ \mathcal{D}\mathbf{IC}^\bullet_{\bar{n}}(X)[n] & \xrightarrow{\mathcal{D}\beta[n]} & \mathcal{D}\mathbf{L}^\bullet[n] \end{array}$$

(where d is given by (SD4)) commutes. Hence, the objects of $SD(X)$ are precisely the self-dual interpolations between $\mathbf{IC}^\bullet_{\bar{m}}(X)$ and $\mathbf{IC}^\bullet_{\bar{n}}(X)$.

9.1.3 Lagrangian Structures

We are interested in understanding the structure of sheaves in $SD(X)$. The main tool in this analysis and indeed the building blocks for objects in $SD(X)$ are certain locally constant sheaves called *Lagrangian structures* which we will regard as complexes in the derived category concentrated in one dimension. Fix some k odd, $k \geq 2$. We state the definition first and then provide further discussion and intuition.

Definition 9.1.3 *Let* $\mathbf{A}^\bullet \in SD(U_k)$. *Recall that the lifting obstruction* $\mathcal{O}(\mathbf{A}^\bullet)$ *is self-dual. A* Lagrangian structure *is a morphism* $\mathcal{L} \longrightarrow \mathcal{O}(\mathbf{A}^\bullet)$, $\mathcal{L} \in D^b_c(U_{k+1})$, *which induces injections on stalks and has the property that some distinguished triangle on* $\mathcal{L} \longrightarrow \mathcal{O}(\mathbf{A}^\bullet)$ *is a nullbordism (see Definition* 8.1.11) *for the self-dual lifting obstruction.*

This means precisely the following: First of all, some triangle on $\phi : \mathcal{L} \longrightarrow \mathcal{O}(\mathbf{A}^\bullet)$ has to be of the form

$$\begin{array}{ccc} \mathcal{L} & \xrightarrow{\phi} & \mathcal{O}(\mathbf{A}^\bullet) \\ {}_{\gamma}\nwarrow^{[1]} & & \swarrow_{\phi'} \\ & \mathcal{D}\mathcal{L}[n+1] & \end{array} \tag{9.2}$$

and we require $\mathcal{D}\gamma[n+1] = \pm\gamma[-1]$.

Let $\mathcal{L}_A \to \mathcal{O}(\mathbf{A}^\bullet)$, $\mathcal{L}_B \to \mathcal{O}(\mathbf{B}^\bullet)$ be Lagrangian structures. A *morphism of Lagrangian structures* is a commutative diagram in $D^b_c(U_{k+1})$

$$\begin{array}{ccc} \mathcal{L}_A & \longrightarrow & \mathcal{O}(\mathbf{A}^\bullet) \\ \downarrow & & \downarrow \mathcal{O}(f) \\ \mathcal{L}_B & \longrightarrow & \mathcal{O}(\mathbf{B}^\bullet) \end{array}$$

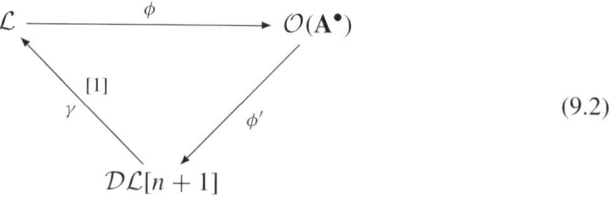

for some $f : \mathbf{A}^\bullet \to \mathbf{B}^\bullet$. It follows from the fact that the lifting obstruction is a functor that the composition of morphisms of Lagrangian structures is well-defined. Thus Lagrangian structures form a category

$$\mathrm{Lag}(U_{k+1} - U_k).$$

The notation $U_{k+1} - U_k$ in the argument is chosen to reflect $\mathrm{supp}(\mathcal{L}) \subset U_{k+1} - U_k$ for a Lagrangian structure $\mathcal{L} \to \mathcal{O}(\mathbf{A}^\bullet)$, $\mathbf{A}^\bullet \in SD(U_k)$.

Let us explain the terminology "Lagrangian" structure. Given a locally constant sheaf $\mathbf{H}$ on a manifold M, together with a nonsingular pairing

$$\phi : \mathbf{H} \otimes \mathbf{H} \longrightarrow \mathbb{R}_M,$$

let us call a subsheaf $\mathbf{E} \subset \mathbf{H}$ *Lagrangian* if, for every $x \in M$, the stalk $\mathbf{E}_x$ is a maximally self-annihilating (i.e. Lagrangian) subspace of $\mathbf{H}_x$ with respect to ϕ_x. Now let $\mathbf{A}^\bullet \in SD(U_k)$, and $i : U_k \hookrightarrow U_{k+1}$, $j : U_{k+1} - U_k \hookrightarrow U_{k+1}$ be inclusions. Put $\Sigma = U_{k+1} - U_k$ and $\mathbf{H} = \mathbf{H}^s(j^* R i_* \mathbf{A}^\bullet)$ (with $s = \bar{n}(k) - n$). Then $\mathbf{H}$ is a local system on the open stratum Σ. The self-duality isomorphism $\delta : \mathcal{D}\mathcal{O}(\mathbf{A}^\bullet)[n+1] \xrightarrow{\cong} \mathcal{O}(\mathbf{A}^\bullet)$ induces a nonsingular pairing

$$\delta' : \mathbf{H} \otimes \mathbf{H} \longrightarrow \mathbb{R}_\Sigma.$$

We shall show that a Lagrangian subsheaf $\mathbf{E} \subset \mathbf{H}$ gives rise to a Lagrangian structure $\mathcal{L} \to \mathcal{O}(\mathbf{A}^\bullet)$. Consider the exact sequence

$$0 \longrightarrow \mathbf{E} \xrightarrow{\alpha} \mathbf{H} \xrightarrow{\beta} \mathbf{Q} \longrightarrow 0 \tag{9.3}$$

where $\mathbf{Q} = \mathbf{H}/\mathbf{E}$, α is the inclusion and β the quotient map. As $\mathbf{E}$ is Lagrangian, δ' induces a nonsingular pairing $\mathbf{E} \otimes \mathbf{Q} \to \mathbb{R}_\Sigma$ such that

$$\begin{array}{ccc}
\mathbf{E} & \xrightarrow{\alpha} & \mathbf{H} \\
\otimes & & \otimes \\
\mathbf{Q} & \xleftarrow{\beta} & \mathbf{H} \\
\downarrow & & \downarrow{\scriptstyle \delta'} \\
\mathbb{R}_\Sigma & & \mathbb{R}_\Sigma
\end{array} \tag{9.4}$$

commutes. Let $\mathcal{L} = \mathbf{E}[-s]$, $\mathcal{Q} = \mathbf{Q}[-s]$, regarded as objects in the derived category and concentrated in dimension s. Then α and β induce unique morphisms $\mathcal{L} \to \mathcal{O}(\mathbf{A}^\bullet)$, $\mathcal{O}(\mathbf{A}^\bullet) \to \mathcal{Q}$ respectively, which we shall again denote by α, β. Now if $\mathcal{A}$ is any local system on a manifold M, then $\mathcal{D}\mathcal{A}[-\dim M] = \mathcal{A}^* \otimes or_M$, where $\mathcal{A}^*$ is the local system with stalks $\mathcal{A}^*_x = \mathrm{Hom}(\mathcal{A}_x, \mathbb{R})$ and or_M is the orientation sheaf, see [B$^+$84, V, 7.10(4)]. In the present case it follows that $\mathcal{D}\mathcal{L}[n+1] = \mathcal{L}^*$, $\mathcal{D}\mathcal{O}(\mathbf{A}^\bullet)[n+1] = \mathcal{O}(\mathbf{A}^\bullet)^*$ and $\mathcal{D}\mathcal{Q}[n+1] = \mathcal{Q}^*$, as Σ is oriented. The pairings in (9.4) induce isomorphisms $\mathcal{O}^* \xrightarrow{\cong} \mathcal{O}$, $\mathcal{Q}^* \xrightarrow{\cong} \mathcal{L}$ such that

$$\begin{array}{ccc}
\mathcal{L} & \xrightarrow{\alpha} & \mathcal{O} \\
\cong\uparrow & & \uparrow\cong \\
\mathcal{Q}^* & \xrightarrow{\beta^*} & \mathcal{O}^*
\end{array}$$

commutes. The short exact sequence (9.3) gives rise to a standard triangle in the derived category and we obtain an isomorphism of triangles

$$\begin{array}{ccccccc}
\mathcal{L} & \xrightarrow{\alpha} & \mathcal{O} & \xrightarrow{\beta} & \mathcal{Q} & \xrightarrow{+1} \\
\cong\uparrow\delta_0 & & \cong\uparrow\delta & & \cong\uparrow\mathcal{D}\delta_0[n+1] & \\
\mathcal{DQ}[n+1] & \xrightarrow{\mathcal{D}\beta[n+1]} & \mathcal{DO}[n+1] & \xrightarrow{(-1)^{n+1}\mathcal{D}\alpha[n+1]} & \mathcal{DL}[n+1] & \xrightarrow{+1}
\end{array}$$

Using $\mathcal{D}\delta_0[n+1]$, we finally arrive at a triangle

$$\mathcal{L} \xrightarrow{\alpha} \mathcal{O} \xrightarrow{\alpha'} \mathcal{DL}[n+1] \xrightarrow{+1}$$

and a diagram

$$\begin{array}{ccccccc}
\mathcal{L} & \xrightarrow{\alpha} & \mathcal{O} & \xrightarrow{\alpha'} & \mathcal{DL}[n+1] & \xrightarrow{+1} \\
\| & & \cong\uparrow\delta & & \| & \\
\mathcal{L} & \xrightarrow{\mathcal{D}\alpha'[n+1]} & \mathcal{DO}[n+1] & \xrightarrow{(-1)^{n+1}\mathcal{D}\alpha[n+1]} & \mathcal{DL}[n+1] & \xrightarrow{+1}
\end{array}$$

which defines the desired Lagrangian structure. Conversely, given a Lagrangian structure it is clear that it gives rise to a Lagrangian subsheaf for the relevant local system. We prefer to use the definition of Lagrangian structures as stated, because it is better adapted to the triangulation of the derived category and to Verdier duality.

9.1.4 Extracting Lagrangian Structures from Self-Dual Sheaves

We demonstrate that every self-dual complex of sheaves in $SD(X)$ has naturally associated Lagrangian structures. More precisely, we will construct a covariant functor

$$\Lambda : SD(U_{k+1}) \longrightarrow \mathrm{Lag}(U_{k+1} - U_k).$$

Suppose we are given $\mathbf{L}^\bullet \in SD(U_{k+1})$. Put

$$\mathcal{L} = \mathbf{H}^s(\mathbf{L}^\bullet)[-s] \in D^b_c(U_{k+1}).$$

Then $\mathcal{L}$ is a mapping cone on the morphism

$${}^{\bar{m}}(\mathbf{L}^\bullet|_{U_k}) \xrightarrow{a} \mathbf{L}^\bullet$$

so that we have a distinguished triangle

$$\begin{array}{ccc} {}^{\bar m}\mathbf{L}^\bullet| & \xrightarrow{\quad a \quad} & \mathbf{L}^\bullet \\ & {\nwarrow}^{[1]} \qquad \swarrow & \\ & \mathcal{L} & \end{array} \tag{9.5}$$

As the square

$$\begin{array}{ccc} \mathcal{D}\mathbf{L}^\bullet[n] & \xrightarrow{\mathcal{D}a[n]} & \mathcal{D}^{\bar m}\mathbf{L}^\bullet|[n] \\ d\downarrow\cong & & \cong\downarrow \\ \mathbf{L}^\bullet & \xrightarrow{\quad b \quad} & {}^{\bar n}\mathbf{L}^\bullet| \end{array}$$

commutes, we see that the triangle obtained from dualizing triangle (9.5) is isomorphic to

$$\begin{array}{ccc} \mathbf{L}^\bullet & \xrightarrow{\quad b \quad} & {}^{\bar n}\mathbf{L}^\bullet| \\ & {\nwarrow}^{[1]} \qquad \swarrow & \\ & \mathcal{D}\mathcal{L}[n+1] & \end{array}$$

Since *ba* is a factorization of the canonical morphism ${}^{\bar m}\mathbf{L}^\bullet| \to {}^{\bar n}\mathbf{L}^\bullet|$, we have a triangle

$$\begin{array}{ccc} {}^{\bar m}\mathbf{L}^\bullet| & \xrightarrow{\quad ba \quad} & {}^{\bar n}\mathbf{L}^\bullet| \\ & {\nwarrow}^{[1]}_{\omega'} \qquad \swarrow_{\omega} & \\ & \mathcal{O}(\mathbf{L}^\bullet|) & \end{array} \tag{9.6}$$

Applying the octahedral axiom to the configuration of triangles consisting of (9.6) and

$$\begin{array}{ccccc} {}^{\bar m}\mathbf{L}^\bullet| & \xrightarrow{\quad a \quad} & \mathbf{L}^\bullet & \xrightarrow{\quad b \quad} & {}^{\bar n}\mathbf{L}^\bullet| \\ & {\nwarrow}^{[1]} \quad \swarrow & & {\nwarrow}^{[1]} \quad \swarrow & \\ & \mathcal{L} & & \mathcal{D}\mathcal{L}[n+1] & \end{array}$$

produces a distinguished triangle

$$\begin{array}{ccc} \mathcal{L} & \xrightarrow{\quad \phi \quad} & \mathcal{O}(\mathbf{L}^\bullet|) \\ & {\nwarrow}^{[1]}_{\gamma} \qquad \swarrow_{\phi'} & \\ & \mathcal{D}\mathcal{L}[n+1] & \end{array} \tag{9.7}$$

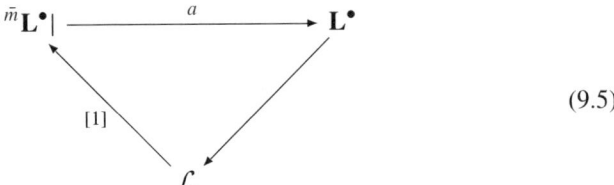

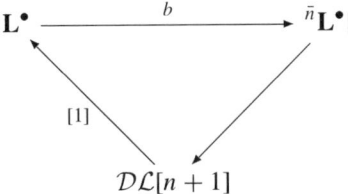

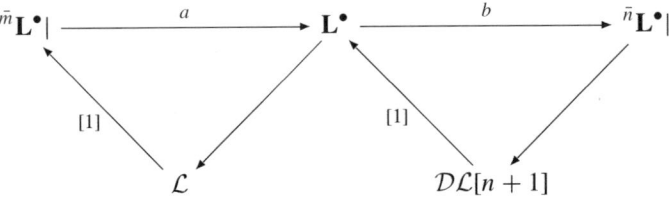

It can be shown (see Theorem 2.4 of [Ban02]) that

$$\mathcal{L} \stackrel{\phi}{\longrightarrow} \mathcal{O}(\mathbf{L}^\bullet|)$$

is a Lagrangian structure. We set

$$\Lambda(\mathbf{L}^\bullet) := (\mathcal{L} \stackrel{\phi}{\longrightarrow} \mathcal{O}(\mathbf{L}^\bullet|)) \in \mathrm{Lag}(U_{k+1} - U_k).$$

Let $f : \mathbf{L}_1^\bullet \to \mathbf{L}_2^\bullet$ be a morphism, $\mathbf{L}_1^\bullet, \mathbf{L}_2^\bullet \in SD(U_{k+1})$. Then we have a diagram

$$\begin{array}{ccccc}
\mathbf{L}_1^\bullet & \xrightarrow{\cong} & \tau_{\leq \bar{n}(k)-n}\mathbf{L}_1^\bullet & \xrightarrow{\tau_\leq(\mathrm{adj})} & \tau_{\leq \bar{n}(k)-n} Ri_{k*}i_k^*\mathbf{L}_1^\bullet = {}^{\bar{n}}\mathbf{L}_1^\bullet| \\
{\scriptstyle f}\downarrow & & \downarrow & & \downarrow{\scriptstyle {}^{\bar{n}}f|} \\
\mathbf{L}_2^\bullet & \xrightarrow{\cong} & \tau_{\leq \bar{n}(k)-n}\mathbf{L}_2^\bullet & \xrightarrow{\tau_\leq(\mathrm{adj})} & \tau_{\leq \bar{n}(k)-n} Ri_{k*}i_k^*\mathbf{L}_2^\bullet = {}^{\bar{n}}\mathbf{L}_2^\bullet|
\end{array}$$

where the upper horizontal composition is $b_1 : \mathbf{L}_1^\bullet \to {}^{\bar{n}}\mathbf{L}_1^\bullet|$ and the lower horizontal composition is $b_2 : \mathbf{L}_2^\bullet \to {}^{\bar{n}}\mathbf{L}_2^\bullet|$. As $1 \stackrel{\mathrm{adj}}{\longrightarrow} Ri_{k*}i_k^*$ is a natural morphism of functors, the above diagram commutes. Hence we have a commutative square on derived sheaves

$$\begin{array}{ccc}
\mathbf{H}^s(\mathbf{L}_1^\bullet) & \xrightarrow{\mathbf{H}^s(b_1)} & \mathbf{H}^s({}^{\bar{n}}\mathbf{L}_1^\bullet|) \\
{\scriptstyle \mathbf{H}^s(f)}\downarrow & & \downarrow{\scriptstyle \mathbf{H}^s({}^{\bar{n}}f|)} \\
\mathbf{H}^s(\mathbf{L}_2^\bullet) & \xrightarrow{\mathbf{H}^s(b_2)} & \mathbf{H}^s({}^{\bar{n}}\mathbf{L}_2^\bullet|)
\end{array} \tag{9.8}$$

that is, a morphism

$$\begin{array}{ccc}
\mathcal{L}_1 & \xrightarrow{\phi_1} & \mathcal{O}(\mathbf{L}_1^\bullet|) \\
\downarrow & & \downarrow{\scriptstyle \mathcal{O}(f|)} \\
\mathcal{L}_2 & \xrightarrow{\phi_2} & \mathcal{O}(\mathbf{L}_2^\bullet|)
\end{array}$$

of Lagrangian structures, which we define to be

$$\Lambda(f) \in \mathrm{Hom}_{\mathrm{Lag}(U_{k+1}-U_k)}(\Lambda(\mathbf{L}_1^\bullet), \Lambda(\mathbf{L}_2^\bullet)).$$

This completes the construction of the functor Λ.

9.1.5 Lagrangian Structures as Building Blocks for Self-Dual Sheaves

We will explain how a given self-dual sheaf on U_k, and a Lagrangian structure along the adjacent lower dimensional pure stratum $U_{k+1} - U_k$, naturally give rise to a self-dual sheaf on the union U_{k+1}.

Let us write $MS(X)$ for the monomorphism category associated with the category $Sh(X)$ of sheaves on X, that is, objects of $MS(X)$ are injective sheaf maps

$$\mathbf{E} \xrightarrow{\phi} \mathbf{F},$$

$\mathbf{E}, \mathbf{F} \in Sh(X)$, and morphisms in $MS(X)$ are commutative squares

$$\begin{array}{ccc} \mathbf{E}_1 & \xrightarrow{\phi_1} & \mathbf{F}_1 \\ {\scriptstyle g_1}\downarrow & & \downarrow{\scriptstyle g_2} \\ \mathbf{E}_2 & \xrightarrow{\phi_2} & \mathbf{F}_2 \end{array}$$

Fix an integer p. Let $D_c^b(X) \rtimes MS(X)$ denote the twisted product category whose objects are pairs

$$(\mathbf{A}^\bullet, \mathbf{E} \xrightarrow{\phi} \mathbf{F})$$

with $\mathbf{A}^\bullet \in D_c^b(X)$, ϕ injective and $\mathbf{F} = \mathbf{H}^p(\mathbf{A}^\bullet)$, and whose morphisms are pairs

$$(f, (g_1, g_2))$$

with $f : \mathbf{A}^\bullet \to \mathbf{B}^\bullet$ a morphism in $D_c^b(X)$, $g_2 = \mathbf{H}^p(f)$ and the diagram

$$\begin{array}{ccc} \mathbf{E}_1 & \xrightarrow{\phi_1} & \mathbf{H}^p(\mathbf{A}^\bullet) \\ {\scriptstyle g_1}\downarrow & & \downarrow{\scriptstyle g_2 = \mathbf{H}^p(f)} \\ \mathbf{E}_2 & \xrightarrow{\phi_2} & \mathbf{H}^p(\mathbf{B}^\bullet) \end{array}$$

commutes. There is a modified truncation functor

$$\tau_{\leq p}(\cdot, \cdot) : D_c^b(X) \rtimes MS(X) \longrightarrow D_c^b(X)$$

with the following properties:

1. There are canonical morphisms

$$\tau_{\leq p-1} \longrightarrow \tau_{\leq p}(\cdot, \cdot) \longrightarrow \tau_{\leq p}$$

   which factor the canonical $\tau_{\leq p-1} \to \tau_{\leq p}$ and are quasi-isomorphisms in dimensions $< p$.
2. There exists a canonical isomorphism

$$\mathbf{H}^p(\tau_{\leq p}(\mathbf{A}^\bullet, \mathbf{E})) \xrightarrow{\cong} \mathbf{E}$$

   such that

$$\begin{array}{ccc} \mathbf{H}^p(\tau_{\leq p}(\mathbf{A}^\bullet, \mathbf{E})) & \longrightarrow & \mathbf{H}^p(\tau_{\leq p}\mathbf{A}^\bullet) \\ \downarrow{\scriptstyle \cong} & & \| \\ \mathbf{E} & \xrightarrow{\phi} & \mathbf{H}^p(\mathbf{A}^\bullet) \end{array}$$

   commutes.

Thus, $\tau_{\le p}(\cdot,\cdot)$ can be viewed as an "interpolation" between $\tau_{\le p-1}$ and $\tau_{\le p}$, taking a specified subsheaf of the cohomology sheaf at the cut-off dimension into account. We will apply this notion of truncation in constructing a functor $\boxplus$ which associates to a self-dual sheaf $\mathbf{A}^\bullet \in SD(U_k)$ and a Lagrangian structure $\phi : \mathcal{A} \to \mathcal{O}(\mathbf{A}^\bullet)$ a self-dual sheaf $\mathbf{A}^\bullet \boxplus \mathcal{A} \in SD(U_{k+1})$. Let $SD(U_k) \rtimes \mathrm{Lag}(U_{k+1} - U_k)$ denote the twisted product of categories whose objects are pairs

$$(\mathbf{A}^\bullet, \mathcal{A} \xrightarrow{\phi} \mathcal{O}(\mathbf{A}^\bullet)),$$

$\mathbf{A}^\bullet \in SD(U_k)$, $\phi \in \mathrm{Lag}(U_{k+1} - U_k)$, and whose morphisms are pairs with first component a morphism $f \in \mathrm{Hom}_{D^b_c(U_k)}(\mathbf{A}^\bullet, \mathbf{B}^\bullet)$ and second component a commutative square

$$\begin{array}{ccc} \mathcal{A} & \xrightarrow{\phi_A} & \mathcal{O}(\mathbf{A}^\bullet) \\ {\scriptstyle l}\downarrow & & \downarrow{\scriptstyle \mathcal{O}(f)} \\ \mathcal{B} & \xrightarrow{\phi_B} & \mathcal{O}(\mathbf{B}^\bullet) \end{array}$$

(We will usually denote such a pair simply by (f, l).)

Define

$$\mathbf{A}^\bullet \boxplus \mathcal{A} := \tau_{\le s}(Ri_{k*}\mathbf{A}^\bullet, \mathbf{H}^s(\mathcal{A}) \xrightarrow{\mathbf{H}^s(\phi)} \mathbf{H}^s(Ri_{k*}\mathbf{A}^\bullet)).$$

(Recall $s = \bar{n}(k) - n$; in the notation $\mathbf{A}^\bullet \boxplus \mathcal{A}$, we suppress the map ϕ.)

Also, given $f : \mathbf{A}^\bullet \to \mathbf{B}^\bullet$ and

$$\begin{array}{ccc} \mathcal{A} & \xrightarrow{\phi_A} & \mathcal{O}(\mathbf{A}^\bullet) \\ {\scriptstyle l}\downarrow & & \downarrow{\scriptstyle \mathcal{O}(f)} \\ \mathcal{B} & \xrightarrow{\phi_B} & \mathcal{O}(\mathbf{B}^\bullet) \end{array}$$

let

$$f \boxplus l := \tau_{\le s}(Ri_{k*}f, \mathbf{H}^s(l))$$

where we wrote $\mathbf{H}^s(l)$ to abbreviate the commutative square

$$\begin{array}{ccc} \mathbf{H}^s(\mathcal{A}) & \longrightarrow & \mathbf{H}^s(\mathcal{O}(\mathbf{A}^\bullet)) \\ {\scriptstyle \mathbf{H}^s(l)}\downarrow & & \downarrow{\scriptstyle \mathbf{H}^s(Ri_{k*}f)} \\ \mathbf{H}^s(\mathcal{B}) & \longrightarrow & \mathbf{H}^s(\mathcal{O}(\mathbf{B}^\bullet)) \end{array}$$

By the properties of the modified truncation functor, there are canonical morphisms

$$^{\bar{m}}\mathbf{A}^\bullet \longrightarrow \mathbf{A}^\bullet \boxplus \mathcal{A} \longrightarrow {}^{\bar{n}}\mathbf{A}^\bullet$$

which factor the canonical $^{\bar{m}}\mathbf{A}^\bullet \to {}^{\bar{n}}\mathbf{A}^\bullet$. Moreover, there is a canonical isomorphism

$$\mathbf{H}^s(\mathbf{A}^\bullet \boxplus \mathcal{A}) \xrightarrow{\cong} \mathbf{H}^s(\mathcal{A})$$

which is induced by a morphism

$$\mathbf{A}^\bullet \boxplus \mathcal{A} \longrightarrow \mathcal{A}$$

such that the diagram

$$\begin{array}{ccc} \mathbf{A}^\bullet \boxplus \mathcal{A} & \longrightarrow & {}^{\bar{n}}\mathbf{A}^\bullet \\ \downarrow & & \downarrow \\ \mathcal{A} & \xrightarrow{\phi} & \mathcal{O}(\mathbf{A}^\bullet) \end{array}$$

commutes. Evidently, $(\mathbf{A}^\bullet \boxplus \mathcal{A})|_{U_k} \cong \mathbf{A}^\bullet$. It can be shown that $\mathbf{A}^\bullet \boxplus \mathcal{A}$ fits into a distinguished triangle

$$\begin{array}{ccc} {}^{\bar{m}}\mathbf{A}^\bullet & \longrightarrow & \mathbf{A}^\bullet \boxplus \mathcal{A} \\ & {\scriptstyle [1]}\nwarrow \quad \swarrow & \\ & \mathcal{A} & \end{array}$$

One then checks that $\mathbf{A}^\bullet \boxplus \mathcal{A} \in SD(U_{k+1})$: Axioms (SD1)–(SD3) being clearly satisfied, the main point is to verify (SD4). That is, one has to construct a self-duality isomorphism

$$d : \mathcal{D}(\mathbf{A}^\bullet \boxplus \mathcal{A})[n] \xrightarrow{\cong} \mathbf{A}^\bullet \boxplus \mathcal{A}.$$

To do this, apply the octahedral axiom to the following three triangles: The triangle on ϕ, which contains $\mathcal{D}\mathcal{A}[n+1]$ as the third term (cf. the definition of a Lagrangian structure), the triangle on the canonical ${}^{\bar{m}}\mathbf{A}^\bullet \to {}^{\bar{n}}\mathbf{A}^\bullet$, which contains $\mathcal{O}(\mathbf{A}^\bullet)$ as the third term, and the above triangle containing $\mathbf{A}^\bullet \boxplus \mathcal{A}$. The Verdier dual of this last triangle is then isomorphic to the triangle produced by the octahedral axiom. This isomorphism induces d. Thus, $\boxplus$ is a covariant functor

$$\boxplus : SD(U_k) \rtimes \mathrm{Lag}(U_{k+1} - U_k) \longrightarrow SD(U_{k+1}).$$

9.1.6 A Postnikov System

We discuss here the main result concerning the algebraic structure of the category $SD(X)$ of self-dual sheaves compatible with intersection homology theory on non-Witt spaces. The functors Λ and $\boxplus$ will be used in constructing an equivalence of categories between $SD(X)$ and a Postnikov system whose fibers are categories of Lagrangian structures along the various pure singular strata of odd codimension.

Consider the functor

$$(i_k^*, \Lambda) : SD(U_{k+1}) \longrightarrow SD(U_k) \rtimes \mathrm{Lag}(U_{k+1} - U_k).$$

One can show (Theorem 2.8 of [Ban02]) that there is an isomorphism of functors

$$\boxplus(i_k^*, \Lambda) \cong 1_{SD(U_{k+1})}.$$

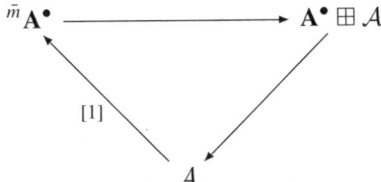

Conversely (Theorem 2.9 of [Ban02]), there is an isomorphism of functors

$$(i_k^*, \Lambda)\boxplus \cong 1_{SD(U_k)\rtimes \mathrm{Lag}(U_{k+1}-U_k)}.$$

Let Const(U_2) denote the category whose single object is the constant sheaf $\mathbb{R}_{U_2}$ on U_2 and whose morphisms are sheaf maps

$$\mathbb{R}_{U_2} \to \mathbb{R}_{U_2}.$$

An induction on the codimension k then proves the following Postnikov-type decomposition of the category $SD(X)$:

Theorem 9.1.4 *For $n = \dim X$ even, there is an equivalence of categories*

$$SD(X) \simeq \mathrm{Lag}(X_1 - X_0) \rtimes \mathrm{Lag}(X_3 - X_2) \rtimes \cdots \rtimes \mathrm{Lag}(X_{n-3} - X_{n-4}) \rtimes Const(U_2).$$

For $n = \dim X$ odd, there is an equivalence of categories

$$SD(X) \simeq \mathrm{Lag}(X_0) \rtimes \mathrm{Lag}(X_2 - X_1) \rtimes \cdots \rtimes \mathrm{Lag}(X_{n-3} - X_{n-4}) \rtimes Const(U_2).$$

The use of real coefficients is not essential, one could work over any field $\Bbbk$ of characteristic $\neq 2$. Of course for general $\Bbbk$ (e.g. $\Bbbk = \mathbb{C}$ the complex numbers), the signature of a bilinear ± 1-symmetric pairing is not available, but all we need to construct self-dual sheaves is the notion of a Lagrangian subspace for such a pairing on a $\Bbbk$-vector space. The use of field coefficients, however, is essential. One reason is related to Borel–Moore–Verdier duality (we need identifications such as $\mathcal{H}^k(X; \mathcal{D}\mathbf{A}^\bullet) \cong \mathrm{Hom}(\mathcal{H}^{-k}(X; \mathbf{A}^\bullet), \Bbbk))$, another is related to the very definition of a self-dual sheaf: If $\Bbbk$ is a field of characteristic $\neq 2$, we may define self-duality as the existence of a quasi-isomorphism $d : \mathbf{A}^\bullet \to \mathcal{D}\mathbf{A}^\bullet[n]$ together with a homotopy from d to $\pm\mathcal{D}d[n]$. For more general coefficient rings, one would have to provide higher coherencies (a homotopy from the previous homotopy to its dual, etc.), see [Wei94].

Further obvious modifications of the axiomatics given include the use of twisted coefficients on the top stratum. For simplicity, we normalize objects of $SD(X)$ to be the constant sheaf of rank one on the top stratum, cf. Definition 9.1.2, axiom (SD1). Replacing (SD1) by a relaxed version permitting arbitrary local systems $\mathcal{S}$ (over $\Bbbk$, of finite rank) with a nonsingular pairing $\mathcal{S} \otimes \mathcal{S} \longrightarrow \Bbbk$ (i.e. Poincaré local systems), one obtains a more general category $\widetilde{SD}(X)$ for which Theorem 9.1.4 is valid (with Const(U_2) changed to allow for twisting).

9.2 L-Classes of Non-Witt Spaces

We have seen in the previous section that a closed, oriented topological pseudomanifold X^n, not necessarily a Witt space, has generalized Poincaré self-duality if, and only if, it possesses Lagrangian structures along all strata of odd codimension. The latter condition is equivalent to $SD(X) \neq \varnothing$, where $SD(X)$ is the category of self-dual sheaves interpolating between $\mathbf{IC}^\bullet_{\bar{m}}(X)$ and $\mathbf{IC}^\bullet_{\bar{n}}(X)$. In the present section we will construct L-classes

$$L_k(X) \in H_k(X; \mathbb{Q})$$

for non-Witt spaces X with generalized Poincaré self-duality, i.e. with $SD(X) \neq \varnothing$. Let X be such a space and choose any $\mathbf{IC}^\bullet_{\mathcal{L}} \in SD(X)$, given by a collection of Lagrangian structures $\mathcal{L}$. Assume that X is Whitney stratified. Using the construction of Sect. 8.2.3, the self-dual sheaf $\mathbf{IC}^\bullet_{\mathcal{L}}$ has L-classes

$$L_k(\mathbf{IC}^\bullet_{\mathcal{L}}) \in H_k(X; \mathbb{Q}).$$

In [Ban06b] we prove:

Theorem 9.2.1 *The L-classes*

$$L_k(X) = L_k(\mathbf{IC}^\bullet_{\mathcal{L}}) \in H_k(X; \mathbb{Q}),$$

$\mathbf{IC}^\bullet_{\mathcal{L}} \in SD(X)$, *are independent of the choice of Lagrangian structure* $\mathcal{L}$.

Thus a non-Witt space has a well-defined L-class $L(X)$, provided $SD(X) \neq \varnothing$. Since L_0 is the signature, we have in particular:

Corollary 9.2.2 *The signature*

$$\sigma(X) = \sigma(\mathbf{IC}^\bullet_{\mathcal{L}}),$$

$\mathbf{IC}^\bullet_{\mathcal{L}} \in SD(X)$, *is independent of the choice of Lagrangian structure* $\mathcal{L}$.

We will sketch the proof of Theorem 9.2.1 for the basic case of a two strata space $X^n \supset \Sigma^s$, $X - \Sigma$ is an n-dimensional manifold and Σ^s an s-dimensional manifold, n even, s odd. (For the general case, we ask the reader to consult [Ban06b].) Given $\mathbf{IC}^\bullet_{\mathcal{L}_0}, \mathbf{IC}^\bullet_{\mathcal{L}_1} \in SD(X)$, determined by Lagrangian structures $\mathcal{L}_0, \mathcal{L}_1$, respectively, along Σ, the central problem is to prove equality of the signatures $\sigma(\mathbf{IC}^\bullet_{\mathcal{L}_0}) = \sigma(\mathbf{IC}^\bullet_{\mathcal{L}_1})$, since then the result on L-classes will follow from the fact that they are determined uniquely by the collection of signatures of subvarieties with normally nonsingular embedding and trivial normal bundle, see Proposition 8.2.11. To prove equality of the signatures, we use bordism theory: We construct a geometric bordism Y^{n+1} from X to $-X$ and cover its interior with a self-dual sheaf complex $\mathbf{S}^\bullet$, which, when pushed to the boundary, restricts to $\mathbf{IC}^\bullet_{\mathcal{L}_0}$ on X, and restricts to $\mathbf{IC}^\bullet_{\mathcal{L}_1}$ on $-X$. One then uses the techniques of Sect. 8.2.2, particularly Proposition 8.2.8, to deduce equality of the signatures. A topologically trivial h-cobordism $Y^{n+1} = X \times [0, 1]$ already works, but of course not with the natural stratification. Our idea is to "cut" the odd-codimensional stratum at $\frac{1}{2}$, which enables us to "decouple" Lagrangian structures because the stratum of odd codimension then consists of two disjoint connected components. This forces the introduction of a new stratum at $\frac{1}{2}$, but its codimension is even and presents no problem. The stratification of Y *with cuts at* $\frac{1}{2}$ is thus defined by the filtration $Y^{n+1} \supset Y_{s+1} \supset Y_s$, where $Y_{s+1} - Y_s = \Sigma^s \times [0, \frac{1}{2}) \sqcup \Sigma^s \times (\frac{1}{2}, 1]$ and $Y_s = \Sigma^s \times \{\frac{1}{2}\}$. The sheaf $\mathbf{S}^\bullet$ will be constructible with respect to this stratification. On $Y - Y_{s+1}$, $\mathbf{S}^\bullet$ is $\mathbb{R}_{Y-Y_{s+1}}[n+1]$, the constant real sheaf in degree $-n-1$. To extend to $Y_{s+1} - Y_s$, we use the Postnikov system Theorem 9.1.4, and the Lagrangian structure whose restriction to $\Sigma^s \times [0, \frac{1}{2})$ is the pullback of $\mathcal{L}_0$ under the first factor projection and whose restriction to $\Sigma^s \times (\frac{1}{2}, 1]$ is the pullback of $\mathcal{L}_1$ under the first factor projection. Finally, we extend to Y_s by the Deligne-step (pushforward and middle perversity truncation), which produces a self-dual sheaf $\mathbf{S}^\bullet$, since Y_s is of even codimension.

9.3 Stratified Maps

In Sect. 8.2.5, we have used the sheaf-theoretic Cappell–Shaneson decomposition Theorem 8.1.7 to obtain a formula that computes the pushforward of L-classes under a stratified map, where the target space of the map has only strata of *even codimension*. Let us recall that formula: If $f : Y^m \to X^n$ is a stratified map such that $m - n$ is even, Y compact, X has only even-codimensional strata and $\mathbf{S}^\bullet \in D^b_c(Y)$ is a self-dual sheaf, then

$$f_* L_k(\mathbf{S}^\bullet) = \sum_{V \in \mathcal{X}} j_* L_k(\overline{V}; \mathfrak{H}_V(\mathbf{S}^\bullet)),$$

where $\mathcal{X}$ is the collection of all connected components of pure strata of X and the $\mathfrak{H}_V(\mathbf{S}^\bullet)$ are local coefficient systems over the components V of pure strata. These local systems have been described in detail in Sect. 8.1.2.

Example 9.3.1 *Here is a simple example that illustrates that nonzero contributions in the above sum can indeed arise from strata of positive codimension. Take* $Y = \mathbb{C}P^2$, $X = \mathbb{C}P^2/\mathbb{C}P^1 = S^4$, *and let* $f : Y \to X$ *be the quotient map. We stratify* Y *as* $Y_4 = Y \supset Y_2 = \mathbb{C}P^1$ *and* X *as* $X_4 = X \supset X_0 = \{x_0\}$, *where* x_0 *is the image of* $\mathbb{C}P^1$ *under* f. *It follows that* f *is a stratified map. We take* $\mathbf{S}^\bullet$ *to be the constant sheaf* $\mathbb{R}[4]$, *which is self-dual on* $\mathbb{C}P^2$. *For* $k = 0$, *the above formula computes the signature*:

$$1 = \sigma(Y) = \sum_{V \in \mathcal{X}} \sigma(\overline{V}; \mathfrak{H}_V) = \sigma(S^4; \mathfrak{H}_{S^4 - \{x_0\}}) + \sigma(\{x_0\}; \mathfrak{H}_{\{x_0\}}).$$

The local system $\mathfrak{H}_{S^4-\{x_0\}}$ *is untwisted because* $S^4 - \{x_0\} \cong \mathbb{R}^4$ *is simply-connected. Thus, by Proposition* 8.2.19,

$$\sigma(S^4; \mathfrak{H}_{S^4-\{x_0\}}) = \sigma(S^4)\sigma((\mathfrak{H}_{S^4-\{x_0\}})_{pt}) = 0 \cdot \sigma((\mathfrak{H}_{S^4-\{x_0\}})_{pt}) = 0.$$

Consequently, the contribution comes from the 0*-dimensional stratum of* X,

$$\sigma(\{x_0\}; \mathfrak{H}_{\{x_0\}}) = 1.$$

The present section investigates target spaces X that are allowed to have strata of odd codimension. We will for instance prove:

Theorem 9.3.2 *Let* X^n *and* Y^m *be closed, oriented, Whitney stratified pseudomanifolds such that* $m - n$ *is even and* $SD(Y) \neq \varnothing$ *(for instance* Y *a Witt space). If*

$f : Y \to X$ is a stratified map and X has only strata of odd codimension (except for the top stratum), then

$$f_* L_i(Y) = L_i(X; \mathcal{S}), \tag{9.9}$$

where $\mathcal{S}$ is a local coefficient system on the top stratum of X.

(See Definition 9.1.2 for the category $SD(Y)$.) This formula then asserts, in sharp contrast to the situation where X has only strata of even codimension, that the strata of odd codimension never contribute terms to the pushforward of the L-class of Y.

We shall write $D(X) = D_c^b(X)$ for the derived category of bounded complexes of sheaves that are constructible with respect to the stratification of X. Given any function $\bar{p} : \mathcal{X} \to \mathbb{Z}$, $({}^{\bar{p}}D^{\leq 0}(X), {}^{\bar{p}}D^{\geq 0}(X))$ is the $\bar{p}$-perverse t-structure on X, ${}^{\bar{p}}\tau^{\leq 0} : D(X) \to {}^{\bar{p}}D^{\leq 0}(X)$ and ${}^{\bar{p}}\tau^{\geq 0} : D(X) \to {}^{\bar{p}}D^{\geq 0}(X)$ are the corresponding t-structure truncations, and ${}^{\bar{p}}P(X)$ is the heart (the abelian category of $\bar{p}$-perverse sheaves on X). The dual perversity $\bar{q}$ is

$$\bar{q}(S) = -\bar{p}(S) - \dim(S), \quad S \in \mathcal{X}.$$

It is pointed out in [BBD82] that the Verdier dualizing functor $\mathcal{D}$ restricts as

$$\mathcal{D} : {}^{\bar{p}}D^{\leq 0}(X) \longrightarrow {}^{\bar{q}}D^{\geq 0}(X), \qquad \mathcal{D} : {}^{\bar{p}}D^{\geq 0}(X) \longrightarrow {}^{\bar{q}}D^{\leq 0}(X),$$

and there is an isomorphism of functors

$$\mathcal{D}\,{}^{\bar{p}}H^0 \cong {}^{\bar{q}}H^0\mathcal{D},$$

with ${}^{\bar{p}}H^0 : D(X) \to {}^{\bar{p}}P(X)$ denoting $\bar{p}$-perverse cohomology. In fact, one has the following precise statement:

Lemma 9.3.3 *In $D(X)$, there are isomorphisms*

$$\mathcal{D}\,{}^{\bar{p}}\tau^{\leq k} \cong {}^{\bar{q}}\tau^{\geq -k}\mathcal{D} \quad \textit{and} \quad \mathcal{D}\,{}^{\bar{p}}\tau^{\geq k} \cong {}^{\bar{q}}\tau^{\leq -k}\mathcal{D}.$$

9.3.1 The Category of Equiperverse Sheaves

We will define a category $EP(X)$ of equiperverse sheaves on X^n (possibly not Witt) which is big enough so that every self-dual sheaf on X is bordant to an equiperverse sheaf, yet small enough so that on the class of non-Witt spaces to be considered in this section (see Sect. 9.3.2), every self-dual equiperverse sheaf is given by Lagrangian structures.

Let X be a stratified, oriented, topological pseudomanifold without boundary. It is natural to ask the following question: What is the set of perversities $\bar{r}$ such that every object of $SD(X)$ is $\bar{r}$-perverse? In other words, for which $\bar{r}$ is $SD(X) \subset {}^{\bar{r}}P(X)$? For $S \in \mathcal{X}$ with $k = \operatorname{codim} S$ even, the stalk condition (SD3) implies $\bar{r}(S) \geq \bar{n}(k) - n$ and the costalk condition (using self-duality) implies $\bar{r}(S) \leq \bar{n}(k) - n + 2$. Thus $\bar{r}(S)$ may have three values,

$$\bar{r}(S) \in \{\bar{n}(k) - n,\ \bar{n}(k) - n + 1,\ \bar{n}(k) - n + 2\}.$$

When one speaks of *the* category of perverse sheaves on a space with only even-codimensional strata, then one refers to ${}^{\bar{r}}P(X)$ with the middle value

$$\bar{r}(S) = \bar{n}(k) - n + 1 = \tfrac{1}{2}\operatorname{codim} S - \dim X,$$

since this value yields a self-dual perversity. The situation for $S \in \mathcal{X}$ with $k = \operatorname{codim} S$ odd is quite different. The stalk condition gives the same lower bound as before, but the costalk condition produces $\bar{r}(S) \leq \bar{n}(k) - n + 1$ as an upper bound, so that $\bar{r}(S)$ may have precisely the two values

$$\bar{r}(S) \in \{\bar{n}(k) - n,\ \bar{n}(k) - n + 1\}.$$

Definition 9.3.4 *Throughout the remainder of the present Sect.* 9.3, $\bar{p}, \bar{q} : \mathcal{X} \to \mathbb{Z}$ *will denote the perversity functions given by*

$$\bar{p}(S) = \begin{cases} \bar{n}(\operatorname{codim} S) - n + 1, & \operatorname{codim} S \textit{ even}, \\ \bar{n}(\operatorname{codim} S) - n, & \operatorname{codim} S \textit{ odd}, \end{cases}$$
$$\bar{q}(S) = \bar{n}(\operatorname{codim} S) - n + 1$$

if $S \in \mathcal{X}$ *is not contained in the top stratum, and* $\bar{p}(S) = \bar{q}(S) = -n$ *for* S *a component of the top stratum.*

We note that $\bar{p}$ and $\bar{q}$ are dual, since for $S \in \mathcal{X}$, $k = \operatorname{codim} S$ odd, the dual value is

$$\begin{aligned} \bar{p}^*(S) &= -\bar{p}(S) - \dim S - \dim X \\ &= -(\tfrac{1}{2}(k-1) - n) - (n-k) - n \\ &= \bar{n}(k) - n + 1 \\ &= \bar{q}(S). \end{aligned}$$

By construction, we have

$$SD(X) \subset {}^{\bar{p}}P(X) \cap {}^{\bar{q}}P(X). \tag{9.10}$$

Definition 9.3.5 *The category of* equiperverse sheaves *on* X *is the full subcategory*

$$EP(X) = {}^{\bar{p}}D^{\leq 0}(X) \cap {}^{\bar{q}}D^{\geq 0}(X) \subset D(X).$$

If X is a space with only even-codimensional strata, for example a complex algebraic variety, then $\bar{p} = \bar{q}$ and $EP(X) = {}^{\bar{p}}P(X) = {}^{\bar{q}}P(X)$ is the usual category of perverse sheaves. There is an obvious inclusion ${}^{\bar{p}}P(X) \cap {}^{\bar{q}}P(X) \subset EP(X)$. The converse inclusion is implied by ${}^{\bar{q}}D^{\geq 0} \subset {}^{\bar{p}}D^{\geq 0}$ and ${}^{\bar{p}}D^{\leq 0} \subset {}^{\bar{q}}D^{\leq 0}$. Thus we can represent equiperverse sheaves as

$$EP(X) = {}^{\bar{p}}P(X) \cap {}^{\bar{q}}P(X). \tag{9.11}$$

Dually to Definition 9.3.5, we set $EP^*(X) = {}^{\bar{p}}D^{\geq 0}(X) \cap {}^{\bar{q}}D^{\leq 0}(X)$. Verdier duality defines an equivalence of categories $\mathcal{D}[n] : EP(X) \simeq EP^*(X)$. The obvious inclusion ${}^{\bar{p}}P(X) \cap {}^{\bar{q}}P(X) \subset EP^*(X)$ is not strict, as we have $\mathcal{D}[n] : EP^*(X) \to {}^{\bar{p}}P(X) \cap {}^{\bar{q}}P(X)$ and $\mathcal{D}[n]$ preserves the subcategory ${}^{\bar{p}}P(X) \cap {}^{\bar{q}}P(X)$. Hence

$$EP(X) = EP^*(X).$$

The category EP is additive, since ${}^{\bar{p}}P$ and ${}^{\bar{q}}P$ are.

9.3.2 Parity-Separated Spaces

Definition 9.3.6 *A* parity-separation *on a stratified space X is a decomposition*

$$X = {}^{o}X \cup {}^{e}X$$

into open subsets ${}^{o}X, {}^{e}X \subset X$ such that ${}^{o}X$ with the induced stratification has only strata of odd codimension (except for the top stratum), and ${}^{e}X$ with the induced stratification has only strata of even codimension.

Alternatively, X has a parity-separation iff the closure $\overline{S}$ of any stratum $S \in \mathcal{X}$, different from the top stratum, contains only strata of its own kind with respect to the parity of codimension. If $X = {}^{o}X \cup {}^{e}X$ is a parity-separation, then ${}^{o}\mathcal{X}$ will denote the set of components of pure strata of ${}^{o}X$ and ${}^{e}\mathcal{X}$ will denote the set of components of pure strata of ${}^{e}X$.

Example 9.3.7 *Let K be a real quadratic number field and $\mathcal{O}_K$ the ring of algebraic integers in K. The Hilbert modular group $\Gamma = PSL_2(\mathcal{O}_K)$ acts on the product $H \times H$ of two upper half planes by*

$$(z, w) \mapsto \left(\frac{az+b}{cz+d}, \frac{\bar{a}w+\bar{b}}{\bar{c}w+\bar{d}}\right),$$

where $\bar{}$ denotes the nontrivial element of the Galois group $Gal(K/\mathbb{Q})$. The Hilbert modular surface *of K is the orbit space $X^4 = (H \times H)/\Gamma$. It is noncompact and has finitely many singular points, the cusps. Various compactifications have been studied: The reductive Borel–Serre compactification $\overline{X}$ has the advantage that the Hecke operators extend to this compactification. In constructing $\overline{X}$, one adjoins certain boundary circles to X. One obtains a stratification $\overline{X}_4 \supset \overline{X}_1 \supset \overline{X}_0 \supset \varnothing$ by taking $\overline{X}_1 - \overline{X}_0$ to be the disjoint union of the boundary circles and $\overline{X}_0$ to be the cusps. The links of the components of $\overline{X}_1 - \overline{X}_0$ are tori; hence $\overline{X}^4$ is not a Witt space. However, it possesses a parity-separation: Take ${}^{o}\overline{X} = \overline{X}_4 - \overline{X}_0$ and take ${}^{e}\overline{X}$ to be the union of small open neighborhoods of the cusps in X. It is shown in [BK04] that $SD(\overline{X}) \neq \varnothing$.*

Let $D^{\mathcal{D}}(X)$ denote the full subcategory of $D(X)$ containing all objects invariant under the action of $\mathcal{D}[n]$, i.e. $D^{\mathcal{D}}(X)$ contains all self-dual sheaves on X (without any stalk restrictions).

Proposition 9.3.8 *If X possesses a parity-separation, then*

$$D^{\mathcal{D}}({}^oX) \cap EP({}^oX) = SD({}^oX),$$

i.e. any self-dual equiperverse sheaf on a space with only odd-codimensional strata (excluding the top stratum) is isomorphic to a sheaf $\mathbf{IC}^\bullet_{\mathcal{L}}(X; \mathcal{S})$ *determined by Lagrangian structures* $\mathcal{L}$ *(see Theorem* 9.1.4*) and given on the top stratum by a local coefficient system* $\mathcal{S}$.

Proof. Any object of $SD({}^oX)$ is in $EP({}^oX)$ by (9.10) in Sect. 9.3.1 and using $EP = {}^{\bar{p}}P \cap {}^{\bar{q}}P$ (see (9.11)). Any object of $SD({}^oX)$ is in $D^{\mathcal{D}}({}^oX)$ by (SD4). Let $\mathbf{S}^\bullet \in D^{\mathcal{D}}({}^oX) \cap EP({}^oX)$. Since $\mathbf{S}^\bullet \in {}^{\bar{p}}D^{\leq 0}({}^oX)$, we have $\mathbf{H}^i(\mathbf{S}^\bullet|_{U_2 \cap {}^oX}) = 0$ for $i > \bar{p}(U_2 \cap {}^oX) = -n$, and since $\mathbf{S}^\bullet \in {}^{\bar{q}}D^{\geq 0}({}^oX)$, we have $\mathbf{H}^i(\mathbf{S}^\bullet|_{U_2 \cap {}^oX}) = 0$ for $i < \bar{q}(U_2 \cap {}^oX) = -n$, establishing (SD1). We prove the lower bound axiom (SD2) by induction on the codimension k of strata. The induction start is furnished by (SD1). Suppose that $\mathbf{H}^i(\mathbf{S}^\bullet|_{U_k})$ vanishes for $i < -n$. If j denotes the open inclusion $j : U_k \hookrightarrow U_{k+1}$ and $i : U_{k+1} - U_k \hookrightarrow U_{k+1}$ denotes the inclusion of the complement, then the distinguished triangle

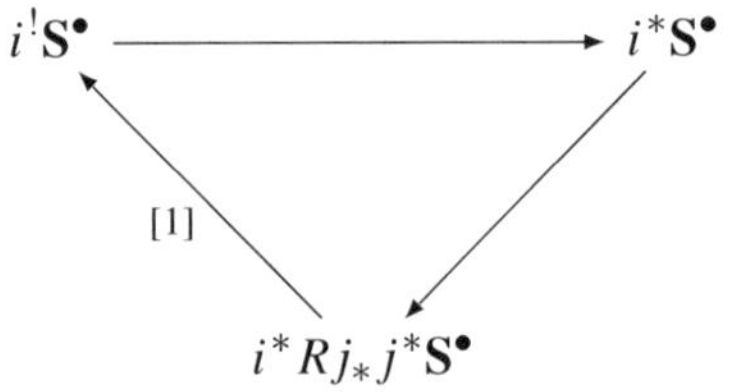

induces an exact sequence

$$\mathbf{H}^m(i^!\mathbf{S}^\bullet) \longrightarrow \mathbf{H}^m(i^*\mathbf{S}^\bullet) \longrightarrow \mathbf{H}^m(i^*Rj_*j^*\mathbf{S}^\bullet).$$

If $m < -n$, then $\mathbf{H}^m(i^!\mathbf{S}^\bullet) = 0$ since $\mathbf{S}^\bullet \in {}^{\bar{q}}D^{\geq 0}$ and the value of $\bar{q}$ on a component of $U_{k+1} - U_k$ is larger than $-n$. The term $\mathbf{H}^m(i^*Rj_*j^*\mathbf{S}^\bullet)$ vanishes when $m < -n$ by the induction hypothesis. Thus $\mathbf{H}^m(\mathbf{S}^\bullet|_{U_{k+1}}) = 0$ for $m < -n$. The stalk condition (SD3) is verified by observing that since ${}^oX$ has only strata of odd codimension, the value of $\bar{p}$ on any component of a stratum of ${}^oX$ is given by $\bar{n}(k) - n$, where k is the codimension (see Definition 9.3.4). Finally, (SD4) is equivalent to $\mathbf{S}^\bullet \in D^{\mathcal{D}}$.

Remark 9.3.9 *The inclusion $SD(X) \subset D^{\mathcal{D}}(X) \cap EP(X)$ holds for any pseudomanifold X, but it is generally strict if X has singular strata of even codimension: Let X^n be a two-strata space with singular stratum Σ, a manifold of dimension $n - 2c$. The sheaf*

$$\mathbf{S}^\bullet = j_*\mathbb{R}_\Sigma[n - c],$$

where $j : \Sigma \hookrightarrow X$ is the closed inclusion, is self-dual on X,

$$\begin{aligned}\mathcal{D}_X\mathbf{S}^\bullet[n] &= \mathcal{D}_X(j_*\mathbb{R}_\Sigma[n-c])[n] \cong j_!(\mathcal{D}_\Sigma\mathbb{R}_\Sigma)[c]\\ &\cong j_*\mathbb{R}_\Sigma[\dim \Sigma][c] \cong j_*\mathbb{R}_\Sigma[n-2c][c]\\ &\cong \mathbf{S}^\bullet,\end{aligned}$$

and is equiperverse as $\mathbf{S}^\bullet|_{X-\Sigma} = 0$ *and* $\mathbf{H}^i(\mathbf{S}^\bullet|_\Sigma) = 0$ *for* $i > \bar{p}(\Sigma) = c - n$. *However,* $\mathbf{S}^\bullet \notin SD(X)$ *since* $\mathbf{H}^i(\mathbf{S}^\bullet|_\Sigma) \cong \mathbb{R}_\Sigma \neq 0$ *for* $i = c - n$, *violating axiom* (SD3).

9.3.3 An Algebraic Bordism Construction

Let $\Omega(X)$ be the abelian group of algebraic bordism classes of self-dual sheaves on X. (Bordism of self-dual sheaves was defined in Definition 8.1.11.) The bordism class of $\mathbf{S}^\bullet \in D^{\mathcal{D}}(X)$ will be written as $[\mathbf{S}^\bullet]$ (thus the self-duality isomorphism will be suppressed in the notation).

Theorem 9.3.10 *Let* X^n *be a topological stratified pseudomanifold and* $\mathbf{S}^\bullet$ *any self-dual sheaf in* $D(X)$. *Then*

$$[\mathbf{S}^\bullet] = [\mathbf{P}^\bullet] \in \Omega(X)$$

for some equiperverse sheaf $\mathbf{P}^\bullet$.

Proof. Given $\mathbf{S}^\bullet \in D^{\mathcal{D}}(X)$, we embed the canonical morphisms

$$^{\bar{p}}\tau^{\leq 0}\mathbf{S}^\bullet \longrightarrow \mathbf{S}^\bullet \longrightarrow {}^{\bar{q}}\tau^{\geq 0}\mathbf{S}^\bullet$$

in distinguished triangles:

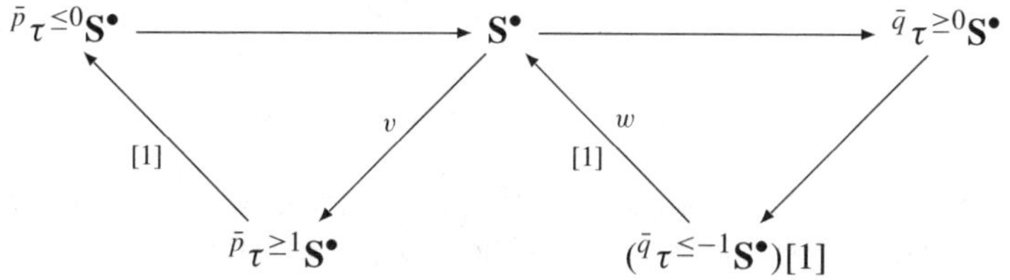

We claim that

$$^{\bar{q}}\tau^{\leq -1}\mathbf{S}^\bullet \xrightarrow{u} \mathbf{S}^\bullet \xrightarrow{v} {}^{\bar{p}}\tau^{\geq 1}\mathbf{S}^\bullet, \tag{9.12}$$

where $u = w[-1]$, sets up an elementary bordism. If D is any triangulated category and $(D^{\leq 0}, D^{\geq 0})$ a t-structure on it, then $\mathrm{Hom}_D(X, Y) = 0$ whenever $X \in D^{\leq 0}$ and $Y \in D^{\geq 1}$. Now $^{\bar{q}}\tau^{\leq -1}\mathbf{S}^\bullet \in {}^{\bar{q}}D^{\leq -1} \subset {}^{\bar{p}}D^{\leq 0}$, $^{\bar{p}}\tau^{\geq 1}\mathbf{S}^\bullet \in {}^{\bar{p}}D^{\geq 1}$ and so

$$\mathrm{Hom}_{D(X)}(^{\bar{q}}\tau^{\leq -1}\mathbf{S}^\bullet, {}^{\bar{p}}\tau^{\geq 1}\mathbf{S}^\bullet) = 0,$$

which implies $vu = 0$. The duality $\mathcal{D}v[n] = u$ follows from Lemma 9.3.3. Thus (9.12) induces a bordism from $\mathbf{S}^\bullet$ to a self-dual sheaf $\mathbf{C}^\bullet_{u,v}$,

$$[\mathbf{S}^\bullet] = [\mathbf{C}^\bullet_{u,v}] \in \Omega(X).$$

The remaining task is to show that $\mathbf{C}^\bullet_{u,v}$ is equiperverse. Since $vu = 0$, there exists a lift u' of u,

$$\begin{array}{ccc} & & {}^{\bar{p}}\tau^{\leq 0}\mathbf{S}^{\bullet} \\ & \overset{u'}{\nearrow} & \downarrow \\ {}^{\bar{q}}\tau^{\leq -1}\mathbf{S}^{\bullet} & \xrightarrow{\ u\ } & \mathbf{S}^{\bullet} \end{array}$$

and $\mathbf{C}^{\bullet}_{u,v}$ is isomorphic to the mapping cone $\mathbf{C}^{\bullet}_{u'}$ of u'. The lift is unique as $\mathrm{Hom}({}^{\bar{q}}\tau^{\leq -1}\mathbf{S}^{\bullet}, ({}^{\bar{p}}\tau^{\geq 1}\mathbf{S}^{\bullet})[-1]) = 0$. The relation ${}^{\bar{q}}D^{\leq -1} \subset {}^{\bar{p}}D^{\leq 0}$ implies an isomorphism of functors

$${}^{\bar{q}}\tau^{\leq -1} \circ {}^{\bar{p}}\tau^{\leq 0} \cong {}^{\bar{q}}\tau^{\leq -1}.$$

Under this identification, we recognize u' as the canonical morphism

$${}^{\bar{q}}\tau^{\leq -1}({}^{\bar{p}}\tau^{\leq 0}\mathbf{S}^{\bullet}) \xrightarrow{u'} {}^{\bar{p}}\tau^{\leq 0}\mathbf{S}^{\bullet}.$$

From the distinguished triangle

$$\begin{array}{ccc} {}^{\bar{q}}\tau^{\leq -1}({}^{\bar{p}}\tau^{\leq 0}\mathbf{S}^{\bullet}) & \longrightarrow & {}^{\bar{p}}\tau^{\leq 0}\mathbf{S}^{\bullet} \\ \quad \nwarrow {\scriptstyle [1]} & & \swarrow \quad \\ & {}^{\bar{q}}\tau^{\geq 0}({}^{\bar{p}}\tau^{\leq 0}\mathbf{S}^{\bullet}) & \end{array}$$

we conclude

$$\mathbf{C}^{\bullet}_{u,v} \cong \mathbf{C}^{\bullet}_{u'} \cong {}^{\bar{q}}\tau^{\geq 0}\,{}^{\bar{p}}\tau^{\leq 0}\mathbf{S}^{\bullet}.$$

Clearly, $\mathbf{P}^{\bullet} = {}^{\bar{q}}\tau^{\geq 0}\,{}^{\bar{p}}\tau^{\leq 0}\mathbf{S}^{\bullet}$ is in ${}^{\bar{q}}D^{\geq 0}$. Since $\mathbf{P}^{\bullet}$ is self-dual and $\mathcal{D}[n]$ maps ${}^{\bar{q}}D^{\geq 0} \to {}^{\bar{p}}D^{\leq 0}$, $\mathbf{P}^{\bullet}$ is also in ${}^{\bar{p}}D^{\leq 0}$ and hence equiperverse. □

9.3.4 Characteristic Classes and Stratified Maps

Theorem 9.3.11 *Let X^n be a stratified pseudomanifold with parity-separation $X = {}^{o}X \cup {}^{e}X$ and $\mathbf{S}^{\bullet} \in D(X)$ a self-dual sheaf. Then*

$$[\mathbf{S}^{\bullet}] = [\mathbf{P}^{\bullet}] \in \Omega(X)$$

with

$$\mathbf{P}^{\bullet}|_{{}^{o}X} \cong \mathbf{IC}^{\bullet}_{\mathcal{L}}({}^{o}X; {}^{o}\mathcal{S}) \in SD({}^{o}X)$$

and

$$\mathbf{P}^{\bullet}|_{{}^{e}X} \cong \bigoplus_{Z \in {}^{e}\mathcal{X}} j_{*}\mathbf{IC}^{\bullet}_{\bar{m}}(\overline{Z}; \mathcal{S}^{Z})[\tfrac{1}{2}\operatorname{codim} Z].$$

Here, ${}^{o}\mathcal{S}$ is a local coefficient system on the top stratum of ${}^{o}X$, $\mathbf{IC}^{\bullet}_{\mathcal{L}}$ is a self-dual intersection chain sheaf on ${}^{o}X$ determined by a Lagrangian structure $\mathcal{L}$ (see Theorem 9.1.4), $\overline{Z}$ is the closure of Z, $j : \overline{Z} \hookrightarrow {}^{e}X$ is the inclusion, $\mathcal{S}^{Z}$ is a local coefficient system on Z, and $\mathbf{IC}^{\bullet}_{\bar{m}}$ is the middle perversity intersection chain sheaf.

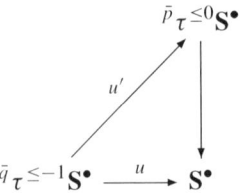

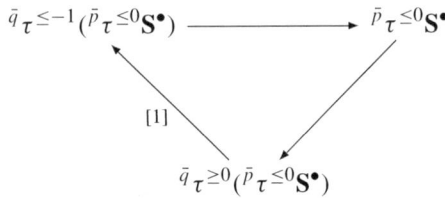

To paraphrase: The result says that only strata of even codimension can contribute terms in the decomposition, while the strata of odd codimension will *never contribute any terms*.

Proof. (of Theorem 9.3.11) By Theorem 9.3.10, $[\mathbf{S}^\bullet] = [\mathbf{T}^\bullet] \in \Omega(X)$ with $\mathbf{T}^\bullet \in EP(X)$. The latter sheaf has restrictions $\mathbf{U}^\bullet = \mathbf{T}^\bullet|_{^oX} \in EP(^oX) \cap D^{\mathcal{D}}(^oX)$ and $\mathbf{V}^\bullet = \mathbf{T}^\bullet|_{^eX} \in EP(^eX) \cap D^{\mathcal{D}}(^eX)$. According to Proposition 9.3.8, $\mathbf{U}^\bullet \in SD(^oX)$, and by Theorem 9.1.4

$$\mathbf{U}^\bullet \cong \mathbf{IC}^\bullet_{\mathcal{L}}(^oX; {}^o\mathcal{S})$$

for some Lagrangian structure $\mathcal{L}$ and local coefficient system $^o\mathcal{S}$. For the piece containing only strata of even codimension, $EP(^eX) = {}^{\bar{q}}P(^eX)$, and the decomposition

$$[\mathbf{V}^\bullet] = [\mathbf{V}^\bullet_0] \in \Omega(^eX),$$

$$\mathbf{V}^\bullet_0 \cong \bigoplus_{Z \in {}^e\mathcal{X}} j_*\mathbf{IC}^\bullet_{\bar{m}}(\overline{Z}; \mathcal{S}^Z)[\tfrac{1}{2}\operatorname{codim} Z]$$

is the central result of [CS91]. The overlap $^oX \cap {}^eX$ is contained in the top stratum, so that

$$\mathbf{U}^\bullet|_{^oX\cap^eX} \cong \mathbf{V}^\bullet_0|_{^oX\cap^eX}.$$

Now ${}^{\bar{q}}P(X)$ is a stack. Therefore there exists a sheaf $\mathbf{P}^\bullet$ on X such that $\mathbf{P}^\bullet|_{^oX} \cong \mathbf{U}^\bullet$ and $\mathbf{P}^\bullet|_{^eX} \cong \mathbf{V}^\bullet_0$. □

Theorem 9.3.12 *Let X^n and Y^m be closed, oriented, Whitney stratified pseudomanifolds such that $m - n$ is even and $SD(Y) \neq \varnothing$ (for instance Y a Witt space). If $f : Y \to X$ is a stratified map and X has only strata of odd codimension (except for the top stratum), then*

$$f_*L(Y) = L(X; \mathcal{S}^{X-\Sigma}).$$

Proof. Let $\mathbf{S}^\bullet \in SD(Y)$ be any sheaf on Y such that $L(Y) = L(\mathbf{S}^\bullet)$ (e.g. $\mathbf{S}^\bullet = \mathbf{IC}^\bullet_{\bar{m}}(Y)$ if Y is Witt). The pushforward $Rf_*\mathbf{S}^\bullet[\frac{1}{2}(n-m)]$ is again self-dual on X, since f is proper. As X has only strata of odd codimension, it admits a parity-separation $X = {}^oX \cup {}^eX$ with $^eX = \varnothing$. By Theorem 9.3.11, $Rf_*\mathbf{S}^\bullet[\frac{1}{2}(n-m)]$ is bordant to $\mathbf{IC}^\bullet_{\mathcal{L}}(X; \mathcal{S}^{X-\Sigma})$ for some Lagrangian structure $\mathcal{L}$. Then $f_*L(\mathbf{S}^\bullet) = L(Rf_*\mathbf{S}^\bullet[\frac{1}{2}(n-m)]) = L(\mathbf{IC}^\bullet_{\mathcal{L}}(X; \mathcal{S}^{X-\Sigma})) = L(X; \mathcal{S}^{X-\Sigma})$, since bordism preserves L-classes (Proposition 8.2.12), and using Theorem 9.2.1. □

9.4 The Presence of Monodromy

In Sect. 8.3, we have seen that the signature $\sigma(X; \mathcal{S})$ of a Witt space X with coefficients in a globally defined Poincaré local system $\mathcal{S}$ can be computed as

$$\sigma(X; \mathcal{S}) = \epsilon_*(\widetilde{\operatorname{ch}}([\mathcal{S}]_K) \cap L(X)) \in \mathbb{Z}.$$

(This is Theorem 8.3.4.)

In [Ban06a], the Witt hypothesis on X is removed and it is shown that the above formula continues to hold for an arbitrary Whitney stratified pseudomanifold X, as long as it still possesses an L-class:

Theorem 9.4.1 *Let X^n be a closed, oriented, Whitney stratified pseudomanifold and let $\mathcal{S}$ be a nondegenerate, symmetric local system on X. If X possesses Lagrangian structures along the strata of odd codimension (so that $L(X) \in H_*(X;\mathbb{Q})$ is defined), then*

$$L(X;\mathcal{S}) = \widetilde{\mathrm{ch}}[\mathcal{S}]_K \cap L(X). \tag{9.13}$$

For the special case of the twisted signature $\sigma(X;\mathcal{S}) = L_0(X;\mathcal{S})$, one has therefore

$$\sigma(X;\mathcal{S}) = \langle \widetilde{\mathrm{ch}}[\mathcal{S}]_K, L(X)\rangle. \tag{9.14}$$

In geometric situations such as the ones considered in the previous Sect. 9.3, the local system $\mathcal{S}$ is typically only given on the top stratum $X - \Sigma$. If it extends, as a local system, over the entire space, then (9.13) can be used to compute the twisted L-classes. A necessary, and on normal spaces also sufficient, condition for the existence of a unique extension is that $\mathcal{S}$ is strongly transverse to the singular set Σ, see Definition 8.2.21 and Proposition 8.2.22. This holds automatically on the class of supernormal spaces, for which there exists a well understood classification theory, [CW91] and [Wei94]. Our requirement that the local system be defined on the entire space cannot be eliminated without substitute, because formulae (9.13) and (9.14) will become false in general: In [Ban04], we construct four-dimensional orbifolds with isolated singularities, together with local systems on their top stratum, such that these systems do not extend to the entire space, and the difference between the left- and right-hand side of (9.14) is given by a nonvanishing rho-invariant. These examples are particularly striking, since the underlying spaces have rather weak singularities, being rational homology manifolds. A brief summary of the construction goes like this: Starting out with a lens-space $L = L_p(1,1)$ and classifying map $L \xrightarrow{l} B\mathbb{Z}_p = L_p^\infty$, there exists a smooth, compact 4-manifold $M \xrightarrow{f} L_p^\infty$ whose boundary is a number of copies of $L \xrightarrow{l} L_p^\infty$. Using Atiyah–Patodi–Singer, there exists a character $\mathbb{Z}_p \xrightarrow{\alpha} U(1)$ with rho-invariant $\rho_\alpha(L) \neq 0$. Via f, this pulls back to a local coefficient system $\mathcal{S}$ on M. Since $\rho_\alpha(L) \neq 0$, we have $\sigma(M;\mathcal{S}) \neq \sigma(M)$. Let X be the space obtained from M by attaching cones on each of the boundary components of M. Then X is the global quotient of a manifold by an action of $\mathbb{Z}_p$ and $\sigma(X;\mathcal{S}) \neq \sigma(X) = \langle \widetilde{\mathrm{ch}}[\mathcal{S}]_K, L(X)\rangle$.

We must confine ourselves here to a mere sketch of the proof of Theorem 9.4.1. By the Thom–Pontrjagin construction, the primary objective is to establish (9.14). The basic strategy now is still inspired by the proof of Theorem 8.3.4 for Witt spaces. To obtain the latter theorem, we used Witt bordism. This suggests that here we should use the bordism theory based on pairs $(X, \mathbf{IC}^\bullet_{\mathcal{L}}(X))$, where X is a stratified pseudomanifold and $\mathbf{IC}^\bullet_{\mathcal{L}}(X) \in SD(X)$ is given by a Lagrangian structure $\mathcal{L}$ (cf. Sect. 8.2.2). This bordism theory is called *signature homology* $S_*(-)$ and has been constructed topologically by Minatta [Min04], following a suggestion of Kreck. The coefficients are $S_{4k}(pt) \cong \mathbb{Z}$, given by the signature, and zero otherwise. (These coefficient groups were already introduced in [Ban02], Chap. 4, where they were denoted by Ω_*^{SD}.) The pertinent feature for us is that the canonical map $\Omega_*^{SO}(X) \to S_*(X)$ is at odd primes surjective, allowing us to pull back signature calculations to smooth manifolds, starting with the identity map in $S_*(X)$. The following problem arises: If a stratified non-Witt space X carries perverse self-dual sheaves extending constant coefficients on the top stratum, then it is not automatically clear that it also carries perverse self-dual sheaves extending a nonconstant local system $\mathcal{S}$ on the top stratum. (If X has only strata of even codimension, then this is true: The Deligne formula for the middle perversity yields a self-dual extension, both when applied to constant coefficients and when applied to $\mathcal{S}$.) We avoid these monodromy difficulties altogether by constructing a piecewise linear version $S_*^{PL}(-)$ of topological signature homology. As pointed out in, among other places, [Sul04] (Sect. 2), the simplicial stratification of a PL space has the virtue of rendering all link bundles trivial.

10

L^2 Cohomology

Jeff Cheeger discovered, working independently of Goresky and MacPherson and not being aware of their intersection homology, that Poincaré duality on triangulated pseudomanifolds equipped with a suitable Riemannian metric on the top stratum, can be recovered by using the complex of L^2 differential forms on the top stratum. The connection between his and the work of Goresky and MacPherson was pointed out by Dennis Sullivan in 1976. We shall give a brief introduction to Cheeger's theory based on [Che80]. Further references include [Che79] and [Che83].

Let (M^n, g) be a Riemannian manifold without boundary, where g denotes the metric. Let $\Omega^k(M)$ denote the vector space of smooth differential k-forms on M. The differential

$$d : \Omega^k(M) \longrightarrow \Omega^{k+1}(M)$$

is a linear map such that $d^2 = 0$. This yields a complex $(\Omega^\bullet(M), d)$, whose cohomology spaces

$$H^k_{DR}(M) = H^k(\Omega^\bullet(M))$$

are called the *de Rham cohomology* of M. Now assume that M is oriented. Then the Hodge $*$-operator is a linear map

$$* : \Omega^k(M) \longrightarrow \Omega^{n-k}(M)$$

such that $*^2 = (-1)^{k(n-k)}$. The volume form $dV = *1$, where 1 is the constant function. One then defines

$$\delta = (-1)^{n(k+1)+1} * d* : \Omega^k(M) \longrightarrow \Omega^{k-1}(M).$$

Clearly,

$$\delta^2 = 0, \qquad *\delta = (-1)^k d*, \qquad \delta* = (-1)^{k+1} * d.$$

The Laplacian on forms is the endomorphism

$$\Delta = \delta d + d\delta : \Omega^k(M) \longrightarrow \Omega^k(M).$$

When M is compact, we can define an inner product $(\cdot, \cdot)$ on k-forms by

$$(\omega, \eta) = \int_M \omega \wedge *\eta.$$

With respect to this inner product, $(d\omega, \eta) = (\omega, \delta\eta)$, i.e. δ is the adjoint of d. This characterizes δ, since $(\cdot, \cdot)$ is positive definite. It follows that Δ is self-adjoint. A k-form ω is called *harmonic*, if $\Delta\omega = 0$. For compact M, $\Delta\omega = 0$ implies $d\omega = 0$ and $\delta\omega = 0$. (The converse, of course, is always true.) Let

$$\mathcal{H}^k(M) = \{\omega \in \Omega^k(M) \mid d\omega = \delta\omega = 0\}.$$

Theorem 10.0.1 (*The Hodge decomposition*) *If M is a compact, oriented, Riemannian manifold, then $\mathcal{H}^k(M)$ is finite-dimensional and there is an orthogonal sum decomposition*

$$\Omega^k(M) = \Delta(\Omega^k(M)) \oplus \mathcal{H}^k(M).$$

From this one deduces that there is a unique harmonic form in each de Rham cohomology class, and the k-th de Rham cohomology vector space is isomorphic to $\mathcal{H}^k(M)$.

Let (M, g) be an oriented Riemannian manifold, not necessarily compact.

Definition 10.0.2 *A k-form $\omega \in \Omega^k(M)$ is L^2, if*

$$\int \omega \wedge *\omega < \infty.$$

Let $\Omega^k_{(2)}(M)$ be the vector space of all $\omega \in \Omega^k(M)$ which are L^2 and such that $d\omega$ is L^2 as well. This yields a complex $(\Omega^\bullet_{(2)}(M), d)$, whose cohomology spaces

$$H^k_{(2)}(M) = H^k(\Omega^\bullet_{(2)}(M))$$

are called the L^2 *cohomology* of M. Let

$$\mathcal{H}^k_{(2)}(M) = \{\omega \in \Omega^k_{(2)}(M) \mid d\omega = \delta\omega = 0\}.$$

Inclusion induces a natural map

$$\mathcal{H}^k_{(2)}(M) \longrightarrow H^k_{(2)}(M).$$

Definition 10.0.3 *Let (M, g) and (M', g') be two Riemannian n-manifolds triangulated smoothly by T and T', respectively. A* piecewise smooth quasi-isometry *is a homeomorphism $f : M' \to M$ such that*

1. *there exist smooth subdivisions S of T and S' of T' for which the restriction of f to each closed top-dimensional simplex of S' is a diffeomorphism onto some closed top-dimensional simplex of S, and*
2. *there exists a constant C so that for every closed top-dimensional simplex of S',*

$$\frac{1}{C}g' \leq f^*(g) \leq Cg'.$$

Definition 10.0.4 *A* metric collar *is a smoothly triangulated Riemannian manifold piecewise smoothly quasi-isometric to* $(0, 1) \times L^m$, *where* L^m *is some triangulated Riemannian manifold and* $(0, 1) \times L^m$ *is given the product metric. Define* $\text{pmc}(L) = (0, 1) \times L$ *to be equipped with the metric*

$$dr^2 + r^2 g,$$

where g *is the metric on* L. *A* punctured metric cone *is a smoothly triangulated Riemannian manifold piecewise smoothly quasi-isometric to* $\text{pmc}(L)$. *A* Riemannian i-handle *is a smoothly triangulated Riemannian manifold piecewise smoothly quasi-isometric to* $(0, 1)^{n-i} \times \text{pmc}(L^{i-1})$ *with the product metric*; $(0, 1)^{n-i}$ *is flat.*

Cheeger proceeds to prove Stokes' theorems for metric collars and punctured metric cones which we will not state here. In order to establish an isomorphism of $H^*_{(2)}$ with the linear dual of $IH^{\bar{m}}_*$, one needs suitable Poincaré lemmas for metric collars and punctured metric cones:

Lemma 10.0.5 (*Poincaré lemma for metric collars*)

$$H^i_{(2)}((0, 1) \times L) \cong H^i_{(2)}(L).$$

Lemma 10.0.6 (*Poincaré lemma for punctured metric cones*) *For* $i \leq \frac{m}{2}$,

$$H^i_{(2)}(\text{pmc}(L^m)) \cong H^i_{(2)}(L).$$

For $i \geq \frac{m+1}{2}$,

$$H^i_{(2)}(\text{pmc}(L^m)) = 0.$$

In Sect. 4.1.3, we used Borel–Moore chains, i.e. infinite chains, to define piecewise linear intersection homology, because this is the chain model that sheafifies readily. One may of course use finite PL chains also; we denote the resulting intersection homology groups for the lower middle-perversity $\bar{m}$ by $IH^c_*(X)$. These groups then satisfy

$$IH^c_i((0, 1) \times L) \cong IH^c_i(L)$$

as well as the local vanishing result

$$IH^c_i(c^\circ L^m) = \begin{cases} IH^c_i(L^m), & i \leq \frac{m}{2}, \\ 0, & i \geq \frac{m+1}{2}. \end{cases}$$

At this point, one certainly gets the idea that middle-perversity intersection homology and L^2 cohomology are closely related. The reader may wish to compare this with the local vanishing results for Borel–Moore type middle-perversity intersection homology, e.g. Examples 4.1.14 (4.1) and 4.1.15 (4.2). The next item on the agenda is to get Mayer–Vietoris sequences for L^2 cohomology.

Definition 10.0.7 *Let* $U \subset M$ *be an open subset of a manifold* M *whose closure* $\overline{U}$ *is a manifold with boundary* $\partial\overline{U}$. *We say that* $\partial\overline{U}$ *is* collared, *if there exists an open neighborhood* V *of* $\partial\overline{U}$ *in* M *and a piecewise smooth quasi-isometry between* V *and a metric collar* $(-1, 1) \times L$ *which maps* $\{0\} \times L$ *onto* $\partial\overline{U}$.

Lemma 10.0.8 *Let $U, V \subset M$ be open subsets such that $U \cup V = M$ and the closure of $U \cap V$ in U is a manifold with boundary collared in U. Then the Mayer–Vietoris sequence*

$$\cdots \longrightarrow H^i_{(2)}(M) \longrightarrow H^i_{(2)}(U) \oplus H^i_{(2)}(V) \longrightarrow H^i_{(2)}(U \cap V) \longrightarrow \cdots$$

is exact.

Definition 10.0.9 *A* Riemannian pseudomanifold *is a triple (X^n, d, g), where*

1. *X is a stratified PL-pseudomanifold with smooth top stratum $X - \Sigma$,*
2. *(X, d) is a metric space,*
3. *$(X - \Sigma, g)$ is a Riemannian manifold, and*
4. *$d|_{X-\Sigma}$ is induced by g.*

Example 10.0.10 *Let X^n be a closed, triangulated pseudomanifold. That is, each $(n-1)$-simplex is the face of precisely two n-simplices, and every point is contained in some closed n-simplex. We will describe how any such X can be endowed with the structure of a Riemannian pseudomanifold. Let X_i denote the i-skeleton, i.e. the union of all simplices of dimension $\leq i$. Set $\Sigma = X_{n-2}$. Then $X - \Sigma$ is a manifold. Let K denote the simplicial complex with $|K| \cong X$. Now take every simplex in K to be a* regular *Euclidean simplex, e.g. assign to every simplex a Euclidean metric with all edge lengths equal to 1. Glue these metric simplices back together so that $|K| \cong X$, but this time, use isometries to glue the faces. At this stage, K is a so-called* metric simplicial complex. *Let us define a distance function d such that $(|K|, d)$ is a metric space. A* piecewise geodesic *γ in K is a path $\gamma : [a, b] \to |K|$, where $[a, b]$ is subdivided into a finite number of subintervals so that the restriction of γ to each closed subinterval is a path lying entirely in some closed simplex σ of K and that restricted path is the unique geodesic (in our case, Euclidean straight line) connecting its endpoints in the metric of σ. The* length *of γ is the sum of the lengths of the geodesics into which it has been partitioned. Set*

$$d(x, y) = \inf_{\gamma} length(\gamma),$$

where the infimum is taken over all piecewise geodesics in K with $\gamma(a) = x$, $\gamma(b) = y$. For a general metric simplicial complex, this is only a pseudometric. In our case however, since we have only used finitely many isometry types to build $|K|$, Theorem 7.19 *of [BH99] applies and allows us to conclude that $(|K|, d)$ is indeed a geodesic metric space. It remains to discuss the Riemannian metric g on $|K| - \Sigma$. Every open n-simplex carries the smooth flat Euclidean metric by the construction above. An $(n-1)$-simplex $\tau \in K$ is the face of precisely two n-simplices σ and σ'. We lay out these two regular n-simplices next to each other in $\mathbb{R}^n$ so that they form a bipyramid with base τ, and then restrict the Euclidean metric on $\mathbb{R}^n$ to it. This shows that the Riemannian metrics on the interiors of σ and σ' extend smoothly across τ to give a smooth, flat Riemannian metric on the interior of $\sigma \cup_\tau \sigma'$, which, of course, induces d. Thus, we obtain a smooth, flat Riemannian metric g on $|K| - \Sigma$ which induces $d|_{|K|-\Sigma}$. This method does not extend further, because an $(n-2)$-simplex*

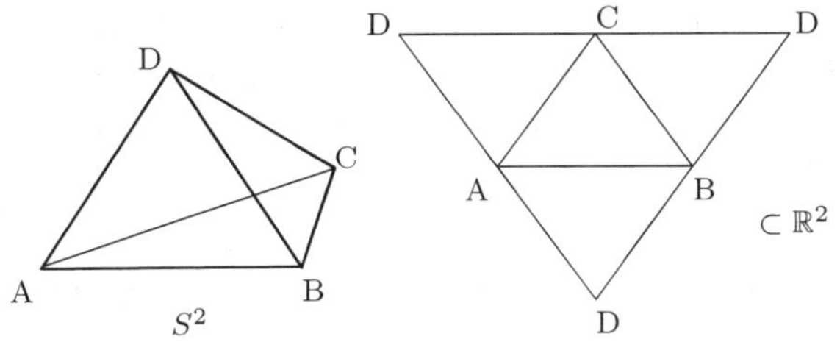

Fig. 10.1. The flat metric on $S^2 - \Sigma$.

may be the face of more than two n-simplices and one cannot necessarily lay out 3 *or more n-simplices in* $\mathbb{R}^n$ *so that their union reflects the way they are glued together in* $|K|$ *at the* $(n-2)$*-simplex.*

More generally, we shall consider metrics piecewise smoothly quasi-isometric to g.

Example 10.0.11 *Figure* 10.1 *illustrates the above process for* $X = S^2$, *triangulated as the boundary of a tetrahedron (left). Here* $\Sigma = \{A, B, C, D\}$. *On the right,* $S^2 - \Sigma$ *is laid out flatly in* $\mathbb{R}^2$ *and edges labeled by the same endpoints are to be identified isometrically. The Euclidean metric on* $\mathbb{R}^2$ *induces the flat metric on* $S^2 - \Sigma$. *Note that this metric cannot be smoothly extended into* Σ *to give a smooth Riemannian metric on* S^2, *because this would contradict Gauss–Bonnet.*

Let (X, d, g) be a Riemannian pseudomanifold with d, g as in Example 10.0.10. Let σ be some simplex in X and let x be an interior point of σ. Then the link $L(\sigma)$ of σ at x is itself a pseudomanifold, and indeed a *Riemannian* pseudomanifold as follows: If $\epsilon > 0$ is sufficiently small, then the union of all line segments in X, emanating from x, of length $\leq \epsilon$ and normal to σ has as its boundary a space $L_\epsilon(x)$ which is PL-homeomorphic to $L(\sigma)$ and which is a union of simplices of curvature 1 with totally geodesic faces. Such a metric is piecewise smoothly quasi-isometric to a piecewise flat metric, and thus $L(\sigma)$ is a Riemannian pseudomanifold. The union of line segments itself is isometric to a metric cone $cL(\sigma)$ on $L(\sigma)$, and x has a neighborhood which is quasi-isometric to $U \times c^\circ L(\sigma)$ with the product metric, where U is an open subset of the interior of σ and has the induced metric.

Definition 10.0.12 *A* handle decomposition *of a Riemannian pseudomanifold* (X, d, g) *is an open cover* $\{U_i\}$ *of* $X - \Sigma$ *such that for every* i, $(U_i, g|_{U_i})$ *is a Riemannian* j*-handle for some* j.

Example 10.0.13 *Let* (X, d, g) *be a Riemannian pseudomanifold with* d, g *as in Example* 10.0.10 *(or piecewise smoothly quasi-isometric to this). Then we have seen that every point* $x \in X$ *has an open neighborhood quasi-isometric to* $U \times c^\circ L(\sigma)$ *with the product metric, where* U *is an open subset of the open simplex that contains* x. *Then*

$$U \times (c^\circ(L(\sigma) - \Sigma) - \{c\})$$

is a Riemannian handle, because $c^\circ(L(\sigma) - \Sigma) - \{c\}$ *is a punctured metric cone* $\mathrm{pmc}(L(\sigma) - \Sigma)$. *Thus we obtain a handle decomposition for* (X, d, g).

Definition 10.0.14 (*Cheeger*) *A Riemannian i-handle* $(0,1)^{n-i} \times \mathrm{pmc}(L^{i-1})$ *is called* admissible, *if either i is even, or if $i = 2k+1$, $H^k_{(2)}(L^{2k}) = 0$. A Riemannian pseudomanifold is called* admissible, *if it possesses a handle decomposition containing only admissible handles.*

This admissibility condition corresponds precisely to the Witt condition of Definition 6.4.2, though Cheeger and Siegel were working independently.

Theorem 10.0.15 (*Cheeger's Hodge Theorem*) *Let X^n be a closed, admissible, Riemannian pseudomanifold. Then $H^i_{(2)}(X-\Sigma)$ is finite-dimensional and the natural map*

$$\mathcal{H}^i_{(2)}(X-\Sigma) \longrightarrow H^i_{(2)}(X-\Sigma)$$

is an isomorphism.

Proof. This is proven by an induction on handles. Surjectivity of the map follows from the finite-dimensionality of $H^*_{(2)}(X-\Sigma)$. The L^2 cohomology of all handles is finite-dimensional by the Poincaré lemmas for metric collars and punctured metric cones. Using the Mayer–Vietoris sequence, $H^*_{(2)}(X-\Sigma)$ is finite-dimensional. Injectivity of the natural map follows if Stokes' theorem holds in the L^2 sense. This in turn is again proven inductively over handles, starting from the Stokes' theorems for metric collars and punctured metric cones. □

Set

$$H^i_{(2)}(X) := H^i_{(2)}(X-\Sigma).$$

As a corollary to the Hodge theorem, one obtains

Corollary 10.0.16 (*Generalized Poincaré duality for admissible pseudomanifolds*) *Let X^n be a closed, admissible, oriented, Riemannian pseudomanifold. The pairing*

$$\begin{aligned} H^i_{(2)}(X) \otimes H^{n-i}_{(2)}(X) &\longrightarrow \mathbb{R} \\ [\omega] \otimes [\eta] &\mapsto \textstyle\int_{X-\Sigma} \omega \wedge \eta \end{aligned}$$

is nonsingular.

Let $\omega \in \Omega^i_{(2)}(X-\Sigma)$. Then $\int_\xi \omega$ is defined for almost all intersection chains $\xi \in IC^c_i(X)$ (though not all) and Stokes' theorem holds. This induces a map

$$H^i_{(2)}(X) \longrightarrow (IH^c_i(X))^*.$$

Theorem 10.0.17 (*Cheeger*) *Let X^n be a closed, admissible, oriented, Riemannian pseudomanifold. The map*

$$H^i_{(2)}(X) \longrightarrow (IH^c_i(X))^*$$

is an isomorphism.

Proof. Induct over the handles using the Mayer–Vietoris sequence and the two Poincaré lemmas. □

Without the admissibility condition, $H^i_{(2)}(X^n)$ and $H^{n-i}_{(2)}(X^n)$ need not have the same rank. Cheeger points out that sometimes Poincaré duality can be restored by imposing $*$-invariant ideal boundary conditions, meaning that in the middle-dimensional space of harmonic L^2 forms on even-dimensional links, one should choose a maximal self-annihilating subspace under the wedge product. This is a precursor of the purely topological notion of a Lagrangian structure, introduced in [Ban02] and discussed in Sect. 9.1 of this book. To the author's knowledge, the problem of somehow using Lagrangian subspaces to restore Poincaré duality occurred to a number of people independently, among them Cheeger, Morgan, Goresky and MacPherson. It was brought to the author's attention by Sylvain Cappell.

References

[Ati69] M. F. Atiyah, *The signature of fibre-bundles*, Global Analysis; Papers in honor of K. Kodaira (D. C. Spencer and S. Iyanayaga, eds.), Princeton Univ. Press, Princeton, 1969, pp. 73–84.

[B+84] A. Borel et al., *Intersection cohomology*, Progr. Math., no. 50, Birkhäuser Verlag, Boston, 1984.

[Ban02] M. Banagl, *Extending intersection homology type invariants to non-Witt spaces*, Memoirs Amer. Math. Soc. **160** (2002), no. 760, 1–83.

[Ban04] M. Banagl, *The signature of partially defined local systems in dimension four*, preprint, 2004.

[Ban06a] M. Banagl, *Computing twisted signatures and L-classes of non-Witt spaces*, Proc. London Math. Soc. (3) **92** (2006), 428–470.

[Ban06b] M. Banagl, *The L-class of non-Witt spaces*, Annals of Math. **163** (2006), no. 3, 743–766.

[BBD82] A. Beilinson, J. Bernstein, and P. Deligne, *Faisceaux pervers, analyse et topologie sur les espaces singuliers*, Astérisque **100** (1982), 1–171.

[BCS03] M. Banagl, S. E. Cappell, and J. L. Shaneson, *Computing twisted signatures and L-classes of stratified spaces*, Math. Ann. **326** (2003), no. 3, 589–623.

[BH99] M. R. Bridson and A. Haefliger, *Metric spaces of non-positive curvature*, Grundlehren Math. Wiss., no. 319, Springer Verlag, 1999.

[BK04] M. Banagl and R. Kulkarni, *Self-dual sheaves on reductive Borel–Serre compactifications of Hilbert modular surfaces*, Geom. Dedicata **105** (2004), 121–141.

[Bre97] G. E. Bredon, *Sheaf theory*, second ed., Grad. Texts in Math., no. 170, Springer Verlag, 1997.

[Bry93] Jean-Luc Brylinski, *Loop spaces, characteristic classes and geometric quantization*, Progr. Math., no. 107, Birkhäuser Verlag, Boston, 1993.

[Che46] S. S. Chern, *Characteristic classes of Hermitian manifolds*, Ann. of Math. **47** (1946), 85–121.

[Che79] J. Cheeger, *On the spectral geometry of spaces with cone-like singularities*, Proc. Natl. Acad. Sci. USA **76** (1979), 2103–2106.

[Che80] J. Cheeger, *On the Hodge theory of Riemannian pseudomanifolds*, Proc. Sympos. Pure Math. **36** (1980), 91–146.

[Che83] J. Cheeger, *Spectral geometry of singular Riemannian spaces*, J. Differential Geom. **18** (1983), 575–657.

[CHS57] S. S. Chern, F. Hirzebruch, and J.-P. Serre, *The index of a fibered manifold*, Proc. Amer. Math. Soc. **8** (1957), 587–596.

[CS91] S. E. Cappell and J. L. Shaneson, *Stratifiable maps and topological invariants*, J. Amer. Math. Soc. **4** (1991), 521–551.

[CSW91] S. E. Cappell, J. L. Shaneson, and S. Weinberger, *Classes topologiques caractéristiques pour les actions de groupes sur les espaces singuliers*, C. R. Acad. Sci. Paris Sér. I Math. **313** (1991), 293–295.

[CW91] S. E. Cappell and S. Weinberger, *Classification de certaines espaces stratifiés*, C. R. Acad. Sci. Paris Sér. I Math. **313** (1991), 399–401.

[Fri03] G. Friedman, *Stratified fibrations and the intersection homology of the regular neighborhoods of bottom strata*, Topology Appl. **134** (2003), 69–109.

[GM80] M. Goresky and R. D. MacPherson, *Intersection homology theory*, Topology **19** (1980), 135–162.

[GM83] M. Goresky and R. D. MacPherson, *Intersection homology II*, Invent. Math. **71** (1983), 77–129.

[GM99] S. I. Gelfand and Yu. I. Manin, *Homological algebra*, Springer Verlag, 1999.

[God58] Roger Godement, *Topologie algébrique et théorie des faisceaux*, Hermann, Paris, 1958.

[Gun66] R. C. Gunning, *Lectures on Riemann surfaces*, Mathematical Notes, Princeton University Press, Princeton, New Jersey, 1966.

[Har66] R. Hartshorne, *Residues and duality*, Lecture Notes in Math., no. 20, Springer Verlag, 1966.

[HBJ92] F. Hirzebruch, Th. Berger, and R. Jung, *Manifolds and modular forms*, Aspects of Math., vol. E20, Vieweg, 1992.

[Hir69] F. Hirzebruch, *The signature of ramified coverings*, Collected Math. Papers in Honor of Kodaira, Tokyo University Press, Tokyo, 1969, pp. 253–265.

[HR96] B. Hughes and A. A. Ranicki, *Ends of complexes*, Cambridge Tracts in Math., no. 123, Cambridge University Press, 1996.

[HW41] W. Hurewicz and H. Wallman, *Dimension theory*, Princeton Univ. Press, Princeton, N.J., 1941.

[Ive86] Birger Iversen, *Cohomology of sheaves*, Universitext, Springer Verlag, 1986.

[Kin85] H. C. King, *Topological invariance of intersection homology without sheaves*, Topology Appl. **20** (1985), 149–160.

[Kle89] S. L. Kleiman, *The development of intersection homology theory*, A century of mathematics in America, part II, Hist. Math., 2, Amer. Math. Soc., Providence, RI, 1989, pp. 543–585.

[Kod67] K. Kodaira, *A certain type of irregular algebraic surfaces*, J. Anal. Math. **19** (1967), 207–215.

[KS90] Masaki Kashiwara and Pierre Schapira, *Sheaves on manifolds*, Grundlehren Math. Wiss., no. 292, Springer Verlag, 1990.

[Lef26] S. Lefschetz, *Intersections and transformations of manifolds*, Trans. Amer. Math. Soc. **28** (1926), 1–29.

[Lüc02] W. Lück, *A basic introduction to surgery theory*, ICTP Lecture Notes Series, vol. 9, Part 1, Abdus Salam International Centre for Theoretical Physics, Trieste, 2002, school "High-dimensional manifold theory" in Trieste, May/June 2001.

[McC75] C. McCrory, *Cone complexes and PL-transversality*, Trans. Amer. Math. Soc. **207** (1975), 1–23.

[Mey72] W. Meyer, *Die Signatur von lokalen Koeffizientensystemen und Faserbündeln*, Bonner Math. Schriften **53** (1972).

[MH73] J. Milnor and D. Husemoller, *Symmetric bilinear forms*, Springer, 1973.

[Min04] A. Minatta, *Hirzebruch homology*, Ph.D. thesis, Ruprecht-Karls Universität Heidelberg, 2004.

[MM79] I. Madsen and J. Milgram, *The classifying spaces for surgery and cobordism of manifolds*, Ann. of Math. Studies, vol. 92, Princeton Univ. Press, Princeton, 1979.

[MS74] J. W. Milnor and J. D. Stasheff, *Characteristic classes*, Ann. of Math. Studies, no. 76, Princeton University Press, 1974.

[Poi95] H. Poincaré, *Analysis situs*, Journal de l'École, Polytechnique **1** (1895), 1–121, Reprinted in *Oeuvres de H. Poincaré VI*, 193–288.

[Poi99] H. Poincaré, *Complément à l'analysis situs*, Rend. Circ. Matem. Palermo **13** (1899), 285–343, Reprinted in *Oeuvres de H. Poincaré VI*, 290–337.

[Pon47] L. S. Pontrjagin, *Characteristic cycles on differentiable manifolds*, Mat. Sb. N.S. **21 (63)** (1947), 233–284, Amer. Math. Soc. Transl. **32** (1950).

[Ran92] A. A. Ranicki, *Algebraic L-theory and topological manifolds*, Cambridge Tracts in Math., no. 102, Cambridge University Press, 1992.

[Ran02] A. A. Ranicki, *Algebraic and geometric surgery*, Oxford Math. Monographs, Oxford University Press, 2002.

[Sie83] P. H. Siegel, *Witt spaces: A geometric cycle theory for KO-homology at odd primes*, Amer. J. Math. **105** (1983), 1067–1105.

[Sti36] E. Stiefel, *Richtungsfelder und Fernparallelismus in n-dimensionalen Mannigfaltigkeiten*, Comment. Math. Helv. **8** (1936), 305–353.

[Sul70a] D. Sullivan, *A general problem*, p. 230, Manifolds, Amsterdam, Lecture Notes in Math., no. 197, Springer Verlag, 1970.

[Sul70b] D. Sullivan, *Geometric topology notes, part I*, MIT, Cambridge, MA, 1970.

[Sul04] D. Sullivan, *René Thom's work on geometric homology and bordism*, Bull. Amer. Math. Soc. **41** (2004), no. 3, 341–350.

[Ver66] J. L. Verdier, *Dualité dans la cohomologie des espaces localement compacts*, vol. 300, Sém. Bourbaki, 1965–66.

[Wei94] S. Weinberger, *The topological classification of stratified spaces*, Chicago Lectures in Math., Univ. of Chicago Press, Chicago, 1994.

[Whi35] H. Whitney, *Sphere-spaces*, Proc. Natl. Acad. Sci. USA **21** (1935), 464–468.

Index

Springer Monographs in *Mathematics*

This series publishes advanced monographs giving well-written presentations of the "state-of-the-art" in fields of mathematical research that have acquired the maturity needed for such a treatment. They are sufficiently self-contained to be accessible to more than just the intimate specialists of the subject, and sufficiently comprehensive to remain valuable references for many years. Besides the current state of knowledge in its field, an SMM volume should also describe its relevance to and interaction with neighbouring fields of mathematics, and give pointers to future directions of research.

Abhyankar, S.S. **Resolution of Singularities of Embedded Algebraic Surfaces** 2nd enlarged ed. 1998
Alexandrov, A.D. **Convex Polyhedra** 2005
Andrievskii, V.V.; Blatt, H.-P. **Discrepancy of Signed Measures and Polynomial Approximation** 2002
Angell, T.S.; Kirsch, A. **Optimization Methods in Electromagnetic Radiation** 2004
Ara, P.; Mathieu, M. **Local Multipliers of C^*-Algebras** 2003
Armitage, D.H.; Gardiner, S.J. **Classical Potential Theory** 2001
Arnold, L. **Random Dynamical Systems** corr. 2nd printing 2003 (1st ed. 1998)
Arveson, W. **Noncommutative Dynamics and E-Semigroups** 2003
Aubin, T. **Some Nonlinear Problems in Riemannian Geometry** 1998
Auslender, A.; Teboulle, M. **Asymptotic Cones and Functions in Optimization and Variational Inequalities** 2003
Banagl, M. **Topological Invariants of Stratified Spaces** 2006
Banasiak, J.; Arlotti, L. **Perturbations of Positive Semigroups with Applications** 2006
Bang-Jensen, J.; Gutin, G. **Digraphs** 2001
Baues, H.-J. **Combinatorial Foundation of Homology and Homotopy** 1999
Böttcher, A.; Silbermann, B. **Analysis of Toeplitz Operators** 2006
Brown, K.S. **Buildings** 3rd printing 2000 (1st ed. 1998)
Chang, K.-C. **Methods in Nonlinear Analysis** 2005
Cherry, W.; Ye, Z. **Nevanlinna's Theory of Value Distribution** 2001
Ching, W.K. **Iterative Methods for Queuing and Manufacturing Systems** 2001
Chudinovich, I. **Variational and Potential Methods for a Class of Linear Hyperbolic Evolutionary Processes** 2005
Coates, J.; Sujatha, R. **Cyclotomic Fields and Zeta Values** 2006
Crabb, M.C.; James, I.M. **Fibrewise Homotopy Theory** 1998
Dineen, S. **Complex Analysis on Infinite Dimensional Spaces** 1999
Dugundji, J.; Granas, A. **Fixed Point Theory** 2003
Ebbinghaus, H.-D.; Flum, J. **Finite Model Theory** 2006
Edmunds, D.E.; Evans, W.D. **Hardy Operators, Function Spaces and Embeddings** 2004
Elstrodt, J.; Grunewald, F.; Mennicke, J. **Groups Acting on Hyperbolic Space** 1998
Engler, A.J.; Prestel, A. **Valued Fields** 2005
Fadell, E.R.; Husseini, S.Y. **Geometry and Topology of Configuration Spaces** 2001
Fedorov, Yu. N.; Kozlov, V.V. **A Memoir on Integrable Systems** 2001
Flenner, H.; O'Carroll, L.; Vogel, W. **Joins and Intersections** 1999
Gelfand, S.I.; Manin, Y.I. **Methods of Homological Algebra** 2nd ed. 2003 (1st ed. 1996)
Griess, R.L. Jr. **Twelve Sporadic Groups** 1998
Gras, G. **Class Field Theory** corr. 2nd printing 2005
Greuel, G.-M.; Lossen, C.; Shustin, E. **Introduction to Singularities and Deformations** 2007
Hida, H. ***p*-Adic Automorphic Forms on Shimura Varieties** 2004
Ischebeck, E; Rao, R.A. **Ideals and Reality** 2005
Ivrii, V. **Microlocal Analysis and Precise Spectral Asymptotics** 1998
Jakimovski, A.; Sharma, A.; Szabados, J. **Walsh Equiconvergence of Complex Interpolating Polynomials** 2006
Jech, T. **Set Theory** (3rd revised edition 2002)
Jorgenson, J.; Lang, S. **Spherical Inversion on SLn (R)** 2001
Kanamori, A. **The Higher Infinite** corr. 2nd printing 2005 (2nd ed. 2003)

Kanovei, V. **Nonstandard Analysis, Axiomatically** 2005
Khoshnevisan, D. **Multiparameter Processes** 2002
Koch, H. **Galois Theory of *p*-Extensions** 2002
Komornik, V. **Fourier Series in Control Theory** 2005
Kozlov, V.; Maz'ya, V. **Differential Equations with Operator Coefficients** 1999
Lam, T.Y. **Serre's Problem on Projective Modules** 2006
Landsman, N.P. **Mathematical Topics between Classical & Quantum Mechanics** 1998
Leach, J.A.; Needham, D.J. **Matched Asymptotic Expansions in Reaction-Diffusion Theory** 2004
Lebedev, L.P.; Vorovich, I.I. **Functional Analysis in Mechanics** 2002
Lemmermeyer, F. **Reciprocity Laws: From Euler to Eisenstein** 2000
Malle, G.; Matzat, B.H. **Inverse Galois Theory** 1999
Mardesic, S. **Strong Shape and Homology** 2000
Margulis, G.A. **On Some Aspects of the Theory of Anosov Systems** 2004
Miyake, T. **Modular Forms** 2006
Murdock, J. **Normal Forms and Unfoldings for Local Dynamical Systems** 2002
Narkiewicz, W. **Elementary and Analytic Theory of Algebraic Numbers** 3rd ed. 2004
Narkiewicz, W. **The Development of Prime Number Theory** 2000
Neittaanmaki, P.; Sprekels, J.; Tiba, D. **Optimization of Elliptic Systems. Theory and Applications** 2006
Onishchik, A.L. **Projective and Cayley–Klein Geometries** 2006
Parker, C.; Rowley, P. **Symplectic Amalgams** 2002
Peller, V. (Ed.) **Hankel Operators and Their Applications** 2003
Prestel, A.; Delzell, C.N. **Positive Polynomials** 2001
Puig, L. **Blocks of Finite Groups** 2002
Ranicki, A. **High-dimensional Knot Theory** 1998
Ribenboim, P. **The Theory of Classical Valuations** 1999
Rowe, E.G.P. **Geometrical Physics in Minkowski Spacetime** 2001
Rudyak, Y.B. **On Thorn Spectra, Orientability and Cobordism** 1998
Ryan, R.A. **Introduction to Tensor Products of Banach Spaces** 2002
Saranen, J.; Vainikko, G. **Periodic Integral and Pseudodifferential Equations with Numerical Approximation** 2002
Schneider, P. **Nonarchimedean Functional Analysis** 2002
Serre, J-P. **Complex Semisimple Lie Algebras** 2001 (reprint of first ed. 1987)
Serre, J-P. **Galois Cohomology** corr. 2nd printing 2002 (1st ed. 1997)
Serre, J-P. **Local Algebra** 2000
Serre, J-P. **Trees** corr. 2nd printing 2003 (1st ed. 1980)
Smirnov, E. **Hausdorff Spectra in Functional Analysis** 2002
Springer, T.A.; Veldkamp, F.D. **Octonions, Jordan Algebras, and Exceptional Groups** 2000
Székelyhidi, L. **Discrete Spectral Synthesis and Its Applications** 2006
Sznitman, A.-S. **Brownian Motion, Obstacles and Random Media** 1998
Taira, K. **Semigroups, Boundary Value Problems and Markov Processes** 2003
Talagrand, M. **The Generic Chaining** 2005
Tauvel, P.; Yu, R.W.T. **Lie Algebras and Algebraic Groups** 2005
Tits, J.; Weiss, R.M. **Moufang Polygons** 2002
Uchiyama, A. **Hardy Spaces on the Euclidean Space** 2001
Üstünel, A.-S.; Zakai, M. **Transformation of Measure on Wiener Space** 2000
Vasconcelos, W. **Integral Closure. Rees Algebras, Multiplicities, Algorithms** 2005
Yang, Y. **Solitons in Field Theory and Nonlinear Analysis** 2001
Zieschang, P.-H. **Theory of Association Schemes** 2005

Printing: Krips bv, Meppel
Binding: Stürtz, Würzburg